T0073362

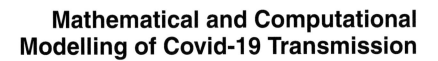

Mathematical and Computational Modelling of Covid-19 Transmission

RIVER PUBLISHERS SERIES IN MATHEMATICAL, STATISTICAL AND COMPUTATIONAL MODELLING FOR ENGINEERING

Series Editors:

Mangey Ram
Graphic Era University, India

Tadashi Dohi
Hiroshima University, Japan

Applied mathematical techniques along with statistical and computational data analysis has become vital skills across the physical sciences. The purpose of this book series is to present novel applications of numerical and computational modelling and data analysis across the applied sciences. We encourage applied mathematicians, statisticians, data scientists and computing engineers working in a comprehensive range of research fields to showcase different techniques and skills, such as differential equations, finite element method, algorithms, discrete mathematics, numerical simulation, machine learning, probability and statistics, fuzzy theory, etc.

Books published in the series include professional research monographs, edited volumes, conference proceedings, handbooks and textbooks, which provide new insights for researchers, specialists in industry, and graduate students.

Topics included in this series are as follows:-

- Discrete mathematics and computation
- Fault diagnosis and fault tolerance
- Finite element method (FEM) modeling/simulation
- Fuzzy and possibility theory
- Fuzzy logic and neuro-fuzzy systems for relevant engineering applications
- Game Theory
- Mathematical concepts and applications
- Modelling in engineering applications
- Numerical simulations
- Optimization and algorithms
- Queueing systems
- Resilience
- Stochastic modelling and statistical inference
- Stochastic Processes
- Structural Mechanics
- Theoretical and applied mechanics

For a list of other books in this series, visit www.riverpublishers.com

Mathematical and Computational Modelling of Covid-19 Transmission

Editors

Mandeep Mittal
Amity University Noida, India

Nita H. Shah
Gujarat University, India

Routledge
Taylor & Francis Group

NEW YORK AND LONDON

Published 2024 by River Publishers

River Publishers

Alsbjergvej 10, 9260 Gistrup, Denmark

www.riverpublishers.com

Distributed exclusively by Routledge

605 Third Avenue, New York, NY 10017, USA

4 Park Square, Milton Park, Abingdon, Oxon OX14 4RN

Mathematical and Computational Modelling of Covid-19 Transmission / by Mandeep Mittal, Nita H. Shah.

Routledge is an imprint of the Taylor & Francis Group, an informa business

ISBN 978-87-7022-831-2 (hardback)

ISBN 978-87-7004-045-7 (paperback)

ISBN 978-10-0380-712-4 (online)

ISBN 978-10-3262-314-6 (master ebook)

While every effort is made to provide dependable information, the publisher, authors, and editors cannot be held responsible for any errors or omissions.

Contents

Preface   xi

List of Figures   xv

List of Tables   xxiii

List of Contributors   xxv

List of Abbreviations   xxix

1 **Mathematical Modelling of COVID-19 Dynamics with Reinfection
   Using Fractional Order**   1
   A. George Maria Selvam, D. Vignesh, and S. Britto Jacob
   1.1 Introduction   2
   1.2 SEIQR Model for COVID-19 Spread   3
   1.3 Fractional-Order SEIQR Model   8
   1.4 Existence of Equilibrium Points   8
   1.5 Stability of the System   10
   1.6 Optimal Control Analysis   11
   1.7 Numerical Simulations   13
        1.7.1 Impact of Reinfected Population Rate (m)   13
        1.7.2 Bifurcation Analysis   14
   1.8 Conclusion   20

2 **Stability Analysis of Caputo–Fabrizio Fractional-order Epidemic
   Model of a Novel Coronavirus (COVID-19)**   25
   A. George Maria Selvam, R. Janagaraj, and R. Dhineshbabu
   2.1 Introduction   26
   2.2 Preliminaries   27
   2.3 Model Formulation   28

2.4   Existence Criteria of Caputo–Fabrizio Model by Picard
       Approximation   30
2.5   Stability Analysis by the Equilibrium States   34
2.6   Numerical Computations by Data Values   36
       2.6.1   COVID-19 Model with Vaccination   42
2.7   Conclusion   47

3   **Impact of Vaccination on COVID-19 to Control its Spread: A
    Case Study of India**   **51**
    Nita H. Shah, Ankush H. Suthar, and Ekta N. Jayswal
    3.1   Introduction   52
    3.2   Mathematical Modelling   54
           3.2.1   Equilibrium Points   57
    3.3   Stability   59
    3.4   Numerical Simulation   60
    3.5   Conclusion   62

4   **A Computational Approach for Regulation of Biomedical
    Waste Expulsion in a Novel Coronavirus Pandemic**   **65**
    Oshin Rawlley, Yatendra Sahu, Rajeev Kumar Gupta,
    Amit Kumar Mishra, Ramakant Bhardwaj, and Satyendra Narayan
    4.1   Introduction   66
    4.2   Literature Review   68
           4.2.1   BMW disposal   68
                    4.2.1.1   Mechanical processes   69
                    4.2.1.2   Thermal processes   69
                    4.2.1.3   Chemical processes   73
                    4.2.1.4   Irradiation processes   73
                    4.2.1.5   Biological processes   73
    4.3   Latest Methodology   73
    4.4   Emission New Rules   75
    4.5   Effects on Nature   76
    4.6   Conclusion   77

5   **Solution for Fractional-order Pneumonia–COVID-19
    Co-infection**   **83**
    Nita H. Shah, Nisha Sheoran, and Ekta Jayswal
    5.1   Introduction   84
    5.2   Formulation of Fractional Order Mathematical Model   85

      5.2.1   Preliminaries of fractional order in Caputo sense . .   85

      5.2.2   Existence of positivity and boundedness of solutions   89

      5.2.3   Equilibrium points and basic reproduction number .   91

  5.3   Stability Analysis   92

  5.4   Numerical Simulation   96

  5.5   Conclusion 104

6   Optimal Control to Curtail the Spread of COVID-19 through Social Gatherings: A Mathematical Model   109

Nita H. Shah, Purvi M. Pandya, and Ankush H. Suthar

  6.1   Introduction 110

  6.2   Mathematical Model 111

  6.3   Optimal Control 115

  6.4   Numerical Simulation 118

  6.5   Conclusion 122

7   Effectiveness of a Booster Dose of COVID-19 Vaccine   125

Nita H. Shah and Moksha H. Satia

  7.1   Introduction 125

  7.2   Mathematical Model 127

      7.2.1   Equilibrium point 129

      7.2.2   Basic reproduction number 133

  7.3   Stability Analysis 135

      7.3.1   Local stability 135

      7.3.2   Global stability 138

  7.4   Numerical Simulation 140

  7.5   Conclusion 146

8   Impact of Post-COVID-19 Pandemic on Quality Education among School and College Students   151

Santosh Kumar, Surya Kant Pal, Isha Sangal,
Khursheed Alam, Mandeep Mittal, Mahesh Kumar Jayaswal

  8.1   Introduction 152

  8.2   Research Gap 153

  8.3   Research Objectives 153

  8.4   Research Hypothesis 154

  8.5   Methodology 154

  8.6   Limitations of the Study 163

  8.7   Future Prospective of Research 164

8.8   Conclusion 164

9   A Mathematical Model for the Dynamics of the Violence Epidemic Spreading due to Misinformation   167
Arindam Kumar Paul and M. Haider Ali Biswas
9.1   Introduction 168
9.2   Previous Studies 169
9.3   An Overview on Misinformation 171
    9.3.1   Intention is not creating violence 172
    9.3.2   Intention is to create violence 172
9.4   Mathematical Model of Spreading Misinformation 172
    9.4.1   Equilibrium analysis 176
    9.4.2   Stability analysis 176
        9.4.2.1   Stability at $E^0\left(S^*, E^*, C^*, I^*, Z^*\right)$ 176
        9.4.2.2   Stability at $\mathrm{E}^*\left(S^*, E^*, C^*, I^*, Z^*\right)$ 177
    9.4.3   Basic reproduction bumber 178
9.5   Numerical Simulations 181
    9.5.1   State Trajectories of Intentional Spread 184
    9.5.2   State Trajectories for Spread with Other Intentions . 186
    9.5.3   Bifurcation analysis 187
9.6   Results and Discussion 189
9.7   Conclusion 192

10   Diseased Predator–Prey Model Incorporating Herd Behaviour in Prey: A Study Under an Alternative Food Source Scenario   197
Sumit Kaur Bhatia, Riju Chaudhary, and Devyanshi Bansal
10.1   Introduction 198
10.2   Mathematical Model Formulation 199
10.3   Qualitative Analysis of the Model 200
    10.3.1   Existence of points of equilibria 201
    10.3.2   Stability analysis 202
        10.3.2.1   The prey-only equilibrium 202
        10.3.2.2   Dynamical behavior near the trivial equilibrium point 203
        10.3.2.3   Disease-free equilibrium point 204
        10.3.2.4   Endemic equilibrium point 205
    10.3.3   Transversality condition for Hopf bifurcation . . . 206
    10.3.4   Global stability analysis 207

10.4 Numerical Simulation 208
10.5 Conclusion 211

11 A COVID-19-related Atangana–Baleanu Fractional Model for Unemployed Youths   215
Albert Shikongo
11.1 Introduction 216
11.2 The Conceptual Model 217
11.3 Mathematical Analysis 221
    11.3.1 Existence of solution 222
    11.3.2 Uniqueness of solution 226
    11.3.3 Positive solution 226
    11.3.4 Equilibrium points 227
    11.3.5 Stability analysis 227
11.4 Derivation of the Numerical Method 228
11.5 Numerical Results and Discussion 233
11.6 Conclusion 236

12 A Fractional-order SVIR Model with Two Infection Classes for COVID-19 in India   241
Nita H. Shah, Kapil Chaudhary, and Ekta Jayswal
12.1 Introduction 242
12.2 Preliminaries 243
12.3 Formulation of Model 245
    12.3.1 Existence and uniqueness of the solution 248
    12.3.2 Equilibrium points 249
12.4 Calculation of Basic Reproduction Number 250
12.5 Stability 251
    12.5.1 251
    12.5.2 Local stability of the endemic equilibrium point E^1 . 253
12.6 Memory Trace and Hereditary Trait 257
12.7 Numerical Simulation 259
12.8 Conclusion 262

13 Forecasting of a COVID-19 Model using LSTM   267
Kopila Chhetri, Himanshi, Sidhanth Karwal, and Sudipa Chauhan
13.1 Introduction 268
13.2 Methodology 274
    13.2.1 Long short-term memory (LSTM) 274

        13.2.2 How does LSTM work   275
    13.3 Data Preparation   278
    13.4 The Proposed LSTM Model   279
    13.5 Root Mean Square Error (RMSE)   279
    13.6 Experimental Results   281
    13.7 Conclusions and Future Work   281

**14 Simulation of COVID-19 Cases in India using AR and ANN
    Models   287**
    Kokila Ramesh, Radha Gupta, and Nethravathi N
    14.1 Introduction   288
    14.2 Literature   289
    14.3 Data   291
    14.4 Methodology   293
    14.5 Results and Discussion   298
    14.6 Conclusions   301

Index   303

About the Editors   305

Preface

The COVID-19 pandemic has been devastating across the globe, from many positive cases to high deaths ever recorded, and has shaken the whole world. The negative impacts are felt in many domains of human life and activities, the global economy, businesses, corporates, and industries, and hospitals have all been under immense stress. In trying to sustain and recover from the rage, scientists have worked day and night to bring about effective and preventive solution approaches to help control the spread up to a certain extent with vaccines and other adoptive healthy social behaviors. Multiple studies have also emerged and addressed both the analysis of the current situation phase to analyse for greater attention and actions and forecast models for prevention and proactive measures. The following chapters below depict models and solutions approaches to different subjects of matter pertaining to the COVID-19 pandemic and methods to improve human life aspects and avoid as possible any further unprecedented disease of such kind. These comprehensive references come to a greater advantage for practitioners, educators, officials in various fields, etc. to explore and gain extraordinary insights. Many subjects are addressed, from examining the impacts and trends of the pandemic to predicting future situations.

Chapter 1 analyses the dynamics and spread of COVID-19 through mathematical modelling in epidemiology and that to understand its mechanism and influence over the spread of the disease, and even predict future situations. Achieves a fractional-order epidemic model taking into account the impact of quarantine and reinfection to examine the dynamics of the COVID-19 outbreak.

Chapter 2 investigates the dynamic behavior of the COVID-19 disease using the non-integer Caputo-Fabrizio derivative to study the coronavirus sickness. Examines the stability of the suggested fractional epidemic model of the novel coronavirus (COVID-19) and provides a parametric rule for the essential reproduction number ratio.

Chapter 3 develops a fractional order compartmental model to study the effect of vaccination on the spread of coronavirus disease 2019 (COVID-19). Studies a particular case of the outbreak in Indian states. The model explores the special characteristics of vaccination on moderately and severely infected classes separately.

Chapter 4 examines sophisticated methods for waste disposal to achieve minimal damage to nature. A high priority requirement for proper management of waste dispensing and avoiding further unprecedented diseases that may cause greater social upheaval.

Chapter 5 proposes and analyses a Caputo fractional order mathematical model to study the dynamics of pneumonia COVID-19 co-infection. The model entertains four equilibrium points namely, disease-free, pneumonia-free, COVID-19-free, and the endemic point where both diseases co-exist. An increment is observed in the recovered population when the memory effect is introduced by considering the fractional order model.

Chapter 6 designs a systematic pathway to scrutinize the COVID-19 spread pattern of individuals by transmission due to gatherings. Formulates a non-linear differential equation and calculates the threshold to study the impact of the viral pandemic due to social gatherings. Also, applies optimal controls to help limit the disastrous outcomes up to a certain level.

Chapter 7 investigates the effectiveness of the two vaccine doses and one booster dose based on the advancement of COVID-19 through a system of non-linear differential equations. Two different parts of the population are found, the exposed individuals and the unexposed population. The stability of the model braces the ongoing situation.

Chapter 8 develops a regression model on Post-COVID-19 data collected via a primary source and measures the accuracy of corona-positive patients as well as the worst-case scenario in India. Based on the outcomes, COVID-19 is expected to continue to grow with an unpredictable factor and fear of uncertainty for the environment and several health issues among individuals.

Chapter 9 represents and analyses the internal mechanism of the misinformation of the epidemic spread that causes violence, disturbance, etc. Mathematical modelling in the form of a system of five non-linear ordinary differential equations is developed to deal with the problem.

Chapter 10 studies and analyses the model of a prey-predator that exhibits herb behavior by prey population. Boundedness details of the solution, condition for the existence of points of equilibria, and their stability are investigated. The biomass conversion parameter is shown to play a crucial impact in the system's dynamics.

Chapter 11 suggests a system of ordinary differential equations (ODEs) modelling the dynamics of youth unemployment in our society. Due to the COVID-19 negative impact on global economic activities, the dynamics extend to a system of Atangana-Baleanu fractional (ABF) dynamics. A numerical method is developed for discretizing the Atangana-Baleanu fractional derivative in the Caputo sense.

Chapter 12 proposes an SVIR model with two infection classes corresponding to the Delta and Omicron COVID-19 variants and describes their fractional-order transmission dynamics. Numerical simulations and relevant findings are exhibited for a better insight into the model.

Chapter 13 proposes an AI (Artificial Intelligence) based forecasting model for COVID-19 propagation using available time-series data. Focuses on a few selected countries for the model and forecasts COVID-19 confirmed, active, recovered, and deaths, based on Long Short-Term Memory (LSTM) with one input layer, three LSTM layers, and ten hidden units (neurons).

Chapter 14 conducts a statistical analysis of India's COVID-19 confirmed, active, recovered, and death cases for a five months period (January 2022 – June 2022), and develops a statistical model that predicts the number of cases for July month. The data obtained helps understand the inter and intra relationships between cases. The prediction model is entertained using the ANN (Artificial neuron network) model with the help of a backpropagation algorithm. The model intends to inform for prevention and proactive measures.

List of Figures

Figure 1.1   Schematic representation of the SEIQR model.   . .   4

Figure 1.2   Time plots of susceptible and exposed population with different fractional-order values.   14

Figure 1.3   Time plots of infected and exposed population with different fractional-order values.   14

Figure 1.4   Time plots of recovered population with different fractional-order values.   15

Figure 1.5   Time plots of susceptible and infected population varying reinfection rate m.   15

Figure 1.6   Time plots of quarantined and recovered population varying reinfection rate m.   15

Figure 1.7   Bifurcations of susceptible and exposed population.   16

Figure 1.8   Bifurcations of infected and quarantined population.   16

Figure 1.9   Bifurcations of recovered population.   17

Figure 1.10   Susceptible population for step size $h = 2.5, 3.0$ and 3.5.   17

Figure 1.11   Exposed population for step size $h = 2.5, 3.0$ and 3.5.   18

Figure 1.12   Infected population for step size $h = 2.5, 3.0$ and 3.5.   18

Figure 1.13   Quarantined population for step size $h = 2.5,$ 3.0 and 3.5.   19

Figure 1.14   Recovered population for step size $h = 2.5, 3.0$ and 3.5.   19

Figure 2.1   Schematic diagram for the SEIHR model of (2.1).   . .   29

Figure 2.2   Behaviour of susceptible state $S(\kappa)$ for various fractional orders μ in India.   38

Figure 2.3   Behaviour of exposed state $E(\kappa)$ for various fractional orders μ in India.   39

Figure 2.4   Behaviour of infected state $I(\kappa)$ for various fractional orders μ in India.   39

Figure 2.5   Behaviour of hospitalized state $H(\kappa)$ for various fractional orders μ in India.   40

Figure 2.6   Behaviour of recovered state $R(\kappa)$ for various fractional orders μ in India.   40

Figure 2.7   Behaviour of susceptible state $S(\kappa)$ for various fractional orders μ in Tamil Nadu.   41

Figure 2.8   Behaviour of exposed state $E(\kappa)$ for various fractional orders μ in Tamil Nadu.   41

Figure 2.9   Behaviour of infected state $I(\kappa)$ for various fractional orders μ in Tamil Nadu.   42

Figure 2.10   Behaviour of hospitalized state $H(\kappa)$ for various fractional orders μ in Tamil Nadu.   43

Figure 2.11   Behaviour of recovered state $R(\kappa)$ for various fractional orders μ in Tamil Nadu.   43

Figure 2.12   Behaviour of $S(\kappa)$ for $\mu = 0.95$ with and without vaccination in India.   44

Figure 2.13   Behaviour of $I(\kappa)$ for $\mu = 0.95$ with and without vaccination in India.   44

Figure 2.14   Behaviour of $R(\kappa)$ for $\mu = 0.95$ with and without vaccination in India.   45

Figure 2.15   Behaviour of $S(\kappa)$ for $\mu = 0.95$ with and without vaccination in Tamil Nadu.   45

Figure 2.16   Behaviour of $I(\kappa)$ for $\mu = 0.95$ with and without vaccination in Tamil Nadu.   46

Figure 2.17   Behaviour of $R(\kappa)$ for $\mu = 0.95$ with and without vaccination in Tamil Nadu.   46

Figure 3.1   Partial vaccination rate in India on 9 August 2021. .   53

Figure 3.2   Complete vaccination rate in India on 9 August 2021.   54

Figure 3.3   Schematic diagram of the model.   55

Figure 3.4   Spread of COVID-19 in Indian states after complete vaccination.   60

Figure 3.5   Variation in the model with time.   61

Figure 3.6   Bifurcation on non-vaccinated (a) and partially vaccinated (b) classes with respect to R_0.   62

Figure 3.7   Sensitivity of R_0 with respect to the parameters related to vaccination.   62

Figure 4.1   History of pandemics.   67

Figure 4.2   Vehicle for medical waste transportation.   68

Figure 4.3   Waste distribution of BMW.   69

Figure 4.4 Various thermal techniques. 70

Figure 4.5 Emission limits before. 72

Figure 4.6 Incineration after segregation. 74

Figure 4.7 Production and treatment trend. 75

Figure 4.8 (a) Delhi regained its breath. With the onset of the pandemic Delhi saw a positive change in its nature. The animals started floating on the ground, river Yamuna was taking good breath also the pollution levels were gone down. This picture displays the same where the clear skies are flowing with drastic reduced pollution levels. (b) PM 2.5 levels in India's atmosphere compared between 30 March 2020 Italy also observed dips in the NO_2 levels seen in the satellite images. The NO_2 levels saw a great fall due to the reduced human intervention in the nature. The satellite images were able to capture these pictures showing a positive impact on the nature. Clearly indicating the fact that how we have destroyed the nature in various forms and how the pandemic showed the improved levels in air quality index. . . 78

Figure 4.9 Surface concentrations of nitrogen dioxide over northern Italy, 31 January versus 15 March 2020. Copernicus Atmosphere Monitoring Service (CAMS); ECMWF. These are the dense surface concentrations of NO_2 which has shown a reduction trend after the pandemic. These pictures confirm a gradual decrease in the pollutants level for the Northern Italy. 79

Figure 5.1 Transmission plot of population among different classes. 87

Figure 5.2 Solution of susceptible population. 96

Figure 5.3 Solution of individuals exposed to COVID-19. . . . 97

Figure 5.4 Solution of population exposed to pneumonia. . . . 97

Figure 5.5 Solution of population infected by COVID-19. . . . 98

Figure 5.6 Solution of population exposed to co-infection. . . 99

Figure 5.7 Solution of pneumonia infectives. 99

Figure 5.8 Solution of co-infected class. 100

Figure 5.9 Solution of recovered class. 100

Figure 5.10 Directional plot among the susceptible class, COVID-19 exposed class and pneumonia exposed class. 101

Figure 5.11 Directional flow of population among the susceptible class, class exposed to pneumonia and pneumonia infected class. 101

Figure 5.12 Directional plot of classes exposed to co-infection, co-infected and recover. 102

Figure 5.13 Forward bifurcation plot. 102

Figure 5.14 Behaviour of infected class considered in the model, i.e. COVID-19 infectives, pneumonia infected and co-infected. 103

Figure 5.15 Recovery plotted for infected class. 103

Figure 5.16 Plot of recovery and infectives of population of COVID-19, pneumonia and its co-infection. 104

Figure 6.1 Mathematical model. 112

Figure 6.2 Model with control variables. 115

Figure 6.3 Transmission pattern. 118

Figure 6.4 Directional flows of state variables. (a) Behaviour of exposed individuals towards recovery and (b) Behaviour of infected individuals towards recovery. 118

Figure 6.5 Trajectory field and solution curve. 119

Figure 6.6 Effect on compartment through oscillations. 120

Figure 6.7 Social gatherings and its observed impact. 121

Figure 6.8 Effect of controls on respective compartments. . . . 121

Figure 7.1 Flow diagram of individual. 128

Figure 7.2 Transmission of individual. 142

Figure 7.3 Individuals having dose I of vaccination. 142

Figure 7.4 Individuals having dose II of vaccination. 143

Figure 7.5 Individuals having booster dose of vaccination. . . 143

Figure 7.6 Behaviour of infected individual after each dose of vaccination. 144

Figure 7.7 Behaviour of individuals after each dose of vaccination. 144

Figure 7.8 Behaviour of recovered individuals and critically ill individuals after booster dose. 145

Figure 7.9 Cyclic behaviour of individuals in critical condition. 146

Figure 8.1 The closure of gyms and other outdoor activities due to lockdown affected your physique. 155

Figure 8.2   Online mode of learning put an adverse effect on
                   your physical health.   156
Figure 8.3   Monotonous daily routine, causing depression and
                   anxiety in the lives of the students.   157
Figure 8.4   Gender/lockdown has proven to be beneficial to
                   maintaining your physical health (male vs. female).   158
Figure 8.5   Currently pursuing/completely satisfied with the
                   delivery of information and clearance of doubts
                   (school students vs. UG students vs. PG students). .   159
Figure 8.6   Currently pursuing/facing a lack of confidence due
                   to the virtual learning platform (school students vs.
                   UG students vs. PG students).   159
Figure 8.7   Online mode of learning put an adverse effect on
                   your physical health.   160
Figure 8.8   Screen time increment which is causing mental
                   stress.   161
Figure 9.1   Diagram of misinformation spread model.   175
Figure 9.2   Basic reproduction number for β.   179
Figure 9.3   Basic reproduction number for r.   180
Figure 9.4   Basic reproduction number for γ.   180
Figure 9.5   State trajectories when spread leads to violence. . .   183
Figure 9.6   State trajectories when spread does not lead to
                   violence.   183
Figure 9.7   Different state trajectories for variation of α.   184
Figure 9.8   Different state trajectories for variation of l.   185
Figure 9.9   Different state trajectories for variation of γ.   186
Figure 9.10   Different state trajectories for variation of b.   187
Figure 9.11   Different state trajectories for variation of b.   188
Figure 9.12   Different state trajectories for variation of β.   189
Figure 9.13   Changes of state trajectories with time for different
                   parameters.   190
Figure 9.14   Changes of state trajectories with time for different
                   parameters.   191
Figure 9.15   Changes of state trajectories with time for $\alpha = 0 \rightarrow$
                   0.2.   192
Figure 9.16   Changes of state trajectories with time for $\mu = 0 \rightarrow$
                   0.5.   193
Figure 10.1 $a = 0.02, b = 0.85, c = 0.5, r_n = 0.2$.   209
Figure 10.2 $a = 0.02, b = 0.89, c = 0.8, r_n = 0.1890$.   210

Figure 10.3 $a = 0.02, b = 0.89, c = 0.8, r_n = 0.1890.$ 210
Figure 10.4 $a = 0.8, b = 0.79, r_n = 0.00049, c = 0.07.$ 210
Figure 10.5 $a = 0.8, b = 0.79, r_n = 0.00049, c = 0.0005.$ 211
Figure 10.6 $a = 0.02, b = 0.89, c = 0.8, r_n = 0.$ 211
Figure 11.1   Graphical representation of the flow of youths among the compartments in eqn (11.2.1).   219
Figure 11.2   Numerical solution presenting the dynamics for (a) learners, (b) employed, (c) SMEs, (d) tertiary, (e) corporate and (f) unemployed youths under the impact of COVID-19.   234
Figure 11.3   Numerical solution presenting the dynamics for (a) learners, (b) employed, (c) SMEs, (d) tertiary, (e) corporate, and (f) unemployed youths, with no COVID-19.   235
Figure 12.1   The flow diagram of the proposed model.   246
Figure 12.2   Variation in S, V, I_0 and I_δ over a period of time in days.   259
Figure 12.3   Changes in I_δ for different values of ϕ over a period of time in days.   260
Figure 12.4   Changes in I_0 for different values of ϕ over a period of time in days.   260
Figure 12.5   Variation in I_δ over V   261
Figure 12.6   Variations in V with respect to I_0.   261
Figure 12.7   Quiver of I_δ, I_0 with respect to R.   262
Figure 13.1   Top 10 hotspot countries of COVID-19 cases   . . .   271
Figure 13.2   Active, Recovered and Deaths in Hotspot Countries as of 4 June 2022.   272
Figure 13.3   LSTM block.   276
Figure 13.4   Train and test RMSE with R^2.   280
Figure 13.5   Prediction for next day, blue is the true number, orange is the train number and green is the forecast number. (a) China, (b) Germany, (c) Iran, (d) Italy, (e) Spain, (f) United States, (g) France, (h) Turkey, (i) United Kingdom and (j) India.   283
Figure 14.1   Observed number of confirmed, active, recovered and death cases in India.   292
Figure 14.2   Relative frequency histogram plot of COVID-19 confirmed cases, active cases, recovered cases and death cases at all India level.   292

Figure 14.3   Scatter plot between the confirmed, active, recovered and death cases to observe the relationship between the variables.   293

Figure 14.4   The autocorrelation function plot for all India regions to understand the intra-relationship within the variables.   294

Figure 14.5   Active cases of all India from 11 January to 1 June have been compared between observed (blue line) and model fit (red line).   296

Figure 14.6   Model comparison with the observed data using ANN model in the training period.   297

List of Tables

Table 3.1   A description of the model's input parameters.   56

Table 4.1   Incineration emissions range.   72

Table 4.2   Methods used for red bag disposal.   74

Table 4.3   Colour coding and type of containers for BMW. . . .   75

Table 5.1   Parametric values with description used to model
equations.   86

Table 6.1   Notation with parametric values.   112

Table 7.1   Notation and its description.   141

Table 8.1   Interpretation of data.   162

Table 8.2   Interpretation of data.   163

Table 9.1   Parameter values.   182

Table 11.1 Parameter values for equation in (11.4.20) for
Figure 11.2.   236

Table 11.2 Parameter values in equation (11.4.20) for
Figure 11.3.   236

Table 12.1 Compartments used in the proposed model.   245

Table 12.2 Values and meaning of parameters used in the proposed model.   247

Table 14.1 The descriptive statistics of all the four cases, such as
death, recovered, active and confirmed at India level. .   291

Table 14.2 The unknowns of the regression (1) for all India. . . .   295

Table 14.3 The parameters depicting the model performance with
the observed data for the training period from 11
January 2022 to 1 June 2022.   295

Table 14.4 Model comparison with the observed data statistics
during the training period.   297

Table 14.5 Day wise comparison of the testing of a model for the
period of 30 days (1 June to 30 June 2022) using AR
and ANN methods with observed data and forecast for
the month of July.   299

Table 14.6 Model performance during the testing period of 30 days (from 1 June to 30 June 2022) using the parameters, such as RMSE, CC and PP, between the observed data and the forecast. 300

List of Contributors

Ali Biswas, M. Haider, *Mathematics Discipline, Science Engineering and Technology School, Khulna University, India*

Bansal, Devyanshi, *Amity School of Engineering and Technology, Amity University, Uttar Pradesh, India*

Bhardwaj, Ramakant, *Department of Mathematics, Amity University, India*

Bhatia, Sumit Kaur, *Amity Institute of Applied Sciences, Amity University, Uttar Pradesh, India*

Britto Jacob, S., *Department of Mathematics, Kristu Jayanti College (Autonomous), India*

Chaudhary, Kapil, *Department of Mathematics, Gujarat University, India*

Chaudhary, Riju, *Amity Institute of Applied Sciences, Amity University, Uttar Pradesh, India*

Chauhan, Sudipa, *Amity Institute of Applied Science, Amity University, India*

Chetri, Kopila, *Amity Institute of Applied Science, Amity University, India*

Dhineshbabu, R., *Department of Mathematics, Sacred Heart College (Autonomous), India, Department of Mathematics, Faculty of Engineering, Karpagam Academy of Higher Education, India, and Department of Mathematics, Sri Venkateswara College of Engineering and Technology (Autonomous), India*

Gupta, Radha, *Department of Mathematics, Dayananda Sagar College of Engineering, India*

Gupta, Rajeev Kumar, *Pandit Deendayal Energy University, India*

Himanshi, *Amity Institute of Applied Science, Amity University, India*

Janagaraj, R., *Department of Mathematics, Sacred Heart College (Autonomous), India, Department of Mathematics, Faculty of Engineering, Karpagam Academy of Higher Education, India, and Department of Mathematics, Sri Venkateswara College of Engineering and Technology (Autonomous), India*

Jayswal, Ekta N., *Department of Mathematics, Gujarat University, India*

Jayswal, Ekta, *Department of Mathematics, Gujarat University, India*

Karwal, Siddhant, *Amity Institute of Science and Technology, Amity University, India*

Mishra, Amit Kumar, *Amity School of Engineering and Technology, Amity University, India*

Narayan, Satyendra, *Department of Applied Computing, Sheridan Institute of Technology and Advanced Learning in Oakville, Canada*

Nethravathi, N., *Department of Mathematics, Dayananda Sagar College of Engineering, India*

Pandya, Purvi M., *Department of Mathematics, Gujarat University, India*

Paul, Arindam Kumar, *Mathematics Discipline, Science Engineering and Technology School, Khulna University, India*

Ramesh, Kokila, *Department of Mathematics, FET, Jain (Deemed-to-be-University), India*

Rawlley, Oshin, *Vellore Institute of Technology, India*

Sahu, Yatendra, *Indian Institute of Information Technology and Management, India*

Satia, Moksha H., *Department of Mathematics, Aditya Silver Oak Institute of Technology, Silver Oak University, India*

Selvam, A. George Maria, *Department of Mathematics, Sacred Heart College (Autonomous), India; Department of Mathematics, Faculty of Engineering, Karpagam Academy of Higher Education, India, and; Department of Mathematics, Sri Venkateswara College of Engineering and Technology (Autonomous), India*

Shah, Nita H., *Department of Mathematics, Gujarat University, India*

Sheoran, Nisha, *Department of Mathematics, Gujarat University, India*

Shikongo, Albert, *Mechanical and Metallurgical Engineering, School of Engineering and the Built Environment, JEDS Campus, University of Namibia, Namibia*

Suthar, Ankush H., Department of Mathematics, Gujarat University, India

Vignesh, D., Department of Mathematics, School of Advanced Sciences, Kalasalingam, Academy of Research and Education, Krishnankoil, Srivilliputhur, India

List of Abbreviations

ABC	Atangana–Baleanu–Caputo
ABF	Atangana–Baleanu fractional
AI	Artificial intelligence
ANN	Artificial neural network
ANOVA	Analysis of variance
AQI	Air quality index
ARIMA	Autoregressive integrated moving average
BMW	Biomedical waste
CNN	Convolutions neural networks
COVID	Coronavirus disease
GHQ	Global health questionnaire
HCW	Health care workers
ICU	Intensive care unit
IVPs	Initial valued problems
LSTM	Long short-term memory
MERS	Middle east respiratory syndrome
MLF	Mittag–Leffler function
NSFD	Non-standard finite difference
ODEs	Ordinary differential equations
PCL-C	PTSD Checklist-civilian version
PTSD	Post-traumatic stress disorder
RMSE	Root mean square error
RNN	Recurrent neural network
RP	Reverse polymerization
SDG	Sustainable development goals
SEIR	Susceptible exposed infectious recovered
SEIRD	Susceptible exposed infected recovered deceased
SME	Small- and mid-size enterprises
TDP	Thermal depolymerization
VOC	Variant of concern
WHO	World health organization

1

Mathematical Modelling of COVID-19 Dynamics with Reinfection Using Fractional Order

A. George Maria Selvam[1], D. Vignesh[2], and S. Britto Jacob[3]

[1]Department of Mathematics, Sacred Heart College (Autonomous), India
[2]Department of Mathematics, School of Advanced Sciences, Kalasalingam Academy of Research and Education, India
[3]Department of Mathematics, Kristu Jayanti College (Autonomous), India
E-mail: agmshc@gmail.com; dvignesh260@gmail.com; brittojacob21@gmail.com

Abstract

Mathematical modelling in epidemiology is used to analyse the dynamical behaviour and spread of diseases caused by different micro-organisms. Also, it provides the understanding of the underlying mechanisms that influence the spread of disease and help to predict the future situation and even control the pandemic. Epidemic diseases like influenza, plague, cholera, MERS, measles, and Ebola have been affecting individuals as well as human society in many ways. This chapter aims at formulating a fractional-order epidemic model taking into account the impact of quarantine and reinfection to examine the dynamics of the outbreak of COVID-19. The non-local behaviour and hysteresis effect of the fractional-order models is quite significant in understanding the epidemics of disease. The model is initially constructed with integer order differential equations which are then converted to its fractional-order form. The discrete version of the corresponding system is then obtained by employing discretization process. The value estimated for the basic reproduction number is evaluated analytically. Stability analysis and optimal control analysis are performed for the considered SEIQR model.

The dynamic changes that occur in the system due to some parameters are identified and their impact are presented graphically for suitable values. The chaotic behaviour exhibited by the system is studied with bifurcation diagrams and the impact of varying step size on the dynamics of the different population groups is demonstrated with time-varying plots.

Keywords: Fractional Order, Stability, COVID-19 Model, Bifurcation.

MSC Subject Classification 2010: 26A33, 35B35, 44A10, 92D30.

1.1 Introduction

Mathematical biology is a field of mathematics that provides insight into the understanding of the interactions among the biological populations, disease spread among and within the species, physical and chemical systems and so on. The mechanism influencing the spread of any disease could be mathematically described in the form of system of equations. Analysing the system of equations thus developed could provide a wide range of properties and behavioural aspects of the biological systems. The epidemiology in partnership with mathematics has been of greater importance as it provides biologists a new path for experimentation [5, 6, 16, 23, 28]. The study of infections in humans due to viruses and spread among the human population has been studied for a long time. The virus once entering a host body replicates and then is released to attack the target cells.

Some of the most common infections that are induced to human by viruses include cold, chicken pox, influenza, cold sores, AIDS and most recently COVID-19 caused by corona-virus. Not all the diseases that are caused by viruses are fatal and around 21 different disease-causing virus families were identified. Some very serious infections caused by certain family of viruses are influenza, herpes simplex, SARS, hepatitis, measles, SARS and so on. The number of human deaths due to respiratory diseases mostly due to viruses in 1997 is about 3.5 million which has increased to 4.4 million deaths in 2007. Most of these deaths are observed in developing and under-developing countries.

Generalization of classical differentiation and integration with arbitrary real order has attracted scientists and researchers of various fields for construction of model due to its rate of accuracy and memory factor. In terms of ecosystem and biological system modelling the hysteresis effect also known as the memory effect is vital in bringing out the past memory to analyse their equivalent future state. Classical and arbitrary-order calculus have originated during the same period of time. Lack of geometrical and physical interpretation of arbitrary order calculus has led to development of the field.

The importance of hysteresis effect on modelling of epidemics can be understand from the fact that past experiences on spread of particular disease can alert the human immune system to act on similar situation that occurs in future state. Vaccines are regarded as best examples for long-term memory effect helping human bodies to stimulate immune memory against diseases. Thus, fractional-order systems receive additional attraction than the classical calculus models. Some recent literature on analysis of fractional-order models include [11, 17, 19, 24, 25, 29].

Here in this chapter, we attempt to illustrate the hysteresis effect of fractional-order COVID-19 model. The additional advantage of employing fractional calculus is possibility of using different orders to match the real-time behaviour of the spread.

1.2 SEIQR Model for COVID-19 Spread

The single-stranded enveloped RNA virus called corona-viruse is not new to biologist and medical researchers. Existence of this type of viruses were discovered in 1965 and their distribution is widely observed in birds and mammals. The corona-viruse causes respiratory infections that are mild and some viruses like SARS in China and MERS in South Korea and Saudi Arabia are very fatal.

A cluster of patients with pneumonia were reported in the city of Wuhan, Hubei Province, China in December 2019. Later, the reason behind the cause was confirmed due to a new type of coronavirus (2019-nCoV). The transmission rate is very high for the is particular type of virus among humans. The outbreak was later declared as a pandemic by WHO on 30 January 2020. The authors in [1, 3, 4, 7–10, 12–15, 18, 21, 22, 26, 27, 30] discussed the dynamics of COVID-19.

The primary source of spread are saliva droplets, discharges from nose while coughing or sneezing. The Epidemic model representing the COVID-19 spread is developed as SEIQR model with consideration of possible reinfection of the recovered individuals. The equilibrium state of the model under infection free and positive equilibrium are discussed with identification of the basic reproduction number and further study the global stability of the model.

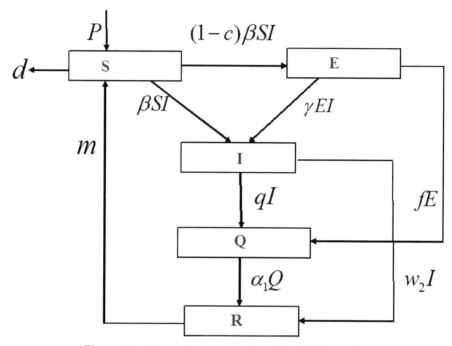

Figure 1.1   Schematic representation of the SEIQR model.

Susceptible Population $(S(\zeta))$ The susceptible population is fraction of individuals who are able to contract disease. A constant growth of susceptible populations are considered with the introduction of P. The decrease in the population at the rate β occurs when a susceptible individual interacts with the infected population and the fraction of individuals at a rate d who are subjected to develop immunity against COVID-19 are considered as a protected group of people and are removed from the population. The population that recovered from infection are subjected to get infected again which tends to increase the exposed and infected population:

$$S'(\zeta) = P + mR(\zeta) - \beta S(\zeta)I(\zeta) - dS(\zeta).$$

Exposed Population $(E(\zeta))$ The compartment considers the fraction of population who are around the infected people but not yet infectious. The population increases at rate $'1 - c'$ by interaction between the susceptible and infected population. The interaction with infectious individuals decreases the population at the rate γ. The population of the compartment also decreases when a fraction f of population who are quarantined.

$$E'(\zeta) = (1-c)\beta S(\zeta)I(\zeta) - fE((\zeta) - \gamma E(\zeta)I(\zeta).$$

Infected Population $(I(\zeta))$ The population who test positive for the infection by corona-virus belong to this compartment. The increase in the infections is from fraction (c) of susceptible population and exposed population who test positive at the rate γ. Part of infected individuals tends to develop immunity against the virus and individuals after suitable treatments recover at the rate w_2. A group of individuals on infection whose recovery time is larger than expected are quarantined at rate q.

$$I'(\zeta) = c\beta S(\zeta)I(\zeta) + \gamma E(\zeta)I(\zeta) - qI(\zeta) - w_2 I(\zeta).$$

Quarantined Population $(Q(\zeta))$ The group of individuals who tested positive for COVID-19 act as a carrier and spread the disease to the person with whom they come in contact. In order to control the increasing infection, the people who are exposed at a rate f and infected individuals at the rate q are quarantined. The compartment deals with population after certain period getting recovered at rate α_1:

$$Q'(\zeta) = qI(\zeta) - \alpha_1 Q(\zeta) + fE(\zeta).$$

Recovered Population $(R(\zeta))$ The population that are infected are assumed to recover at the rate w_2 and α_1 belong to this compartment and are also assumed that the fading away of immunity in these individuals at the rate of m increases the probability of getting infected again resulting in decreasing population:

$$R'(\zeta) = \alpha_1 Q(\zeta) + w_2 I(\zeta) - mR(\zeta).$$

Hence the system of SEIQR COVID model is

$$\frac{dS(\zeta)}{d\zeta} = P - \beta S(\zeta)I(\zeta) + mR(\zeta) - dS(\zeta),$$

$$\frac{dE(\zeta)}{d\zeta} = (1-c)\beta S(\zeta)I(\zeta) - \alpha E(\zeta) - \delta E(\zeta)I(\zeta),$$

$$\frac{dI(\zeta)}{d\zeta} = c\beta S(\zeta)I(\zeta) + \delta E(\zeta)I(\zeta) - qI(\zeta) - w_2 I(\zeta),$$

$$\frac{dQ(\zeta)}{d\zeta} = qI(\zeta) - \alpha_1 Q(\zeta) + \alpha E(\zeta),$$

$$\frac{dR(\zeta)}{d\zeta} = \alpha_1 Q(\zeta) + w_2 I(\zeta) - mR(\zeta). \tag{1.1}$$

Theorem 1. Positivity: If the initial conditions $S(0) > 0, E(0) > 0, I(0) > 0, Q(0) > 0, R(0) > 0$ and $\zeta_0 > 0$, then for all $\zeta \in [0, \zeta_0]$. $S(\zeta), E(\zeta), I(\zeta), Q(\zeta), R(\zeta)$ remain positive in R_+^5.

Proof: It is assumed that all the parameters used in the model are positive and hence we can set a lower bound for each of the equations in (1.1) as follows: $\dot{S} > -(\beta I + d)S, \dot{E} > -(\alpha + \delta I)E, \dot{I} > (c\beta S + \delta E - q - w_2)I, \dot{Q} > -\alpha_1 Q, \dot{R} > -mR$.

Solving the above inequalities respectively leads to

$$S(\zeta) > S(0) \exp\left(-d\zeta - \int (\beta I)\zeta d\zeta\right) > 0,$$

$$E(\zeta) > E(0) \exp\left(-\alpha\zeta + \int (\delta I)\zeta d\zeta\right) > 0,$$

$$I(\zeta) > I(0) \exp\left(c \int (\beta S)\zeta d\zeta + \int (\delta E)\zeta d\zeta\right) > 0,$$

$$Q(\zeta) > Q(0) \exp\left(-\alpha_1 \zeta\right) > 0,$$

$$R(\zeta) > R(0) \exp\left(-m\zeta\right) > 0. \tag{1.2}$$

Thus for all $\zeta \in [0, \zeta_0], S(\zeta), E(\zeta), I(\zeta), Q(\zeta)$ and $R(\zeta)$ are positive in R_+^5.

Theorem 2. Boundedness: For the functions $S(\zeta), E(\zeta), I(\zeta), Q(\zeta)$ and $R(\zeta)$ of Eqn (1.1), there exist positive constants $S_\mathcal{M}, E_\mathcal{M}, I_\mathcal{M}, Q_\mathcal{M}, R_\mathcal{M}$ such that

$$\lim_{\zeta \to \infty} \sup S(\zeta) \leq S_\mathcal{M},$$

$$\lim_{\zeta \to \infty} \sup E(\zeta) \leq E_\mathcal{M},$$

$$\lim_{\zeta \to \infty} \sup I(\zeta) \leq I_\mathcal{M}, \tag{1.3}$$

$$\lim_{\zeta \to \infty} \sup Q(\zeta) \leq Q_\mathcal{M},$$

$$\lim_{\zeta \to \infty} \sup R(\zeta) \leq R_\mathcal{M},$$

for all $\zeta \in [0, \zeta_0], \zeta_0 > 0$.

Proof: To show boundedness, we add all the equations in the model (1.1) and obtain $\frac{dN}{d\zeta} \leq P - dS$. It then follows that $\frac{dN}{d\zeta} \leq 0$ if $S \geq \frac{P}{d}$.

Thus, solving $\frac{dN}{d\zeta} \leq P - dS$ by applying Gronwall's Inequality leads to

$$N(\zeta) \leq N(0) \exp \left(\int (P)\zeta d\zeta - \int (dS)\zeta d\zeta \right).$$

In particular,

$$N(\zeta) \leq \lim_{t \to \infty} \sup N(0) \exp \left[\int (P)\zeta d\zeta - \int (dS)\zeta d\zeta \right] = \frac{P}{d}$$

if $N(0) \leq \frac{P}{d}$.

Now, choosing $S_M = E_M = I_M = Q_M = R_M = \frac{P}{d}, \zeta_0 > 0$ it can be concluded that S, E, I, Q, R are all bounded. Since

$$S(\zeta), E(\zeta), I(\zeta), Q(\zeta), R(\zeta) \leq N(\zeta) \leq \frac{P}{d}.$$

Thus the region

$$\Omega = \left\{ (S, E, I, Q, R) \in R_+^5 : S(\zeta) + E(\zeta) + I(\zeta) + Q(\zeta) + R(\zeta) \leq \frac{P}{d} \right\},$$
$$(1.4)$$

is positively invariant region for model (1.1).

Moreover, if $N(0) \geq \frac{P}{d}$, then either the solution of Eqn (1.1) enters Ω in a finite time or $N(\zeta)$ approaches $\frac{P}{d}$ asymptotically. Hence, the region Ω attracts all solution of Eqn (1.1) in R_+^5.

Theorem 3. Existence and Uniqueness: In model (1.1), if the initial conditions $S(0) > 0, E(0) > 0, I(0) > 0, Q(0) > 0, R(0) > 0$ and $\zeta_0 > 0$, then for all $\zeta \in R$ the solutions $S(\zeta), E(\zeta), I(\zeta), Q(\zeta), R(\zeta)$ exist in R_+^5.

Proof: Model (1.1) can be expressed in the form $\dot{x} = f(x)$,

$$\text{where } \dot{x} = \begin{pmatrix} S(\zeta) \\ E(\zeta) \\ I(\zeta) \\ Q(\zeta) \\ R(\zeta) \end{pmatrix}, f(x) = \begin{pmatrix} P - \beta SI + mR - dS \\ (1-c)\beta SI - \alpha E - \delta EI \\ c\beta SI + \delta EI - qI - w_2 I \\ qI - \alpha_1 Q + \alpha E \\ \alpha_1 Q + w_2 I - mR \end{pmatrix}.$$

Since f has a continuous first derivative in R_+^5, it is then locally Lipschitz. As a result, by the well known fundamental existence and uniqueness theorem [20], and Eqn (1.2), (1.3), there exists a unique, positive and bounded solution for the system of differential equation (1.1) in R_+^5.

1.3 Fractional-Order SEIQR Model

Fractional order is used widely in mathematical modelling due to its memory effect. The fractional analogue of model (1.1) is

$$D^\alpha S(\zeta) = P - \beta S(\zeta)I(\zeta) + mR(\zeta) - dS(\zeta),$$

$$D^\alpha E(\zeta) = (1 - c)\beta S(\zeta)I(\zeta) - \alpha E(\zeta) - \delta E(\zeta)I(\zeta),$$

$$D^\alpha I(\zeta) = c\beta S(\zeta)I(\zeta) + \delta E(\zeta)I(\zeta) - qI(\zeta) - w_2 I(\zeta), \qquad (1.5)$$

$$D^\alpha Q(\zeta) = qI(\zeta) - \alpha_1 Q(\zeta) + \alpha E(\zeta),$$

$$D^\alpha R(\zeta) = \alpha_1 Q(\zeta) + w_2 I(\zeta) - mR(\zeta).$$

Here $\zeta > 0$ and α is the fractional order satisfying $\alpha \in (0; 1]$. Now, applying the discretization process of a fractional order system outlined in [2], we obtain the discrete fractional-order system as follows:

$$S(\zeta_{j+1}) = S(\zeta_j) + \frac{h^\alpha}{\Gamma(\alpha + 1)} \left(P - \beta S(\zeta)I(\zeta) + mR(\zeta) - dS(\zeta) \right),$$

$$E(\zeta_{j+1}) = E(\zeta_j) + \frac{h^\alpha}{\Gamma(\alpha + 1)} \left((1 - c)\beta S(\zeta)I(\zeta) - \alpha E(\zeta) - \delta E(\zeta)I(\zeta) \right),$$

$$I(\zeta_{j+1}) = I(\zeta_j) + \frac{h^\alpha}{\Gamma(\alpha + 1)} \left(c\beta S(\zeta)I(\zeta) + \delta E(\zeta)I(\zeta) - qI(\zeta) - w_2 I(\zeta) \right),$$

$$Q(\zeta_{j+1}) = Q(\zeta_j) + \frac{h^\alpha}{\Gamma(\alpha + 1)} \left(qI(\zeta) - \alpha_1 Q(\zeta) + \alpha E(\zeta) \right),$$

$$R(\zeta_{j+1}) = R(\zeta_j) + \frac{h^\alpha}{\Gamma(\alpha + 1)} \left(\alpha_1 Q(\zeta) + w_2 I(\zeta) - mR(\zeta) \right).$$

$$(1.6)$$

1.4 Existence of Equilibrium Points

The system of eqn (1.6) has a disease-free equilibrium point

$$\left(S, E, I, Q, R = \frac{P}{d}, 0, 0, 0, 0 \right).$$

In disease-free situation where there is no infection $I = 0$.

Also the endemic equilibrium (EE) point is $(S^*, E^*, I^*, Q^*, R^*)$

$$S^* = \frac{P}{d},$$

$$E^* = \frac{dq - dw_2 - c\beta P}{\gamma d},$$

$$I^* = \frac{\alpha(dq + dw_2 - c\beta P)}{\gamma(P\beta - dq - dw_2)},$$

$$Q^* = \frac{\alpha(dq + dw_2 - c\beta P)(P\beta - dw_2)}{d\gamma(P\beta - dq - dw_2)\alpha_1},$$

$$R^* = \frac{P\beta\alpha[dq + dw_2 - c\beta P]}{d\gamma(p\beta - dq - dw_2)m}.$$

The transmission rate of the diseases is measured with evaluation of the value of R_0. In other words, number of secondary individuals getting infected from an infected case in a completely susceptible population.

The factors that influence the basic reproduction number include

(1) Contact rate in the considered population.
(2) Transmission rate during contact.
(3) time period of infection.

Depending on the value of R_0, the spread of disease can be classified. For R_0 being greater than 1, the infection rate is high and represents the epidemic in the susceptible population. Decrease in infection is represented by $R_0 < 1$.

R_0 can be determined by next generation matrix (NGM) approach. Hence the basic reproduction number is

$$R_0 = \frac{c\beta P}{d(q + w_2)}.$$

The Jacobian matrix for the corresponding system (1.6) is

$$J(S^*, E^*, I^*, Q^*, R^*) =$$

$$\begin{bmatrix} 1 - \Lambda(\beta I + d) & 0 & \Lambda(-\beta S^*) & 0 & m \\ \Lambda((1 - c)\beta I^*) & 1 - \Lambda(f - \gamma I^*) & \Lambda(1 - c)\beta S^* - \gamma E^* & 0 & 0 \\ \Lambda(c\beta I^*) & \Lambda(\gamma I^*) & 1 - \Lambda(q + w_2 - c\beta S^* - \gamma E^*) & 0 & 0 \\ 0 & \Lambda(f) & \Lambda(q) & 1 - \Lambda(\alpha_1) & 0 \\ 0 & 0 & \Lambda(w_2) & \Lambda(\alpha_1) & 1 - \Lambda(m) \end{bmatrix},$$

$$(1.7)$$

where $\Lambda = \frac{h^\mu}{\Gamma(\mu+1)}$.

1.5 Stability of the System

In this section, we present the stability results of the system (1.6) for disease-free equilibrium (DFE) point and endemic equilibrium(EE) point.

Proposition 4. The DFE point $N_0 = \left(\frac{P}{d}, 0, 0, 0, 0\right)$ is locally asymptotically stable (LAS) if $R_0 < 1$ and unstable if $R_0 > 1$.

Proof:

The Jacobian matrix for eqn (1.6) at DFE is

$$J(S^*, E^*, I^*, Q^*, R^*) =$$

$$\begin{bmatrix} 1 - \Lambda(d) & 0 & -\Lambda\left(\frac{\beta P}{d}\right) & 0 & \Lambda(m) \\ 0 & 1 - \Lambda(\alpha) & \Lambda\left((1-c)\frac{\beta P}{d}\right) & 0 & 0 \\ 0 & 0 & 1 - \Lambda(q + w_2 - \frac{\beta P c}{d}) & 0 & 0 \\ 0 & \Lambda(f) & \Lambda(q) & 1 - \Lambda(\alpha_1) & 0 \\ 0 & 0 & \Lambda(w_2) & \Lambda(\alpha_1) & 1 - \Lambda(m) \end{bmatrix},$$

$$(1.8)$$

The eigenvalues are $\lambda_1 = 1 - \Lambda(d)$, $\lambda_2 = 1 - \Lambda(\alpha)$, $\lambda_3 = 1 - \Lambda(q + w_2 - \frac{\beta P c}{d})$, $\lambda_4 = 1 - \Lambda(\alpha_1)$, $\lambda_5 = 1 - \Lambda(m)$.

Hence the DFE is locally asymptotically stable whenever $R_0 < 1$.

Proposition 5. The EE point $N^* = (S^*, E^*, I^*, Q^*, R^*)$ is locally asymptotically stable iff $R_0 > 1$.

Proof: The Jacobian matrix for eqn (1.6) at EE is

$$J(S^*, E^*, I^*, Q^*, R^*) =$$

$$\begin{bmatrix} 1 + \Lambda\left(\frac{\beta f w}{\gamma} - d\right) & 0 & -\Lambda\left(\frac{\beta P}{d}\right) & 0 & \Lambda(m) \\ -\Lambda\left(\frac{(1-c)\beta f w}{\gamma}\right) & 1 + \Lambda(wf + f) & \Lambda\left(\frac{(1-c)\beta P}{d} + \frac{P\beta c - dq - dw_2}{d}\right) & 0 & 0 \\ -\Lambda\left(\frac{f w c \beta}{\gamma}\right) & -\Lambda(fw) & 1 + \Lambda\left(\frac{c\beta P}{d} - \frac{P\beta c - dq - dw_2}{d} - w_2 - q\right) & 0 & 0 \\ 0 & \Lambda(f) & \Lambda(q) & 1 - \Lambda(\alpha_1) & 0 \\ 0 & 0 & \Lambda(w_2) & \Lambda(\alpha_1) & 1 - \Lambda(m) \end{bmatrix},$$

$$(1.9)$$

The eigenvalues are $\lambda_1 = 1 + \Lambda\left(\frac{\beta f w}{\gamma} - d\right)$, $\lambda_2 = 1 + \Lambda(wf + f)$, $\lambda_3 = 1 + \Lambda\left(\frac{c\beta P}{d} - \frac{P\beta c - dq - dw_2}{d} - w_2 - q\right)$, $\lambda_4 = 1 - \Lambda(\alpha_1)$, $\lambda_5 = 1 - \Lambda(m)$.

Hence the EE point is locally asymptotically stable whenever $R_0 > 1$ and $|\lambda_i| < 1$, $(i = 1, 2, 3, 4, 5)$.

1.6 Optimal Control Analysis

The important aspect of discussing an epidemiological model is to control the infection among the population and provide suitable control strategies that could be implemented among the populations to prevent further spread. Here, in this section, we would like to consider strategies to control the COVID-19 spread with creating awareness among the people by educating them about the pandemic denoted by n_1, basic preventive measures like using sanitizers, wearing face mask, maintaining a reasonable social distance between individuals and so on, that are to be taken by the population with n_2 and n_3 to represent the medical treatments and isolation carried out by public welfare departments and hospitals. Incorporation of the discussed optimal control strategies in eqn (1.1), we have the following system of equation as

$$\begin{aligned}
\dot{S} &= P - \beta SI + mR - dS, \\
\dot{E} &= (1-c)\beta SI - (\alpha + n_1)E - \delta EI, \\
\dot{I} &= c\beta SI + \delta EI - (q + w_2 + n_2)I, \\
\dot{Q} &= (q + n_2)I - (\alpha_1 + n_3)Q + (\alpha + n_1)E, \\
\dot{R} &= (\alpha_1 + n_3)Q + (w_2 + n_2)I - mR.
\end{aligned} \tag{1.10}$$

Now, the function that minimizes the number of exposed cases E, number of infected cases I, and number of quarantined cases Q over a time interval of $[0, T]$ can be defined as

$$\mathcal{J}_{op}(N_i, \Omega) = \int_0^T [\nu_0 E(\varsigma) + \nu_1 I(\varsigma) + \nu_2 Q(\varsigma)] + \left(\frac{\mu_1}{2} n_1^2 + \frac{\mu_2}{2} n_2^2 + \frac{\mu_3}{2} n_3^2 \right) d\varsigma, \tag{1.11}$$

where biologically feasible region is denoted by Ω as in Section 1.2. The balance in factors are ensured by the positive weights ν_0, ν_1, ν_2 and cost and other measures that involves on implementing the strategies are denoted by μ_1, μ_2, μ_3 over $[0, T]$.

Minimizing equation (1.11) provides an optimal control $n_i^*, i = 1, 2, 3$ such that

$$\mathcal{J}_{op}(n_1^*, n_2^*, n_3^*) = \min_{n \in N} \mathcal{J}_{op}(n_1(\varsigma), n_2(\varsigma), n_3(\varsigma)), \tag{1.12}$$

where the control set is given by $N = \{n_i(\varsigma) : 0 \le n_i(\varsigma) \le 1, 0 \le \varsigma \le T, i = 1, 2, 3\}$ subjected to the constraint given by system of differential equation (1.10).

The necessary conditions that need to be satisfied by optimal control called Pontryagin's maximum principle converts eqn (1.10), into a problem of minimizing point-wise a Hamiltonian Hs with respect to $n_i(\zeta)$

$$
\begin{aligned}
Hs\,(y, n_1(\zeta), n_2(\zeta), n_3(\zeta), \lambda_j, \zeta) = &\ (\nu_0 E(\zeta) + \nu_1 I(\zeta) + \nu_2 Q(\zeta)) \\
&+ \left(\frac{\mu_1}{2} n_1^2 + \frac{\mu_2}{2} n_2^2 + \frac{\mu_3}{2} n_3^2\right)(\zeta) \\
&+ \lambda_S h_1(\zeta) + \lambda_E h_2(\zeta) + \lambda_I h_3(\zeta) \\
&+ \lambda_Q h_4(\zeta) + \lambda_R h_5(\zeta), \qquad (1.13)
\end{aligned}
$$

where

$$
h_1 = P - \beta SI + mR - dS,
$$

$$
h_2 = (1 - c)\beta SI - (\alpha + n_1)E - \delta EI,
$$

$$
h_3 = c\beta SI + \delta EI - (q + w_2 + n_2)I,
$$

$$
h_4 = (q + n_2)I - (\alpha_1 + n_3)Q + (\alpha + n_1)E,
$$

$$
h_5 = (\alpha_1 + n_3)Q + (w_2 + n_2)I - mR.
$$

Differentiating the Hamiltonian function with respect to the compartment variables gives the adjoint variables $\lambda_j, j \in \{S, E, I, Q, R\}$ corresponding to the system given as follows

$$
\dot{\lambda}_S = \lambda_S(-d - \beta I) - \lambda_E(1 - c)\beta I,
$$

$$
\dot{\lambda}_E = \lambda_E(\alpha + n_1 - \delta I) - \nu_0 + \lambda_I(\delta E) + \lambda_Q(\alpha + n_1),
$$

$$
\dot{\lambda}_I = \lambda_I(q + w_2 + n_2 + c\beta S + \delta E) - \nu_1 - \lambda_Q(q + n_2) - \lambda_R(w_2 + n_2),
$$

$$
\dot{\lambda}_Q = \lambda_Q(\alpha_1 + n_3) - \nu_2 - \lambda_R(\alpha_1 + n_3),
$$

$$
\dot{\lambda}_R = \lambda_R m,
$$

and $\lambda_S, \lambda_E, \lambda_I, \lambda_Q, \lambda_R$ are the adjoint variables, $\lambda_j = (\lambda_S, \lambda_E, \lambda_I, \lambda_Q, \lambda_R)$, $y = (S, E, I, Q, R)$.

Now, setting the transversality condition

$$
\lambda_j(\zeta) = 0, j \in \{\lambda_S, \lambda_E, \lambda_I, \lambda_Q, \lambda_R\}. \qquad (1.14)
$$

We obtain the optimal controls and the optimality conditions, respectively, as

$$n_1(\zeta) = \left(\frac{\lambda_E + \lambda_Q}{\mu_1}\right) E,$$

$$n_2(\zeta) = \left(\frac{\lambda_I - \lambda_Q - \lambda_R}{\mu_2}\right) I,$$

$$n_3(\zeta) = \left(\frac{\lambda_Q - \lambda_R}{\mu_3}\right) Q. \tag{1.15}$$

and

$$n_1^*(\zeta) = \min\left[\max\left(0, \frac{\lambda_E + \lambda_Q}{\mu_1} E\right), 1\right],$$

$$n_2^*(\zeta) = \min\left[\max\left(0, \frac{\lambda_I - \lambda_Q - \lambda_R}{\mu_2} I\right), 1\right],$$

$$n_3^*(\zeta) = \min\left[\max\left(0, \frac{\lambda_Q - \lambda_R}{\mu_3} Q\right), 1\right]. \tag{1.16}$$

Note that, state eqn (1.10), the adjoint equation (1.11) together with the characterization of the optimal control eqn (1.16) and the transversality condition eqn (1.14) are said to be optimality system. The sections and subsections treated above describe analytical behaviours of the optimal control.

1.7 Numerical Simulations

The birth rate in India per 1000 people is 17.22 and $N = 100$. Then the birth rate is evaluated as $P = \frac{17.22}{1000} \times 100 = 1.722$. The other parameters are assumed to take the following values: $h = 0.01$, $m = 0.03$, $\beta = 0.014$, $d = 0.024$, $c = 0.93$, $f = 0.18$, $\gamma = 0.016$, $w_2 = 0.74$, $\alpha_1 = 0.1$, $q = 0.2$. Fractional Order (μ) is varied to study its impact on the SEIQR model. Initial Population of each compartment are $(S, E, I, Q, R) = (65, 20, 10, 5, 0)$.

1.7.1 Impact of Reinfected Population Rate (m)

For parameter values $\mu = 0.9$, $h = 0.01$, $d = 0.024$, $P = 1.722$, $\beta = 0.014$, $c = 0.93$, $f = 0.18$, $\gamma = 0.016$, $w_2 = 0.74$, $\alpha_1 = 0.1$, $q = 0.2$. Reinfection rate (m) is varied to study its impact on the SEIQR model.

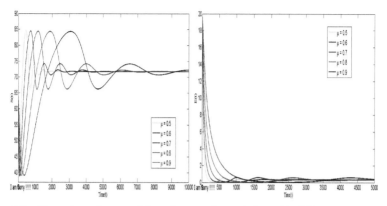

Figure 1.2   Time plots of susceptible and exposed population with different fractional-order values.

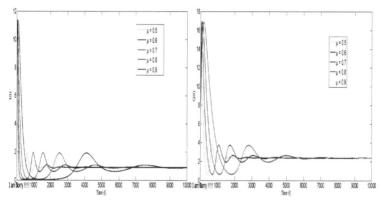

Figure 1.3   Time plots of infected and exposed population with different fractional-order values.

1.7.2 Bifurcation Analysis

Chaos theory in population dynamics is very important for developing the suitable control strategies to bring about harmony in the population. Chaos in real life may be due to small change in regular happening that leads to drastic changes. Such sensitivity to small changes are studied with the help of bifurcation. A bifurcation in the mathematical analysis of dynamical systems represents a sudden change in topological or qualitative behaviour of the system when a parameter value is subject to small variation. Bifurcation in case of a dynamical system representing the physical or chemical networks illustrates the occurrence of chaos from stable state. The transformation from

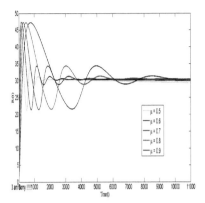

Figure 1.4 Time plots of recovered population with different fractional-order values.

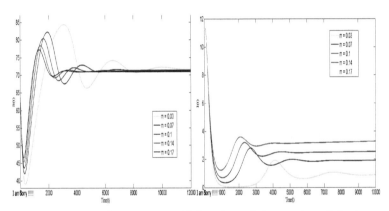

Figure 1.5 Time plots of susceptible and infected population varying reinfection rate m.

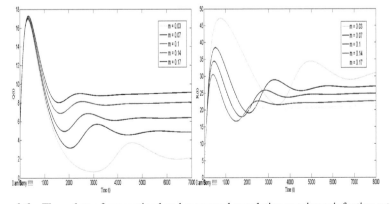

Figure 1.6 Time plots of quarantined and recovered population varying reinfection rate m.

stable state to chaos can be simulated in terms of time plots with orbits which increases periodically leading to unstable behaviour. In biological point of view bifurcation plays a vital part in understanding the functioning of organism and in preserving the extinction of the species in ecosystems. Here in this chapter we would like to understand the impact of the step-size on the qualitative behaviour of the considered discretized fractional-order model of COVID-19 outbreak.

For the bifurcation analysis, we fix the birth rate $P = 1.722$ and varying the step size h from 2.5 to 3.5 along with the feasible parametric values $m = 0.01, \beta = 0.009, d = 0.09, c = 0.033, \alpha = 0.219, \delta = 0.059, q = 0.611, w_2 = 0.499, \alpha_1 = 0.799$ with fractional-order $\mu = 0.5$.

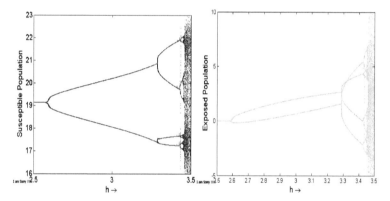

Figure 1.7   Bifurcations of susceptible and exposed population.

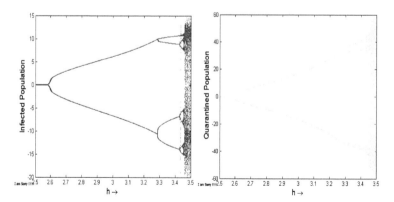

Figure 1.8   Bifurcations of infected and quarantined population.

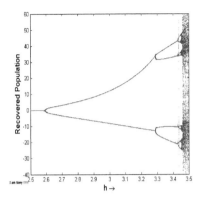

Figure 1.9 Bifurcations of recovered population.

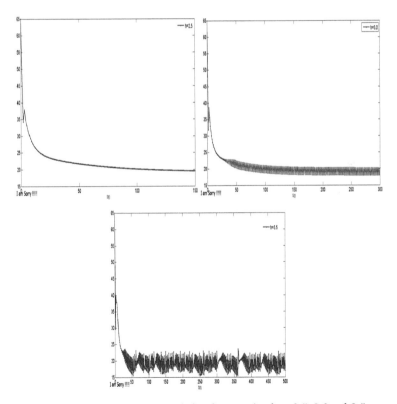

Figure 1.10 Susceptible population for step size $h = 2.5, 3.0$ and 3.5.

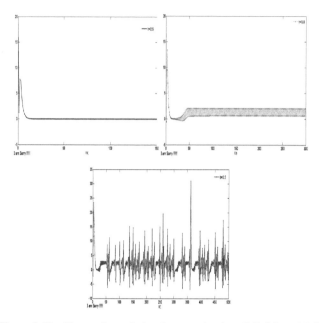

Figure 1.11   Exposed population for step size $h = 2.5, 3.0$ and 3.5.

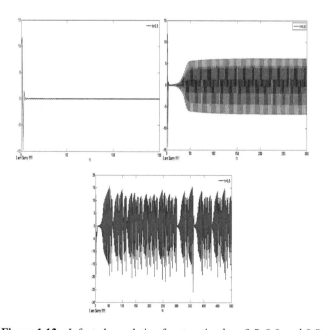

Figure 1.12   Infected population for step size $h = 2.5, 3.0$ and 3.5.

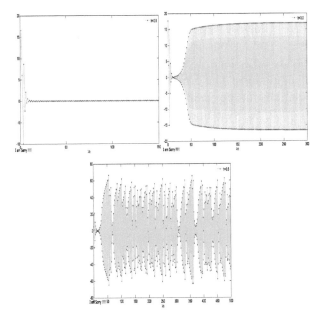

Figure 1.13 Quarantined population for step size $h = 2.5, 3.0$ and 3.5.

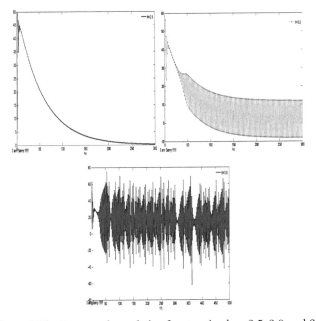

Figure 1.14 Recovered population for step size $h = 2.5, 3.0$ and 3.5.

The transformation of the different populations of the system is illustrated by considering three values of step-size $h = 2.5, 3.0, 3.5$. The time plots for each of the different populations for each step-size values h are presented in Figures 1.10, 1.11, 1.12, 1.13, 1.14. For $h = 2.5$, we observe the stable behaviour of the populations. The occurrence of uniform oscillations among the different populations for increasing time is obtained when $h = 3.0$. When the step- size $h = 3.5$, the system experiences a random and unpredictable motion with increasing time causing chaos in the populations. The chaos in population size can lead to devastating effects which highlight the importance of studying the considered model (1.6) in terms of step size. In our case of an epidemiological problem, unpredictable exposed and infected rises concern over the medical support for the people from the infections. In a similar way, irregularity in dynamics of the recovered population in a way may lead to keeping track and necessary control measures that can be implemented in future. Unstability in the cases of quarantine may lead to severe impact of increase in infection and tragic situation in terms of medical aid. Thus, our study of the bifurcation and time plots under different step sizes will be useful for developing optimal control strategies that could be implement to have superiority over the spread of coronavirus infection.

1.8 Conclusion

A mathematical model of COVID-19 outbreak with quarantine and reinfection factors was developed and mathematical analysis are performed to provide an outline of the system behaviour. The equilibrium states are then used to study the qualitative properties of the considered discretized fractional-order SEIQR model. The control strategies were developed to ensure the reduction in infection among the population. The parameters of the model are used to perform simulations numerically coinciding with the theoretical calculations that were performed. The chapter again ensures the fact the fractional-order epidemiological models better portray the real-life situations. Bifurcation analysis performed with step size provides an insight into chaotic dynamics of the populations. 2-D plots with varying time for different step-size in range of bifurcation diagrams give a clear view on transformation of the system from stable state to unpredictable or chaotic phase.

References

[1] Abraham P, Aggarwal N, Babu GR, Barani S, Bhargava B, Bhatnagar T, Dhama AS, Gangakhedkar RR, Giri S, Gupta N, Kurup KK, Manickam P, Murhekar M, Potdar V, Praharaj I, Rade K, Reddy DCS, Saravanakumar V, Shah N, et al. Laboratory surveillance for SARS-CoV2 in India: performance of testing and descriptive epidemiology of detected COVID-19, January 22–April 30, 2020. Indian J Med Res. 2020;151:424–437. doi: $10.4103/ijmr.IJMR_1896_20$.

[2] Agarwal, Ravi P., Ahmed MA El-Sayed, and Sanaa M. Salman. "Fractional-order Chua's system: discretization, bifurcation and chaos." Advances in Difference Equations 2013, no. 1 (2013): 1-13.

[3] Ahmed, I., Baba, I. A., Yusuf, A., Kumam, P., and Kumam, W. (2020). Analysis of Caputo fractional-order model for COVID-19 with lockdown. Advances in difference equations, 2020(1), 394. https://doi.org/10.1186/s13662-020-02853-0

[4] Akman O, Chauhan S, Ghosh A, et al. The Hard Lessons and Shifting Modeling Trends of COVID-19 Dynamics: Multiresolution Modeling Approach. Bulletin of Mathematical Biology. 2021 Nov;84(1):3. DOI: 10.1007/s11538-021-00959-4. PMID: 34797415; PMCID: PMC8602007.

[5] Baleanu, D., Jajarmi, A., Sajjadi, S.S., Mozyrska, D.: A new fractional model and optimal control of a tumor-immune surveillance with non-singular derivative operator. Chaos, Interdisc. J. Nonlinear Sci. 29(8), 083127 (2019).

[6] Borah, M.J., Hazarika, B., Panda, S.K., Nieto, J.J.: Examining the correlation between the weather conditions and COVID-19 pandemic in India: a mathematical evidence. Results Phys. 19, 103587 (2020).

[7] Deressa, C.T., Mussa, Y.O., Duressa, G.F.: Optimal control and sensitivity analysis for transmission dynamics of coronavirus. Results Phys. (2020).

[8] Deressa, C.T., Duressa, G.F.: Modeling and optimal control analysis of transmission dynamics of COVID-19: the case of Ethiopia. Alex. Eng. J. 60(1), 719-732 (2021).

[9] George Maria Selvam. A, Jehad Alzabut, D. Abraham Vianny, Mary Jacintha and Fatma Bozkurt Yousef, Modeling and stability analysis of the spread of novel coronavirus disease COVID-19, International Journal of Biomathematics, Volume 14, Issue 04 (May 2021).

[10] George Maria Selvam. A, and D.Vignesh, Analysis of the Effects of Social Distancing in Controlling the Spread of COVID-19, Alochana Chakra Journal, Volume IX, Issue IV, April/2020.

[11] George Maria Selvam. A , Janagaraj. R, Dhineshbabu. R, Analysis of Novel Corona Virus (COVID-19) Pandemic with Fractional-Order Caputo–Fabrizio Operator and Impact of Vaccination, Mathematical Analysis for Transmission of COVID-19, April 2021.

[12] Giuliani D, Dickson MM, Espa G, Santi F. Modelling and predicting the spatiotemporal spread of COVID-19 in Italy. BMC Infect Dis 2020;20(1):700.

[13] Gweryina, R. I., Madubueze, C. E., and Kaduna, F. S. (2021). Mathematical assessment of the role of denial on COVID-19 transmission with non-linear incidence and treatment functions. Scientific African, 12, e00811. https://doi.org/10.1016/j.sciaf.2021.e00811

[14] He S, Y Tang S, Rong L. A discrete stochastic model of the COVID-19 outbreak: forecast and control. Math Biosci Eng 2020;17(4):2792-804.

[15] Holm MR, Poland GA. Critical aspects of packaging, storage, preparation, and administration of mRNA and adenovirus-vectored COVID-19 vaccines for optimal efficacy. Vaccine. 2021;39(3):457âĂŞ459. doi: 10.1016/j.vaccine.2020.12.017.

[16] Kermack, W.O., McKendrick, A.G.: Contributions to the mathematical theory of epidemics-I. Bull. Math. Biol. 53(1-2), 33-55 (1991).

[17] Khan, M.A., Atangana, A., Alzahrani, E.: The dynamics of COVID-19 with quarantined and isolation. Adv. Differ. Equ. 2020(1), 425 (2020).

[18] Khan, M.A., Atangana, A.: Modeling the dynamics of novel coronavirus (2019-nCov) with fractional derivative. Alex. Eng. J. 59(4), 2379-2389 (2020).

[19] Kilbas AA, Srivastava HM, Trujillo JJ. Theory and applications of fractional differential equations. North Holland Mathematical Studies, vol 204. Amsterdam, London and New York: Elsevier (North Holland) Science Publishers; 2006.

[20] Lawrence, Perko. "Differential equations and dynamical systems." (1991).

[21] Maria Jones. G, S. Godfrey Winster, A. George Maria Selvam and D.Vignesh, A Mathematical Model and Forecasting of COVID -19 Outbreak in India, Intelligent Computing Applications for COVID -19, 213-234, 2021.

[22] Praveen Agarwal, Juan J. Nieto, Michael Ruzhansky, Delfim F. M. Torres, Analysis of Infectious Disease Problems (Covid-19) and Their

Global Impact, Infosys Science Foundation Series in Mathematical Sciences, Springer, 2021.

[23] Sene, N.: Analysis of the stochastic model for predicting the novel coronavirus disease. Adv. Differ. Equ. 2020(1), 568 (2020).

[24] Srivastava HM. Diabetes and its resulting complications: mathematical modeling via fractional calculus. Lancet 2020;4(3):000163.

[25] Toufik, M., Atangana, A.: New numerical approximation of fractional derivative with non-local and non-singular kernel: application to chaotic models. Eur. Phys. J. Plus 132(10), 144 (2017).

[26] World Health Organization (WHO). https://www.who.int/emergencies/diseases/novel-coronavirus-2019/technical-guidance2020

[27] Yang C, Wang J. A mathematical model for the novel coronavirus epidemic in Wuhan, China. Math Biosci Eng 2020;17(3):2708-24.

[28] Zhang, J., Ma, Z.: Global dynamics of an SEIR epidemic model with saturating contact rate. Math. Biosci. 185(1), 15-32 (2003).

[29] Zhang, Z.: A novel COVID-19 mathematical model with fractional derivatives: singular and nonsingular kernels. Chaos Solitons Fractals 139, 110060 (2020).

[30] Zhang, Z., Zeb, A., Egbelowo, O. F., and Erturk, V. S. (2020). Dynamics of a fractional order mathematical model for COVID-19 epidemic. Advances in difference equations, 2020(1), 420. https://doi.org/10.1186/s13662-020-02873-w

2

Stability Analysis of Caputo–Fabrizio Fractional-order Epidemic Model of a Novel Coronavirus (COVID-19)

A. George Maria Selvam, R. Janagaraj, and R. Dhineshbabu

Department of Mathematics, Sacred Heart College (Autonomous), India
Department of Mathematics, Faculty of Engineering, Karpagam Academy of Higher Education, India
Department of Mathematics, Sri Venkateswara College of Engineering and Technology (Autonomous), India
E-mail: agmshc@gmail.com

Abstract

The coronavirus infection (COVID-19), a highly contagious illness that first appeared in December 2019 in the Chinese city of Wuhan and spread quickly throughout the world, which is a newly discovered infectious disease. This pathogen infected millions of people worldwide and continues to represent a serious threat to human life. In Asian countries, India is the third country to surpass two million coronavirus cases. Numerous mathematical models have traditionally been investigated in order to gain a better understanding of coronavirus infection. Throughout this study, the dynamic behaviour of COVID-19 disease investigates by utilizing the non-integer Caputo–Fabrizio derivative to study the coronavirus sickness. To obtain a generalized model, first design the model in the integer sense and then use the fractional operator. After that, the theoretical findings are reported using the generalized model. The suggested fractional epidemic model of the novel coronavirus (COVID-19) is examined for stability and a parametric rule for the essential reproduction number ratio is provided. Picard approximation is used to demonstrate the fractional model for existence criteria. Finally, numerical simulations are developed to understand the influence of various parameters that determine infection dynamics, and the findings are exhibited as diagrams.

Keywords: SEIHR Modelling, Caputo–Fabrizio Derivative, Coronavirus, Fixed Point, Stability, Numerical Techniques.

2.1 Introduction

Coronaviruses are a type of virus that belongs to the Coronaviridae family. The virus is between 65 and 125 nanometres in size. It is single-stranded RNA with a cell nucleus that is 22–26 kilobases big. The four coronavirus subgroups are alpha, beta, gamma and delta. SARS-COV is one of the most widespread and potentially lethal respiratory illnesses. They are capable of causing pulmonary dysfunction. Since camels, bats and monkeys were the disease's major hosts, it was originally assumed to be an animal-only illness. This one was ultimately spread to humans, resulting in a global catastrophe. The literature [1] discovered the aspects, habitats and propagation of the coronavirus in the human population.

Wuhan, China, has experienced the spread of the deadly coronavirus since December 2019. The number of persons who have passed away is growing tremendously. During the first 50 days of the sickness, 70,000 people were infected. It has also been established that the illness was caused by beta-family coronaviruses. Chinese experts have dubbed the virus Wuhan viral disease or the new Coronavirus 2019. The virus is known as SARS-COV-2, COVID-19, according to the International Red Cross (IRC). Zhao and Chen analyzed the coronavirus transmission in China utilizing mathematical methodologies [2]. As per WHO, the virus has affected approximately thirty million people. Also, 1,44,000 persons are said to have died, while 7,65,000 are said to have been rescued. As a consequence, governments utilized delaying methods to battle the global epidemic, including social isolation, self-quarantine and travel bans. The WHO has now labelled the 2019-nCov epidemic a widespread threat.

The propagation possibility and intensity of the coronavirus were investigated by Shim et al. in South Korea [3]. Kucharski et al. proposed very few control approaches using a mathematical model of coronavirus [4]. Based on the data, Jiang et al. suggested a classification approach for coronavirus forecasts [5]. Fanelli and Piazza used an integer-order mathematical method to study and predict the growth of COVID-19 in Italy, China and France

[6]. The mathematical modelling of the coronavirus paradigm is in excellent agreement also with genuine occurrences thanks to the inclusion of delay effects in the system of differential equations [7, 8]. Delays can be caused by quarantine, segregation or vaccinations, for example. The sickness will probably settle into a stable position in most epidemics if the infection rate is monitored closely. In the current circumstances of 2019-nCov, preventing disease is nearly impossible, thus delaying methods like as social position, quarantine, separation and many others were implemented to prevent the pandemic.

Many epidemiologic modelling studies and research have been carried out to investigate COVID-19 transmission patterns in human populations [9, 10]. Wu et al. used a simple susceptible exposed infectious recovered (SEIR) model to predict COVID-19's transmission capability in China and globally [11]. Several mathematicians employed various analysis tools to look into the present coronavirus pandemic [12–15]. In the literature, there is a diversity of mathematical models for the present epidemic that predict the disease's future. Numerous researchers [16, 17] have proposed different control strategies.

The following is the structure of our present research. In Section 2.2, we explore fundamental concepts and definitions of fractional derivatives utilized in the study. The COVID-19-modified transmission CF fractional order is developed in Section 2.3. The Picard approximation technique is employed in Section 2.4 to investigate the existence and uniqueness of the concerned solution. In Section 2.5, using the next-generation matrix method along with various equilibrium states, we investigate the stability of the model under investigation. We execute numerical simulations in Section 2.6 with the use of available data to corroborate and verify our analytical results. Section 2.7 summarizes the outcome as a conclusion.

2.2 Preliminaries

Fundamental definitions and some simple concepts of fractional derivatives are introduced in this section.

Definition 2.1. ([18, 21]) The fractional derivative of the Caputo-type operator for a function $V \in AC_R^\mu([0, \infty))$ is given by

$$^C D_0^\mu V(\kappa) = \frac{1}{\Gamma(\tau - \mu)} \int_0^\tau (\kappa - \varsigma)^{\tau - \mu - 1} V^{(\tau)}(\varsigma) d\varsigma, \mu \in (\tau - 1, \tau] \text{ and } \tau = \lceil \mu \rceil + 1$$

exists if the value of the is finite.

Definition 2.2. ([19]) Let $\mu \in (0,1), n > m$ and $V \in G^1(n,m)$. Then the C-FFD is

$$^{CF}D_\iota^\mu V(\kappa) = \frac{Z(\mu)}{(1-\mu)} \int^\kappa \exp\left[-\frac{\mu(\iota-\varsigma)}{1-\kappa}\right] V'(\varsigma)d\varsigma, \quad \kappa \geq 0,$$

where $Z(\mu)$ depends on ρ which is a normalization function with $Z(0) = Z(1) = 1$.

Definition 2.3. ([20]) Let V be a function and $\rho \in (0,1)$ be the fractional order, then the integral operator for Caputo–Fabrizio is

$$^{CF}J_0^\mu V(\kappa) = \frac{2(1-\mu)}{(2-\mu)Z(\mu)}V(\kappa) + \frac{2\mu}{(2-\mu)Z(\mu)} \int_0^\iota V(\varsigma)d\varsigma, \quad \kappa \geq 0.$$

If U is a constant, then $^{CF}D_\iota^\rho U(\iota) = 0$.

Definition 2.4. ([22]) Let $\mu \in (0,1)$ be the fractional order and $\theta \in N$, then LT for the C-FFD is

$$L\left(^{CF}D_\kappa^{\theta+\mu}V(\kappa)\right)(\varpi) = \frac{1}{1-\mu}L\left(V^{(\theta+1)}(\kappa)\right)L\left(\exp\left(-\frac{\mu}{1-\mu}\kappa\right)\right)$$
$$= \frac{\varpi^{\theta+1}L(V(\kappa))-\varpi^\theta V(0)-\varpi^{\theta-1}V'(0)-\cdots-V^{(\theta)}(0)}{\varpi+\mu(1-\varpi)}.$$

Specially,

$$L\left(^{CF}D_\kappa^{\mu+\theta}V(\kappa)\right)(\varpi) = \begin{cases} \frac{\varpi L(V(\kappa))-V(0)}{\varpi+\mu(1-\varpi)}, \theta = 0, \\ \frac{\varpi^2 L(V(\kappa))-\varpi V(0)-V'(0)}{\varpi+\mu(1-\varpi)}, \theta = 1. \end{cases}$$

2.3 Model Formulation

Mathematical formulations are of boundless significance in biology and epidemiology. The significant field of epidemiology is concerned with the dynamics of health and illness in populations as well as their outlines [28]. There are a variety of principles in epidemiology strictly connected to infectious diseases, published by Kermack and McKendrick in 1927. These notions demonstrate a crucial function in the structure of mathematical formulation concepts by totalling different types of model. Another traditional remedy model, the SEIR model, includes a compartment of exposed people, *E(t)*, who are infected but not contagious. The complete human population at time k is separated into five parts to analyze the propagation and dissemination of COVID-19 among people. This is done using the SEIHR model. The susceptible population is represented by *S*, exposed by *E*, infected by *I*, hospitalized by H and recovered by R. A majority of these kinds of models are based on integer-order derivatives, which do not account for fading memory and crossover behaviour, which are common in biological processes. The memory and non-localization properties of fractional derivatives make it ideal for forecasting epidemic transitions. Non-singularity, non-locality and exponential form are all properties of the kernel of the fractional Caputo–Fabrizio derivative. In a sense, it is the best choice for simulating illness spread. This research adopts the *SEIR* proposed model with a total population of individuals of size N and an extra class, the Hospitalized (H). [29] A five-dimensional fractional differential equation of the mathematical SEIHR COVID-19 model is

$$
\begin{aligned}
D_\kappa S(\kappa) &= \Theta - \frac{\xi S(\kappa)I(\kappa)}{N-H(\kappa)} - \delta S(\kappa), \\
D_\kappa E(\kappa) &= \frac{\xi S(\kappa)I(\kappa)}{N-H(\kappa)} - \lambda E(\kappa) - \delta E(\kappa), \\
D_\kappa I(\kappa) &= \lambda E(\kappa) - \delta I(\kappa) - \vartheta I(\kappa) - \varepsilon I(\kappa), \\
D_\kappa H(\kappa) &= \varepsilon I(\kappa) - \delta H(\kappa) - \rho H(\kappa) - \chi H(\kappa), \\
D_\kappa R(\kappa) &= \vartheta I(\kappa) + \rho H(\kappa) - \delta R(\kappa),
\end{aligned}
\tag{2.1}
$$

with the initial conditions $(S_0, E_0, I_0, H_0, R_0)$ of the model (2.1). Also, the parameters, Θ is the average recruitment rate, ξ is the transmission rate of the infected population, δ is average life expectancy at birth, λ is the COVID-19 incubation period, ϑ is the recovery rate of the infected population, ε is the average hospitalization rate of the infected population, ρ is the average recovery rate from COVID-19 hospitalization infection and χ is average case fatality rate. The dynamical schematic diagram for the SEIHR model is depicted in Figure 2.1.

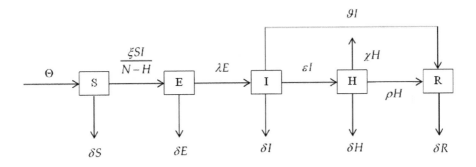

Figure 2.1 Schematic diagram for the SEIHR model of (2.1).

In recent decades, fractional differential equations have garnered immense interest among the community of researchers due to their applicability in the form of mathematical models for processes and systems in various flavours of engineering and scientific fields which are physical, chemical or natural and of date even in social networking. In recent decades, several researchers have been forfeiting attention to the concept of FDs. The concept of the Riemann–Liouville (R–L) and Caputo FDs have some limitations. Caputo and Fabrizio [19] introduced FD with a non-singular kernel in the year 2015 and its properties were investigated by Losada and Nieto [20]. Researching the Caputo–Fabrizio notion together with a non-singular kernel is hence more appropriate. The fractional CF model for $\kappa \geq 0$ and $\mu \in (0,1)$ in the COVID-19 modified transmission has the following form:

$$
\begin{aligned}
{}^{CF}D_\kappa^\mu S(\kappa) &= \Theta - \frac{\xi S(\kappa) I(\kappa)}{N - H(\kappa)} - \delta S(\kappa), \\
{}^{CF}D_\kappa^\mu E(\kappa) &= \frac{\xi S(\kappa) I(\kappa)}{N - H(\kappa)} - \lambda E(\kappa) - \delta E(\kappa), \\
{}^{CF}D_\kappa^\mu I(\kappa) &= \lambda E(\kappa) - \delta I(\kappa) - \vartheta I(\kappa) - \varepsilon I(\kappa), \\
{}^{CF}D_\kappa^\mu H(\kappa) &= \varepsilon I(\kappa) - \delta H(\kappa) - \rho H(\kappa) - \chi H(\kappa), \\
{}^{CF}D_\kappa^\mu R(\kappa) &= \vartheta I(\kappa) + \rho H(\kappa) - \delta R(\kappa),
\end{aligned}
\tag{2.2}
$$

2.4 Existence Criteria of Caputo–Fabrizio Model by Picard Approximation

The existence and unique solution for the CF fractional operator of the COVID-19 transmission model (2.1) discusses by using the Picard approximation technique. From the results of Losada and Nieto [20], the model (2.2) can be written as follows:

$$
\begin{aligned}
S(\kappa) &= S(0) +{}^{CF}J_\kappa^\mu \left[\Theta - \frac{\xi S(\kappa) I(\kappa)}{N - H(\kappa)} - \delta S(\kappa) \right], \\
E(\kappa) &= E(0) +{}^{CF}J_\kappa^\mu \left[\frac{\xi S(\kappa) I(\kappa)}{N - H(\kappa)} - \lambda E(\kappa) - \delta E(\kappa) \right], \\
I(\kappa) &= I(0) +{}^{CF}J_\kappa^\mu \left[\lambda E(\kappa) - \delta I(\kappa) - \vartheta I(\kappa) - \varepsilon I(\kappa) \right], \\
H(\kappa) &= H(0) +{}^{CF}J_\kappa^\mu \left[\varepsilon I(\kappa) - \delta H(\kappa) - \rho H(\kappa) - \chi H(\kappa) \right], \\
R(\kappa) &= R(0) +{}^{CF}J_\kappa^\mu \left[\vartheta I(\kappa) + \rho H(\kappa) - \delta R(\kappa) \right].
\end{aligned}
$$

Utilizing Definition 2.3,

$$
\begin{aligned}
S(\kappa) &= S(0) + \frac{2(1-\mu)}{(2-\mu)Z(\mu)} \left[\Theta - \frac{\xi S(\kappa) I(\kappa)}{N - H(\kappa)} - \delta S(\kappa) \right] \\
&\quad + \frac{2\mu}{(2-\mu)Z(\mu)} \int_0^\kappa \left[\Theta - \frac{\xi S(\varsigma) I(\varsigma)}{N - H(\varsigma)} - \delta S(\varsigma) \right] d\varsigma,
\end{aligned}
$$

$$E(\kappa) = E(0) + \frac{2(1-\mu)}{(2-\mu)Z(\mu)} \left[\frac{\xi S(\kappa)I(\kappa)}{N-H(\kappa)} - \lambda E(\kappa) - \delta E(\kappa) \right]$$
$$+ \frac{2\mu}{(2-\mu)Z(\mu)} \int_0^\kappa \left[\frac{\xi S(\varsigma)I(\varsigma)}{N-H(\varsigma)} - \lambda E(\varsigma) - \delta E(\varsigma) \right] d\varsigma,$$
$$I(\kappa) = I(0) + \frac{2(1-\mu)}{(2-\mu)Z(\mu)} \left[\lambda E(\kappa) - \delta I(\kappa) - \vartheta I(\kappa) - \varepsilon I(\kappa) \right]$$
$$+ \frac{2\mu}{(2-\mu)Z(\mu)} \int_0^\iota \left[\lambda E(\varsigma) - \delta I(\varsigma) - \vartheta I(\varsigma) - \varepsilon I(\varsigma) \right] d\varsigma,$$
$$H(\iota) = H(0) + \frac{2(1-\mu)}{(2-\mu)Z(\mu)} \left[\varepsilon I(\kappa) - \delta H(\kappa) - \rho H(\kappa) - \chi H(\kappa) \right]$$
$$+ \frac{2\mu}{(2-\mu)Z(\mu)} \int_0^\kappa \left[\varepsilon I(\varsigma) - \delta H(\varsigma) - \rho H(\varsigma) - \chi H(\varsigma) \right] d\varsigma,$$
$$R(\iota) = R(0) + \frac{2(1-\mu)}{(2-\mu)Z(\mu)} \left[\vartheta I(\kappa) + \rho H(\kappa) - \delta R(\kappa) \right]$$
$$+ \frac{2\mu}{(2-\mu)Z(\mu)} \int_0^\kappa \left[\vartheta I(\varsigma) + \rho H(\varsigma) - \delta R(\varsigma) \right] d\varsigma.$$

Equivalently,

$$S(\kappa) = S(0) + \frac{2(1-\mu)}{(2-\mu)Z(\mu)} [\Lambda_1(\kappa, S)] + \frac{2\mu}{(2-\mu)Z(\mu)} \int_0^\kappa [\Lambda_1(\varsigma, S)] d\varsigma,$$
$$E(\kappa) = E(0) + \frac{2(1-\mu)}{(2-\mu)Z(\mu)} [\Lambda_2(\kappa, E)] + \frac{2\mu}{(2-\mu)Z(\mu)} \int_0^\kappa [\Lambda_2(\varsigma, E)] d\varsigma,$$
$$I(\kappa) = I(0) + \frac{2(1-\mu)}{(2-\mu)Z(\mu)} [\Lambda_3(\kappa, I)] + \frac{2\mu}{(2-\mu)Z(\mu)} \int_0^\kappa [\Lambda_3(\varsigma, I)] d\varsigma,$$
$$H(\kappa) = H(0) + \frac{2(1-\mu)}{(2-\mu)Z(\mu)} [\Lambda_4(\kappa, H)] + \frac{2\mu}{(2-\mu)Z(\mu)} \int_0^\kappa [\Lambda_4(\varsigma, H)] d\varsigma,$$
$$R(\kappa) = R(0) + \frac{2(1-\mu)}{(2-\mu)Z(\mu)} [\Lambda_5(\kappa, R)] + \frac{2\mu}{(2-\mu)Z(\mu)} \int_0^\kappa [\Lambda_5(\varsigma, R)] d\varsigma.$$

$$(2.3)$$

Here,

$$\Lambda_1(\kappa, S) = \Theta - \frac{\xi S(\kappa)I(\kappa)}{N-H(\kappa)} - \delta S(\kappa),$$
$$\Lambda_2(\kappa, E) = \frac{\xi S(\kappa)I(\kappa)}{N-H(\kappa)} - \lambda E(\kappa) - \delta E(\kappa),$$
$$\Lambda_3(\kappa, I) = \lambda E(\kappa) - \delta I(\kappa) - \vartheta I(\kappa) - \varepsilon I(\kappa),$$
$$\Lambda_4(\kappa, H) = \varepsilon I(\kappa) - \delta H(\kappa) - \rho H(\kappa) - \chi H(\kappa),$$
$$\Lambda_5(\kappa, R) = \vartheta I(\kappa) + \rho H(\kappa) - \delta R(\kappa).$$

Now, taking into account the operations $S(\kappa)$ and $S_*(\kappa)$, we get

$$\|\Lambda_1(\kappa, S(\kappa)) - \Lambda_1(\kappa, S_*(\kappa))\|$$
$$= \left\| \Theta - \frac{\xi S(\kappa)I(\kappa)}{N-H(\kappa)} - \delta S(\kappa) - \Theta + \frac{\xi S_*(\kappa)I(\kappa)}{N-H(\kappa)} + \delta S_*(\kappa) \right\|$$
$$= \left\| -\frac{\xi I(\kappa)}{N-H(\kappa)} [S(\kappa) - S_*(\kappa)] - \delta [S(\iota) - S_*(\iota)] \right\|$$
$$\leq \left(\xi \left\| \frac{I(\kappa)}{N-H(\kappa)} \right\| + \delta \right) \|S(\iota) - S_*(\iota)\|$$
$$\leq \pi_1 \|S(\iota) - S_*(\iota)\|.$$

Here $\pi_1 = \xi \frac{\iota_1}{N-\iota_2} + \delta$ and $\iota_1 = \|I(\kappa)\|, \iota_2 = \|H(\kappa)\|$, which are bounded. Thus, Lipchitz condition fulfils for Λ_1. Additionally, when

$0 < \pi_1 \leq 1$ such that Λ_1 is a contraction. Also,

$$
\begin{aligned}
\|\Lambda_2(\kappa, E(\kappa)) - \Lambda_2(\kappa, E_*(\kappa))\| &\leq \pi_2 \, \|E(\kappa) - E_*(\kappa)\|, \\
\|\Lambda_3(\kappa, I(\kappa)) - \Lambda_3(\kappa, I_*(\kappa))\| &\leq \pi_3 \, \|I(\kappa) - I_*(\kappa)\|, \\
\|\Lambda_4(\kappa, H(\kappa)) - \Lambda_4(\kappa, H_*(\kappa))\| &\leq \pi_4 \, \|H(\kappa) - H_*(\kappa)\|, \\
\|\Lambda_5(\kappa, R(\kappa)) - \Lambda_5(\kappa, R_*(\kappa))\| &\leq \pi_5 \, \|R(\kappa) - R_*(\kappa)\|.
\end{aligned}
$$

Hence, $\Lambda_2, \Lambda_3, \Lambda_4, \Lambda_5$ fulfils the Lipchitz condition. Now consider the recursive formula:

$$
\begin{aligned}
S_l(\kappa) &= \tfrac{2(1-\mu)}{(2-\mu)Z(\mu)} \left[\Lambda_1(\kappa, S_{l-1})\right] + \tfrac{2\mu}{(2-\mu)Z(\mu)} \int_0^\kappa \left[\Lambda_1(\varsigma, S_{l-1})\right] d\varsigma, \\
E_l(\kappa) &= \tfrac{2(1-\mu)}{(2-\mu)Z(\mu)} \left[\Lambda_2(\kappa, E_{l-1})\right] + \tfrac{2\mu}{(2-\mu)Z(\mu)} \int_0^\kappa \left[\Lambda_2(\varsigma, E_{l-1})\right] d\varsigma, \\
I_l(\kappa) &= \tfrac{2(1-\mu)}{(2-\mu)Z(\mu)} \left[\Lambda_3(\kappa, I_{l-1})\right] + \tfrac{2\mu}{(2-\mu)Z(\mu)} \int_0^\kappa \left[\Lambda_3(\varsigma, I_{l-1})\right] d\varsigma, \\
H_l(\kappa) &= \tfrac{2(1-\mu)}{(2-\mu)Z(\mu)} \left[\Lambda_4(\kappa, H_{l-1})\right] + \tfrac{2\mu}{(2-\mu)Z(\mu)} \int_0^\kappa \left[\Lambda_4(\varsigma, H_{l-1})\right] d\varsigma, \\
R_l(\kappa) &= \tfrac{2(1-\mu)}{(2-\mu)Z(\mu)} \left[\Lambda_5(\kappa, R_{l-1})\right] + \tfrac{2\mu}{(2-\mu)Z(\mu)} \int_0^\kappa \left[\Lambda_5(\varsigma, R_{l-1})\right] d\varsigma.
\end{aligned}
\tag{2.4}
$$

From eqn (2.4),

$$
\begin{aligned}
P_{1j} = S_l(\kappa) - S_{l-1}(\kappa) &= \tfrac{2(1-\mu)}{(2-\mu)Z(\mu)} \left[\Lambda_1(\kappa, S_{l-1}) - \Lambda_1(\kappa, S_{l-2})\right] \\
&\quad + \tfrac{2\mu}{(2-\mu)Z(\mu)} \int_0^\kappa \left[\Lambda_1(\varsigma, S_{l-1}) - \Lambda_1(\varsigma, S_{l-2})\right] d\varsigma, \\[4pt]
P_{2j} = E_l(\kappa) - E_{l-1}(\kappa) &= \tfrac{2(1-\mu)}{(2-\mu)Z(\mu)} \left[\Lambda_2(\kappa, E_{l-1}) - \Lambda_2(\kappa, E_{l-2})\right] \\
&\quad + \tfrac{2\mu}{(2-\mu)Z(\mu)} \int_0^\kappa \left[\Lambda_2(\varsigma, E_{l-1}) - \Lambda_2(\varsigma, E_{l-2})\right] d\varsigma, \\[4pt]
P_{3j} = I_l(\kappa) - I_{l-1}(\kappa) &= \tfrac{2(1-\mu)}{(2-\mu)Z(\mu)} \left[\Lambda_3(\kappa, I_{l-1}) - \Lambda_3(\kappa, I_{l-2})\right] \\
&\quad + \tfrac{2\mu}{(2-\mu)Z(\mu)} \int_0^\kappa \left[\Lambda_3(\varsigma, I_{l-1}) - \Lambda_3(\varsigma, I_{l-2})\right] d\varsigma, \\[4pt]
P_{4j} = H_l(\kappa) - H_{l-1}(\kappa) &= \tfrac{2(1-\mu)}{(2-\mu)Z(\mu)} \left[\Lambda_4(\kappa, H_{l-1}) - \Lambda_4(\kappa, H_{l-2})\right] \\
&\quad + \tfrac{2\mu}{(2-\mu)Z(\mu)} \int_0^\kappa \left[\Lambda_4(\varsigma, H_{l-1}) - \Lambda_4(\varsigma, H_{l-2})\right] d\varsigma, \\[4pt]
P_{5j} = R_l(\kappa) - R_{l-1}(\kappa) &= \tfrac{2(1-\mu)}{(2-\mu)Z(\mu)} \left[\Lambda_5(\kappa, R_{l-1}) - \Lambda_5(\kappa, R_{l-2})\right] \\
&\quad + \tfrac{2\mu}{(2-\mu)Z(\mu)} \int_0^\kappa \left[\Lambda_5(\varsigma, R_{l-1}) - \Lambda_5(\varsigma, R_{l-2})\right] d\varsigma.
\end{aligned}
$$

Thus,

$$
\begin{aligned}
&S_l(\kappa) = \textstyle\sum_{q=o}^l P_{1q}(\kappa), \; E_l(\kappa) = \sum_{q=o}^l P_{2q}(\kappa), \; I_l(\kappa) = \sum_{q=o}^l P_{3q}(\kappa), \\
&H_l(\kappa) = \textstyle\sum_{q=o}^l P_{4q}(\kappa), \; R_l(\kappa) = \sum_{q=o}^l P_{5q}(\kappa).
\end{aligned}
\tag{2.5}
$$

Considering triangular inequality to $P'_{1l}s$,

$$
\|P_{1l}(\kappa)\| = \|S_l(\kappa) - S_{l-1}(\kappa)\|
$$

$$= \left\| \begin{matrix} \frac{2(1-\mu)}{(2-\mu)Z(\mu)} [\Lambda_1(\kappa, S_{l-1}) - \Lambda_1(\kappa, S_{l-2})] \\ + \frac{2\mu}{(2-\mu)Z(\mu)} \int_0^\kappa [\Lambda_1(\varsigma, S_{l-1}) - \Lambda_1(\varsigma, S_{l-2})] \, d\varsigma \end{matrix} \right\|,$$

$$\|P_{1l}(\kappa)\| \le \frac{2(1-\mu)}{(2-\mu)Z(\mu)} \|[\Lambda_1(\kappa, S_{l-1}) - \Lambda_1(\kappa, S_{l-2})]\|$$
$$+ \frac{2\mu}{(2-\mu)Z(\mu)} \left\| \int_0^\kappa [\Lambda_1(\varsigma, S_{l-1}) - \Lambda_1(\varsigma, S_{l-2})] \, d\varsigma \right\|.$$

Now, Lipchitz condition is satisfied for Λ_1,

$$\|S_l(\kappa) - S_{l-1}(\kappa)\| \le \frac{2(1-\mu)}{(2-\mu)Z(\mu)} \pi_1 \|S_{l-1} - S_{l-2}\| + \frac{2\mu}{(2-\mu)Z(\mu)}$$

$$\pi_1 \int_0^\kappa \|S_{l-1} - S_{l-2}\| \, d\varsigma.$$

Thus,

$$\|P_{1l}(\kappa)\| \le \frac{2(1-\mu)}{(2-\mu)Z(\mu)} \pi_1 \|P_{1l-1}(\kappa)\| + \frac{2\mu}{(2-\mu)Z(\mu)} \pi_1 \int_0^\kappa \|P_{1l-1}(\varsigma)\| \, d\varsigma. \tag{2.6}$$

In the same way,

$$\|P_{2l}(\kappa)\| \le \frac{2(1-\mu)}{(2-\mu)Z(\mu)} \pi_2 \|P_{2l-1}(\kappa)\| + \frac{2\mu}{(2-\mu)Z(\mu)} \pi_2 \int_0^\kappa \|P_{2l-1}(\varsigma)\| \, d\varsigma,$$
$$\|P_{3l}(\kappa)\| \le \frac{2(1-\mu)}{(2-\mu)Z(\mu)} \pi_3 \|P_{3l-1}(\kappa)\| + \frac{2\mu}{(2-\mu)Z(\mu)} \pi_3 \int_0^\kappa \|P_{3l-1}(\varsigma)\| \, d\varsigma,$$
$$\|P_{4l}(\kappa)\| \le \frac{2(1-\mu)}{(2-\mu)Z(\mu)} \pi_4 \|P_{4l-1}(\kappa)\| + \frac{2\mu}{(2-\mu)Z(\mu)} \pi_4 \int_0^\kappa \|P_{4l-1}(\varsigma)\| \, d\varsigma,$$
$$\|P_{5l}(\kappa)\| \le \frac{2(1-\mu)}{(2-\mu)Z(\mu)} \pi_5 \|P_{5l-1}(\kappa)\| + \frac{2\mu}{(2-\mu)Z(\mu)} \pi_5 \int_0^\kappa \|P_{5l-1}(\varsigma)\| \, d\varsigma. \tag{2.7}$$

Hence model (2.2) has a solution.

Next to prove that eqn (2.3) has unique solution. Let S_1, E_1, I_1, R_1, Q_1 be a new solution of eqn (2.3) such that

$$\|S(\kappa) - S_1(\kappa)\| = \left\| \begin{matrix} \frac{2(1-\mu)}{(2-\mu)Z(\mu)} [\Lambda_1(\kappa, S) - \Lambda_1(\kappa, S_1)] \\ + \frac{2\mu}{(2-\mu)Z(\mu)} \int_0^\kappa [\Lambda_1(\varsigma, S) - \Lambda_1(\varsigma, S_1)] \, d\varsigma \end{matrix} \right\|,$$
$$\le \frac{2(1-\mu)}{(2-\mu)Z(\mu)} \|[\Lambda_1(\kappa, S) - \Lambda_1(\kappa, S_1)]\| + \frac{2\mu}{(2-\mu)Z(\mu)}$$
$$\left\| \int_0^\kappa [\Lambda_1(\varsigma, S) - \Lambda_1(\varsigma, S_1)] \, d\varsigma \right\|.$$

Using Lipchitz condition,

$$\|S(\kappa) - S_1(\kappa)\| \le \frac{2(1-\mu)}{(2-\mu)Z(\mu)} \pi_1 \|S(\kappa) - S_1(\kappa)\|$$

$$+ \frac{2\mu}{(2-\mu)Z(\mu)} \pi_1 \kappa \left\| S(\kappa) - S_1(\kappa) \right\|,$$

thus,

$$\left(1 - \frac{2(1-\mu)}{(2-\mu)Z(\mu)} \pi_1 - \frac{2\mu}{(2-\mu)\psi(\mu)} \pi_1 \kappa \right) \left\| S(\kappa) - S_1(\kappa) \right\| \leq 0. \quad (2.8)$$

Theorem 2.5. If the following assumption holds:

$$\left(1 - \frac{2(1-\mu)}{(2-\mu)Z(\mu)} \pi_1 - \frac{2\mu}{(2-\mu)\psi(\mu)} \pi_1 \kappa \right) \geq 0. \quad (2.9)$$

Then, the solution of the model (2.3) has a unique solution.

Proof. Considering inequalities eqn (2.9) and (2.8) attain at

$$\left(1 - \frac{2(1-\mu)}{(2-\mu)Z(\mu)} \pi_1 - \frac{2\mu}{(2-\mu)\psi(\mu)} \pi_1 \kappa \right) \left\| S(\kappa) - S_1(\kappa) \right\| = 0.$$

Hence $\| S(\kappa) - S_1(\kappa) \| = 0$, this implies $S(\kappa) = S_1(\kappa)$. Also, one can easy to verify

$$E(\kappa) = E_1(\kappa), I(\kappa) = I_1(\kappa), H(\kappa) = H_1(\kappa), R(\kappa) = R_1(\kappa).$$

This completes the proof.

2.5 Stability Analysis by the Equilibrium States

For the COVID-19 model, endemic and disease-free equilibria states are constructed eqn (2.2). Using basic reproduction rate, model (2.4) is investigated. The next-generation matrix technique is applied to calculate R_0. As a result, the equilibria states of endemic and disease-free conditions are both globally (GAS) and locally (LAS) asymptotically stable. The feasible area of eqn (2.2) is

$$\Xi = \left\{ (S, E, I, H, R) | S, E, I, H, R \geq 0, S + E + I + H + R = N \leq \frac{\Theta}{\delta} \right\}.$$

Also, the equilibria states of the model (2.2) under endemic and disease-free conditions are $\Phi_0 = (S_0, E_0, I_0, H_0, R_0) = (\frac{\Theta}{\delta}, 0, 0, 0, 0)$ and $\Phi_* = (S_*, E_*, I_*, H_*, R_*)$, where

$$S_* = \frac{N(\delta + \lambda)(\delta + \varepsilon + \vartheta)(\chi + \delta + \rho) - \Theta \lambda \varepsilon}{\lambda [\xi(\chi + \delta + \rho) - \delta \varepsilon]}, E_* = \frac{\delta + \varepsilon + \vartheta}{\lambda} I_*$$

$$I_* = \frac{[N\delta(\delta + \lambda)(\delta + \varepsilon + \vartheta) - \Theta \lambda \xi](\chi + \delta + \rho)}{(\delta + \lambda)(\delta + \varepsilon + \vartheta)[\delta \varepsilon - \xi(\chi + \delta + \rho)]},$$

$$H_* = \frac{\varepsilon}{(\chi + \delta + \rho)} I_*, \quad R_* = \frac{1}{\delta}\left[\vartheta + \frac{\rho\varepsilon}{(\chi+\delta+\rho)}\right] I_*.$$

Next-generation matrix [22] is applied to compute R_0 for model (2.2). Taking the compartments E, I, H gives

$$Q = \begin{bmatrix} 0 & \frac{\Theta\xi}{N\delta} & 0 \\ 0 & 0 & 0 \\ 0 & 0 & 0 \end{bmatrix}, K = \begin{bmatrix} (\delta+\lambda) & -\frac{\varepsilon\xi}{N\delta} & 0 \\ -\lambda & (\delta+\vartheta+\varepsilon) & 0 \\ 0 & -\varepsilon & (\chi+\delta+\rho) \end{bmatrix}$$

and the next-generation matrix

$$\Xi = QK^{-1} = \begin{bmatrix} \frac{\Theta\xi}{N\delta(\delta+\lambda)} & \frac{\Theta\xi\lambda}{N\delta(\delta+\lambda)(\delta+\varepsilon+\vartheta)} & \frac{\Theta\xi\lambda\varepsilon}{N\delta} \\ 0 & 0 & 0 \\ 0 & 0 & 0 \end{bmatrix}.$$

Hence, $\Omega(\Xi) = \frac{\Theta\xi\lambda}{N\delta(\delta+\lambda)(\delta+\varepsilon+\vartheta)}$. From [29], $R_0 = \frac{\Theta\xi\lambda}{N\delta(\delta+\lambda)(\delta+\varepsilon+\vartheta)}$.

The following results established LAS and GAS for endemic and disease-free equilibria states.

Theorem 2.6. In a feasible area Ξ, $\Phi_0 = (S_0, E_0, I_0, H_0, R_0)$ is LAS when $R_0 < 1$.

Proof. The Jacobian and eigenvalues of the model (2.2) for $\Phi_0 = (S_0, E_0, I_0, H_0, R_0)$ are

$$|J(\Phi_0) - \Psi I| = \begin{vmatrix} -\delta - \Psi & 0 & -\frac{\Theta\xi}{N} & 0 & 0 \\ 0 & -(\delta+\lambda)-\Psi & \frac{\Theta\xi}{N} & 0 & 0 \\ 0 & \lambda & -(\delta+\vartheta+\varepsilon)-\Psi & 0 & 0 \\ 0 & 0 & \varepsilon & -(\chi+\delta+\rho)-\Psi & 0 \\ 0 & 0 & \vartheta & \rho & -\delta-\Psi \end{vmatrix}$$

$=0,$

$$\nu_{1,2} = -\delta, -\delta, \nu_3 = -(\chi+\delta+\rho) \text{ and}$$
$$\nu_{4,5} = -\frac{(2\delta+\lambda+\vartheta+\varepsilon)}{2} \pm \frac{1}{2}\sqrt{(\lambda-\vartheta-\varepsilon)^2 + \frac{4\Theta\xi\lambda}{N}}.$$

Here, the eigenvalues are having negative values by utilizing the Hurwitz criterion [23]. Thus, Φ_0 is LAS when $R_0 < 1$ applying the stability theory [24], completes the proof.

Theorem 2.7. In a feasible area Ξ, $\Phi_0 = (S_0, E_0, I_0, H_0, R_0)$ is GAS when $R_0 < 1$.

Proof. From eqn (2.1), $S'(\kappa) \leq \Theta - \delta S(\kappa)$, after solving the previous equation, we get

$$S(\kappa) \leq \frac{\Theta}{\delta} + \left[S(0) - \frac{\Theta}{\delta}\right]e^{-\delta\kappa}.$$

Now Lyapunov function is defined by

$$Lyp(\kappa) = \lambda E + (\lambda + \delta)I.$$

Thus,

$$
\begin{aligned}
Lyp'(\kappa) &= \lambda E' + (\lambda + \delta)I' \\
&= \lambda \left[\frac{\xi S(\kappa)I(\kappa)}{N - H(\kappa)} - \lambda E(\kappa) - \delta E(\kappa) \right] \\
&\quad + (\lambda + \delta)\left[\lambda E(\kappa) - \delta I(\kappa) - \vartheta I(\kappa) - \varepsilon I(\kappa) \right] \\
&= \lambda \frac{\xi S(\kappa)I(\kappa)}{N} - (\lambda + \delta)(\delta + \vartheta + \varepsilon)I(\kappa) \\
&\leq \left[\frac{\Theta \xi \lambda}{N\delta} - (\lambda + \delta)(\delta + \vartheta + \varepsilon) \right] I(\kappa).
\end{aligned}
$$

Thus, $R_0 = \frac{\Theta \xi \lambda}{N\delta(\delta + \lambda)(\delta + \varepsilon + \vartheta)} < 1$ implies that $L'(\iota) < 0$. Using the LaSalle Invariance principle [30], $\Phi_0 = (S_0, E_0, I_0, H_0, R_0)$ is GAS when $R_0 < 1$. The proof is completed.

Theorem 2.8. In a feasible area Ξ, $\Phi_* = (S_*, E_*, I_*, H_*, R_*)$ is LAS when $R_0 > 1$.

Proof. The Jacobian model (2.2) for $\Phi_* = (S_*, E_*, I_*, H_*, R_*)$ is

$$
|J(\Phi_*) - \Psi I| =
\begin{vmatrix}
-\frac{\xi I_*}{N - H_*} - \delta - \Psi & 0 & -\frac{\xi S_*}{N - H_*} & -\frac{\xi S_* I_*}{(N - H_*)^2} & 0 \\
\frac{\xi I_*}{N - H_*} & -(\delta + \lambda) - \Psi & \frac{\xi S_*}{N - H_*} & \frac{\xi S_* I_*}{(N - H_*)^2} & 0 \\
0 & \lambda & -(\delta + \vartheta + \varepsilon) - \Psi & 0 & 0 \\
0 & 0 & \varepsilon & -(\chi + \delta + \rho) - \Psi & 0 \\
0 & 0 & \vartheta & \rho & -\delta - \Psi
\end{vmatrix}.
$$

Here, solving the Jacobian and the roots having negative values by utilizing the Hurwitz criterion [23]. Thus, Φ_* is LAS when $R_0 > 1$ by applying the stability theory [24].

2.6 Numerical Computations by Data Values

Numerical simulations are presented in this section for the COVID-19 transmission model over a period of κ through fractional order. A fractional-order mathematical formulation of the SEIHR model is given by

$$
\begin{aligned}
D_\kappa^\mu S(\kappa) &= \Theta - \frac{\xi S(\kappa)I(\kappa)}{N - H(\kappa)} - \delta S(\kappa). \\
D_\kappa^\mu E(\kappa) &= \frac{\xi S(\kappa)I(\kappa)}{N - H(\kappa)} - \lambda E(\kappa) - \delta E(\kappa). \\
D_\kappa^\mu I(\kappa) &= \lambda E(\kappa) - \delta I(\kappa) - \vartheta I(\kappa) - \varepsilon I(\kappa). \\
D_\kappa^\mu H(\kappa) &= \varepsilon I(\kappa) - \delta H(\kappa) - \rho H(\kappa) - \chi H(\kappa). \\
D_\kappa^\mu R(\kappa) &= \vartheta I(\kappa) + \rho H(\kappa) - \delta R(\kappa).
\end{aligned}
\tag{2.10}
$$

Applying the numerical scheme from [25], the model (2.10) has been modified as follows:

$$S\left(\kappa_\iota\right) = \left[\Theta - \frac{\xi S(\kappa_{\iota-1})I(\kappa_{\iota-1})}{N - H(\kappa_{\iota-1})} - \delta S(\kappa_{\iota-1})\right] h^\mu - \sum_{r=t}^{\iota} c_r^{(\mu)} S\left(\kappa_{\iota-r}\right),$$

$$E\left(\iota_n\right) = \left[\frac{\xi S(\kappa_{\iota-1})I(\kappa_{\iota-1})}{N - H(\kappa_{\iota-1})} - \lambda E(\kappa_{\iota-1}) - \delta E(\kappa_{\iota-1})\right] h^\mu$$
$$- \sum_{r=t}^{\iota} c_r^{(\mu)} E\left(\kappa_{\iota-r}\right),$$

$$I\left(\iota_n\right) = \left[\lambda E(\kappa_{\iota-1}) - \delta I(\kappa_{\iota-1}) - \vartheta I(\kappa_{\iota-1}) - \varepsilon I(\kappa_{\iota-1})\right] h^\mu$$
$$- \sum_{r=t}^{\iota} c_r^{(\mu)} I\left(\kappa_{\iota-r}\right),$$

$$H\left(\iota_n\right) = \left[\varepsilon I(\kappa_{\iota-1}) - \delta H(\kappa_{\iota-1}) - \rho H(\kappa_{\iota-1}) - \chi H(\kappa_{\iota-1})\right] h^\mu$$
$$- \sum_{r=t}^{\iota} c_r^{(\mu)} H\left(\kappa_{\iota-r}\right),$$

$$R\left(\iota_n\right) = \left[\vartheta I(\kappa_{\iota-1}) + \rho H(\kappa_{\iota-1}) - \delta R(\kappa_{\iota-1})\right] h^\mu - \sum_{r=t}^{\iota} c_r^{(\mu)} R\left(\kappa_{\iota-r}\right),$$

where $\iota = 1, 2, 3, \cdots N$, for $N = [T_{sim}/h]$ and the initial points are $(S(0), E(0), I(0), H(0), R(0))$.

The numerical simulations are calculated for $S(\iota), E(\iota), I(\iota), R(\iota), Q(\iota)$ for non-integers $\mu = 0.92, 0.94, 0.96, 0.98$ and $\mu = 1$ of the model (2.10). Numerical computations of the model (2.10) are presented for two different cases, one from India and another from Tamil Nadu. Both the cases assume that $N = 100$ as the total population [27]. Average life expectancy at birth in India is $\delta = 0.0143$ and for Tamil Nadu is $\delta = 0.0141$, then $\Theta = \delta \times N = 1.43$ for India and for Tamil Nadu Θ is 1.41. From [26], average retrieval rates of hospitalized and reported infection for COVID-19 in India and Tamil Nadu are $\rho = 98.2\% = 0.982$ and the average incident of fatality rate in India and Tamil Nadu are $\rho = 1.3\% = 0.013$, respectively. Take the distinct parameter values for India are $\xi = 2.60, \lambda = 0.08, \vartheta = 0.79, \varepsilon = 0.6$ and consider another set of parameter values for Tamil Nadu as $\xi = 1.47, \lambda = 0.30, \vartheta = 0.60, \varepsilon = 0.56$ from [28] to attain the numerical simulations. Also, for model (2.10), along with $S(0) = 45, E(0) = 30, I(0) = 25, H(0) = 5$ and $R(0) = 0$ from $N = 100$. From the values of the aforementioned set of parameters, the model (2.2) estimate $R_0 = 1.6323$ is determined.

The dynamics of $SEIHR$ populations are displayed in Figures 2.2 and 2.6 for the various fractional orders μ with time κ in the case of India. Similarly for Tamil Nadu, the nature of $SEIHR$ populations is displayed in Figures 2.7 and 2.11 for the various fractional orders μ with time κ. Figures 2.2 and 2.7 exposed the susceptible state decreases and changes to selected values of μ. Figures 2.3 and 2.8 show the exposed state increases and moves to decreases as the order μ with time κ. For various values of μ time κ, the infected state decreases but infection states are decreased when the value μ with time κ increases, see Figures 2.4 and 2.9. From Figures 2.5 and 2.10, the hospitalized state recovers speedily with a variation of μ. Finally, Figures 2.6 and 2.11 exhibits the recovered states, which increase with various values of μ.

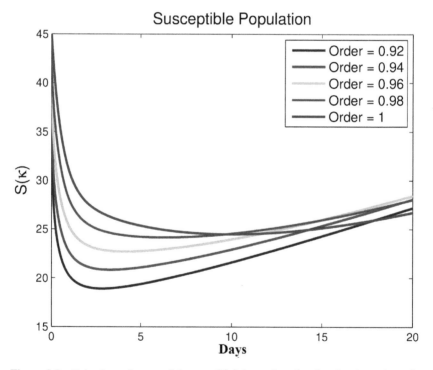

Figure 2.2   Behaviour of susceptible state $S(\kappa)$ for various fractional orders μ in India.

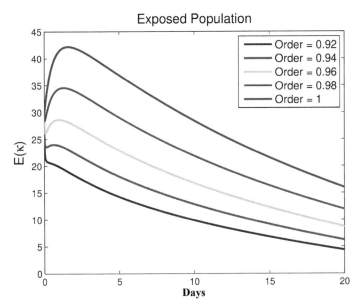

Figure 2.3 Behaviour of exposed state $E(\kappa)$ for various fractional orders μ in India.

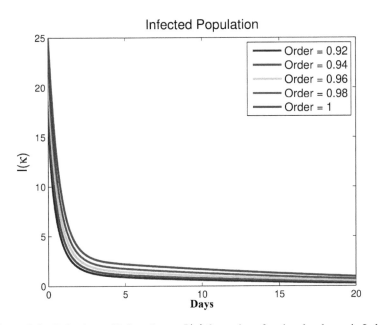

Figure 2.4 Behaviour of infected state $I(\kappa)$ for various fractional orders μ in India.

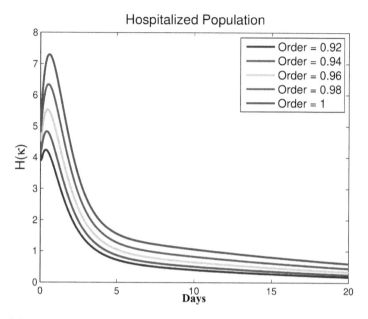

Figure 2.5   Behaviour of hospitalized state $H(\kappa)$ for various fractional orders μ in India.

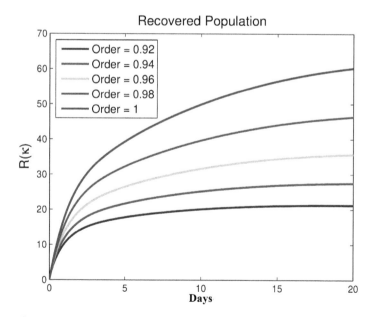

Figure 2.6   Behaviour of recovered state $R(\kappa)$ for various fractional orders μ in India.

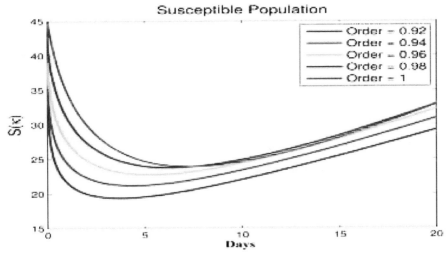

Figure 2.7 Behaviour of susceptible state $S(\kappa)$ for various fractional orders μ in Tamil Nadu.

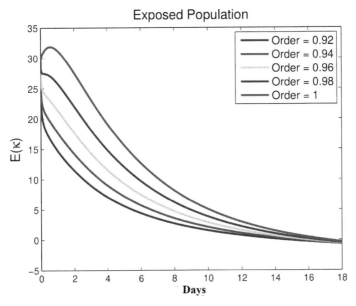

Figure 2.8 Behaviour of exposed state $E(\kappa)$ for various fractional orders μ in Tamil Nadu.

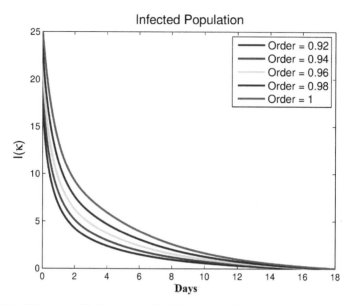

Figure 2.9   Behaviour of infected state $I(\kappa)$ for various fractional orders μ in Tamil Nadu.

2.6.1 COVID-19 Model with Vaccination

The following is a fractional SEIHR model associated with the vaccinated state $[V(\kappa)]$:

$$
\begin{aligned}
D_\kappa^\mu S(\kappa) &= \Theta - \frac{\xi S(\kappa)I(\kappa)}{N-H(\kappa)} - \beta S(\kappa) + \alpha V(\kappa) - \delta S(\kappa),\\
D_\kappa^\mu E(\kappa) &= \frac{\xi S(\kappa)I(\kappa)}{N-H(\kappa)} - \lambda E(\kappa) - \delta E(\kappa),\\
D_\kappa^\mu I(\kappa) &= \lambda E(\kappa) - \delta I(\kappa) - \vartheta I(\kappa) - \varepsilon I(\kappa),\\
D_\kappa^\mu H(\kappa) &= \varepsilon I(\kappa) - \delta H(\kappa) - \rho H(\kappa) - \chi H(\kappa),\\
D_\kappa^\mu R(\kappa) &= \vartheta I(\kappa) + \rho H(\kappa) - \delta R(\kappa),\\
D_\kappa^\mu V(\kappa) &= \beta S(\kappa) - \alpha V(\kappa) - \delta V(\kappa).
\end{aligned}
\tag{2.11}
$$

Here the parameters assume are all positive and have the same significance as in model (2.1). Also α, β are the waning immunity rate of the vaccinated population and rate of vaccination, respectively. Similar parameters are considered in Figures 2.2–2.11 along with $\alpha = 0.001, \beta = 0.3$. Thus, Figures 2.12–2.17 give a picture of $S(\kappa)$, $I(\kappa)$ and $R(\kappa)$ populations of the models (2.10) and (2.11) with and without vaccination through $\mu = 0.95$.

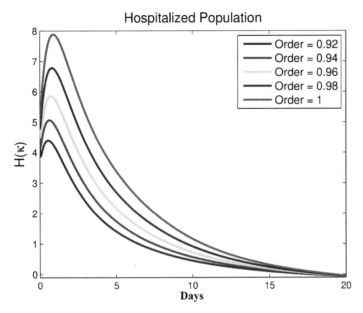

Figure 2.10 Behaviour of hospitalized state $H(\kappa)$ for various fractional orders μ in Tamil Nadu.

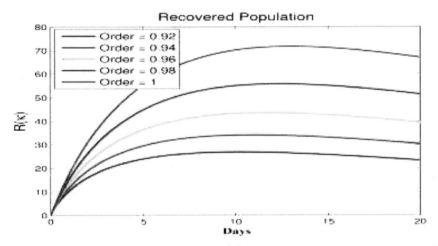

Figure 2.11 Behaviour of recovered state $R(\kappa)$ for various fractional orders μ in Tamil Nadu.

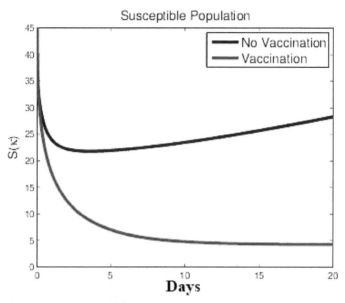

Figure 2.12   Behaviour of $S(\kappa)$ for $\mu = 0.95$ with and without vaccination in India.

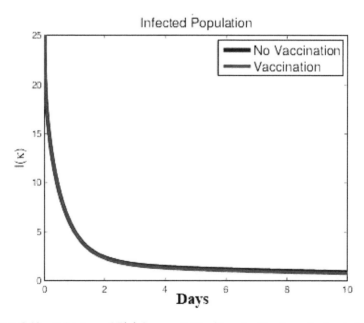

Figure 2.13   Behaviour of $I(\kappa)$ for $\mu = 0.95$ with and without vaccination in India.

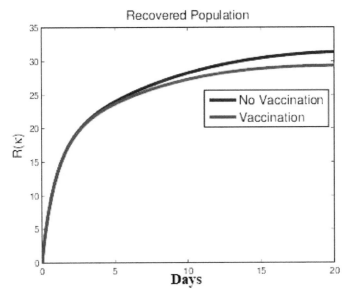

Figure 2.14 Behaviour of $R(\kappa)$ for $\mu = 0.95$ with and without vaccination in India.

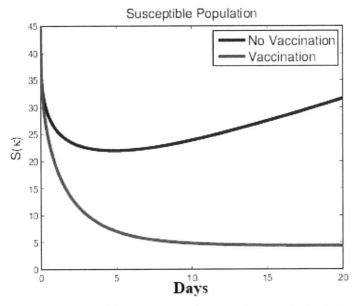

Figure 2.15 Behaviour of $S(\kappa)$ for $\mu = 0.95$ with and without vaccination in Tamil Nadu.

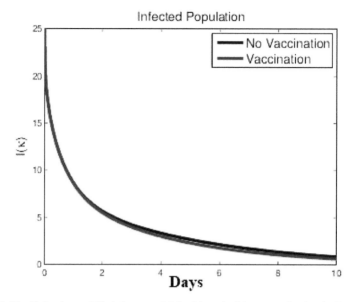

Figure 2.16   Behaviour of $I(\kappa)$ for $\mu = 0.95$ with and without vaccination in Tamil Nadu.

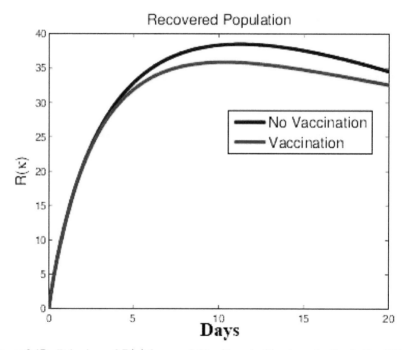

Figure 2.17   Behaviour of $R(\kappa)$ for $\mu = 0.95$ with and without vaccination in Tamil Nadu.

2.7 Conclusion

In this research work, with a newly invented fractional operator in the sense of Caputo–Fabrizio, we developed a novel type of COVID-19 model with five compartments such as $S(\kappa)$, $E(\kappa)$, $I(\kappa)$, $H(\kappa)$ and $R(\kappa)$ classes. The proposed model has been discussed in stability analysis along with various equilibria and the basic reproduction number. The existence and uniqueness of the solutions have been addressed with the help of the Picard approximation technique. We executed numerical computations for our model and quickly analyzed the results for various values of fractional order. The CF fractional provides a more accurate analysis than the traditional integer-order COVID-19 model, which can be seen in our graphical representations. Here, we used several parameters from the Indian country and Tamil Nadu state. We can replicate the same strategy in many other places of the globe in the future, and we can also adapt the same technology to other pandemic diseases.

References

[1] M. D. Shereen, S. Khan, A. Kazmi, N. Bashir, R. Siddique, "COVID-19 infection: Origin, transmission and characteristics of human corona viruses", J. Adv. Res., 24, pp.91–98, 2020.

[2] S. Zhao, H. Chen, "Modelling the epidemic dynamics and control of COVID-19 outbreak in China", Quant. Biol., 11, pp.1–9, 2020.

[3] E. Shim, A. Tariq, W. Choi, Y. Lee, G. Chowell, "Transmission potential and severity of COVID-19 in South Korea", Int. J. Infect. Dis., 93, pp.339–344, 2020.

[4] A. J. Kucharski, Timothy W Russell, Charlie Diamond, Yang Liu, John Edmunds, "Early dynamics of transmission and control of COVID-19: A mathematical modelling study", Lancet Infect. Dis., 20, pp.553–558, 2020.

[5] Xiangao Jiang, Megan Coffee, Anasse Bari, Junzhang Wang, Xinyue Jiang, Jianping Huang, Jichan Shi, Jianyi Dai, Jing Cai, Tianxiao Zhang, Zhengxing Wu, Guiqing He, Yitong Huang, "Towards an artificial intelligence framework for data-driven prediction of corona virus clinical severity", Comput. Mater. Con., 63, pp.537–551, 2020.

[6] D. Fanelli, F. Piazza, "Analysis and forecast of COVID-19 spreading in China, Italy and France", Chaos Soliton. Fract., 134, 109761, 2020. https://doi.org/10.1016/j.chaos.2020.109761

[7] A. Altan, S. Karasu, S, "Recognition of COVID-19 disease from X-ray images by hybrid model consisting of 2D curvelet transform, chaotic slap swarm algorithm and deep learning technique", Chaos Soliton. Fract., 140, 110071, 2020.

[8] Muhammad Naveed, Muhammad Rafiq, Ali Raza, Nauman Ahmed, Ilyas Khan, Kottakkaran Sooppy Nisar, Atif Hassan Soori, "Mathematical analysis of novel corona virus (2019-nCov) delay pandemic model", Comput. Mater.Con., 64, pp.1401–1414, 2020.

[9] T. M. Chen, J. Rui, Q. P. Wang, Z. Y. Zhao, J. A. Cui, L. Yin, "A mathematical model for simulating the phase-based transmissibility of a novel corona virus", Infectious Disease of Poverty, 9(1), pp.1-8, 2020.

[10] S. E. Eikenberry, M. Mancuso, E. Iboi, T. Phan, K. Eikenberry, Y. Kuang, E. Kostelich, A. B. Gumel, "To mask or not to mask: Modeling the potential for face mask use by the general public to curtail the COVID-19 pandemic", Infectious Disease Modelling, 5, pp.293-308, 2020.

[11] Salihu Sabiu Musa, Sania Qureshi, Shi Zhao, Abdullahi Yusuf, Umar Tasiu Mustapha, Daihai He, "Mathematical modeling of COVID-19 epidemic with effect of awareness programs", Infectious Disease Modelling, 6, pp.448-460, 2021.

[12] Mansour A Abdulwasaa, Mohammed S Abdo, Kamal Shah, Taher A Nofal, Satish K Panchal, Sunil V Kawale, Abdel-Haleem Abdel-Aty, "Fractal-fractional mathematical modeling and forecasting of new cases and deaths of COVID-19 epidemic outbreaks in india", Results Phys., 20, 103702, 2021.

[13] Abdullah, Saeed Ahmad, Saud Owyed, Abdel-Haleem Abdel-Aty, Emad E Mahmoud, Kamal Shah, Hussam Alrabaiah, "Mathematical analysis of COVID-19 via new mathematical model", Chaos, Soliton Fract., 143, 110585, 2021.

[14] G. Maria Jones, S. G. Winster, "Prediction of Novel Corona virus (nCOVID-19) Propagation Based on SEIR, ARIMA and Prophet Model.

Predictive and Preventive Measures for Covid-19 Pandemic", Algorithms for Intelligent Systems, Springer, Singapore, 2021.

[15] A. G. M. Selvam, R. Janagaraj, R. Dhineshbabu, "Analysis of Novel Corona Virus (COVID-19) Pandemic with Fractional-Order Caputo-Fabrizio Operator and Impact of Vaccination", Mathematical Analysis for Transmission of COVID-19, Mathematical Engineering, Springer, Singapore, pp.1-28, 2021.

[16] Y. Chayu, J. A. Wang, "Mathematical model for the novel coronavirus epidemic in wuhan, china", Math Biosci Eng., 17(3), 2708, 2020.

[17] Anwarud Din, Tahir Khan, Yongjin Ji, Hassan Tahir, Asaf Khan, "Mathematical analysis of dengue stochastic epidemic model", Results Phys., 20, 103719, 2021.

[18] R. Khalil, M. Al Horani, A. Yousef, M. Sababheh, "A new definition of fractional Derivative", Journal of Computational and Applied Mathematics, 264, pp.65-70, 2014.

[19] M. Caputo, M. Fabrizio, "A new definition of fractional derivative without singular kernel", Progr. Fract. Differ. Appl., 1(2), pp.73-85, 2015.

[20] J. Losada, J. J. Nieto, "Properties of the new fractional derivative without singular kernel", Progr.Fract. Differ. Appl., 1(2), pp.87-92, 2015.

[21] G. Samko, A. Kilbas, O. Marichev, "Fractional integrals and derivatives: Theory and applications", Gordon and Breach, 1993.

[22] A. M. Shaikh, I. N. Shaikh, K. S. Nisar, "A mathematical model of COVID-19 using fractional derivative: outbreak in India with dynamics of transmission and control", Advances in Difference Equations, 2020(373), pp.1-19, 2020.

[23] E. A. Barbashin, "Introduction to the Theory of Stability", Groningen, The Netherlands: Walters-Noordhoff, 1970.

[24] R. C. Robinson, "An Introduction to Dynamical Systems: Continuous and Discrete", Englewood Cliffs, NJ, USA, Prentice-Hall, 2004.

[25] I. Petras, "Fractional-order Nonlinear Systems: Modeling, Analysis and Simulation", Springer, New York, USA, 2011.

[26] https://www.covid19india.org/.

[27] https://knoema.com/atlas/India/Life-expectancy.

[28] L. J. Allen, F. Brauer, P. Van den Driessche, J. Wu, "Mathematical Epidemiology", Berlin, Germany, Springer, 2008.

[29] T. Sardar, S. K. ShahidNadim, S. Rana, J. Chattopadhyay, "Assessment of lockdown effect in some states and overall India: A predictive mathematical study on COVID-19 outbreak", Chaos Solitons Fractals, 139, 110078, 2020, 10.1016/j.chaos.2020.110078.

[30] J. P. LaSalle, "The Stability of Dynamical Systems (Regional Conference Series in Applied Mathematics)", Philadelphia, PA, USA, SIAM, 1976.

3

Impact of Vaccination on COVID-19 to Control its Spread: A Case Study of India

Nita H. Shah, Ankush H. Suthar, and Ekta N. Jayswal

Department of Mathematics, Gujarat University, India
E-mail: nitahshah@gmail.com; ankush.suthar1070@gmail.com;
jayswal.ekta1993@gmail.com

Abstract

The world's largest vaccination program to control COVID-19, which began in India on 16 January 2021 with a goal of first immunizing 300 million people. As of 17 August 2021, more than 55 million people have been vaccinated in India. In this chapter, a fractional order compartmental model is constructed to study the effect of vaccination on the spread of coronavirus disease 2019 (COVID-19). In this compartmental model, the infected class is divided into two subclasses to study special characteristics of vaccination on moderately and severely infected classes separately. In particular, it includes an original approach that estimates the transmission rate of the disease over the ratio of partially and fully vaccinated populations, which provides the effect of the vaccination on transmission of the COVID-19. The chapter includes the study of a particular case of the outbreak in Indian states. The reported data for the disease spread and the rate of partial and complete vaccination in the Indian states are used to identify the model parameters. The stability of the model is proved using mathematical theory. The simulated results of the model show a graphical design of the estimations given by the model. We also study the behaviour of the transmission rate when the whole population is fully vaccinated and the output is plotted on a map using QGIS software. Moreover, the behaviour of the model for different fractional orders is also analysed in the model.

Keywords: COVID-19, Vaccination, Caputo Derivative, Stability, Basic Reproduction Number, Numerical Simulation

3.1 Introduction

The Wuhan district of China identified a novel coronavirus (COVID-19) for the first time in December 2019 [12]. After a short amount of time, the virus spread quickly throughout the world, and at the start of 2020, the World Health Organization (WHO) declared the COVID-19 outbreak to be a pandemic global health disaster [11]. Throughout the outbreak, the world has experienced unexpected disasters including 221 million infected cases and more than 5 million deaths (https://www.worldometers.info/coronavirus/). Till the end of 2021, the world has seen the two-wave pattern of COVID-19. Observed data for both waves show that the characteristics of the infection including diversity in age scale and severity of the disease do vary between both waves [8]. Yet, normalcy has not been declared as many countries are facing the third wave of the COVID-19 outbreak [6].

USA, India and Brazil are the top three countries that are highly affected due to the COVID-19 outbreak. Despite all preventive measures, these countries were not able to manage the transmission of COVID-19. The major reason for the disaster was the unknowing characteristics of the virus and the shortage of effective vaccines for the disease. The first vaccine for the novel coronavirus disease was registered for human clinical testing on 16 March 2020. And at the end of 2020, more than 200 vaccines are developed out of which 52 candidate vaccines are in human trials [1].

In this chapter, a study on the transmission of COVID-19 in India under the influence of the vaccination rate is performed using the mathematical model. It is observed that due to the vast diversity in different regions of India, the distribution of vaccines in the 1.39 billion population of India is a challenging task for the Indian Government. The Indian Government has started the administration of mass vaccination with two types of vaccines, Covishield and Covaxin on 16 January 2021. Both of these vaccines require two doses as researchers found that there was a stronger immune response when a second dose was added after some period of the first dose. As Of 23 September 2021, 45.5% of the Indian population are partially and 15% are fully vaccinated. Reported data from the Ministry of Health and Family

Welfare show that the vaccination rate is different in Indian states depending on the population density and associated administrators in the region. In respective Figures 3.1 and 3.2 the partial and complete vaccination rates in Indian states are plotted using QGIS software for the cumulative coverage report of COVID-19 vaccination provided by the Indian government. We are motivated to estimate the COVID-19 transmission rate for Indian states as a result of the wide range of vaccination rates that we have seen. By comparing the vaccination rate and COVID-19 transmission rate in the area, one may examine the effect of vaccination on the spread of the virus in a particular region. The benefits of vaccination are seen in the simulation of the model, and the model is also stabilized.

The most effective method for preventing the spread of COVID-19 illness is vaccination. An efficient plan is anticipated to vaccinate the sizable population and address the vaccine scarcity. Using an age-stratified mathematical model and optimization algorithms, Matrajt et al. assessed the efficiency of vaccine allocation for a variety of metrics (deaths, symptomatic infections, and maximal non-ICU and ICU hospitalizations) [10]. An SEIR model has been created by Foy et al. to assess the efficacy of plans made for COVID-19 vaccination in India [7]. Their findings are in line with international

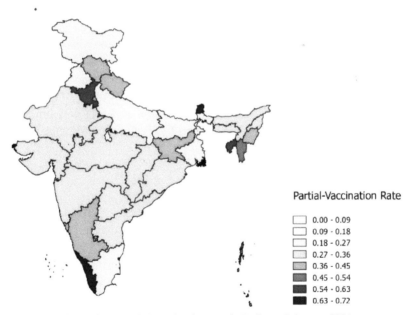

Figure 3.1 Partial vaccination rate in India on 9 August 2021.

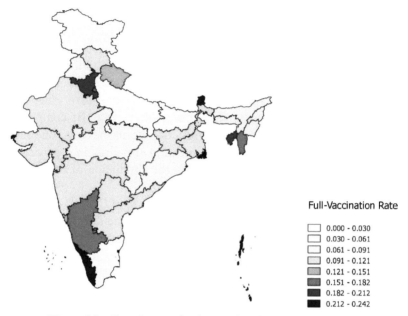

Full-Vaccination Rate

- 0.000 - 0.030
- 0.030 - 0.061
- 0.061 - 0.091
- 0.091 - 0.121
- 0.121 - 0.151
- 0.151 - 0.182
- 0.182 - 0.212
- 0.212 - 0.242

Figure 3.2 Complete vaccination rate in India on 9 August 2021.

suggestions that older age groups receive COVID-19 vaccination allocations first. The beneficial effects of non-pharmacological therapies were evaluated using a compartmental SEIR model created by Chatterjee et al., who also anticipated that hospitalization, intensive care unit (ICU) requirements and fatalities may be reduced by approximately 90% [3]. An age-structured epidemic model simulating COVID-19 transmission in India was built by Mandal et al. to examine the effects of different vaccination regimens on the morbidity and mortality of COVID-19 [9].

3.2 Mathematical Modelling

The mathematical model of the whole population is divided into seven compartments. The first compartment is a class of susceptible individuals (S), susceptible populations are divided into three compartments which are classified based on the vaccination. The second compartment is for non-vaccinated individuals (N_V), the third compartment (F_V) is a class of fully vaccinated individuals, and the fourth compartment contains the population which are partially vaccinated (who took only the first dose of the

COVID-19 vaccine). The infected class is divided into two compartments depending on the intensity of the infection as the compartment containing mild infected cases (I_M) and the compartment contains severe cases (I_S). The last compartment contains the class of recovered individuals (R). Figure 3.3 shows the schematic diagram of the model which shows the transmission of COVID-19.

$$
\begin{aligned}
\frac{dS}{dt} &= B - \beta_1 S N_V - \beta_2 S - \beta_3 S N_V - \mu S, \\
\frac{dN_V}{dt} &= \beta_3 S N_V - \beta_4 N_V - \beta_6 N_V - \beta_7 N_V - \mu N_V, \\
\frac{dP_V}{dt} &= \beta_1 S N_V + \beta_4 N_V - (\beta_5 + \beta_8 + \beta_9) P_V - \mu P_V, \\
\frac{dF_V}{dt} &= \beta_2 S + \beta_5 P_V - \beta_{10} F_V - \mu F_V, \\
\frac{dI_M}{dt} &= \beta_{10} F_V + \beta_8 P_V + \beta_6 N_V - \beta_{11} I_M - \beta_{12} I_M - \mu I_M, \\
\frac{dI_S}{dt} &= \beta_7 N_V + \beta_9 P_V + \beta_{11} I_M - \beta_{13} I_S - \mu \mu I_S, \\
\frac{dR}{dt} &= \beta_{12} I_M + \beta_{13} I_S - \mu R.
\end{aligned}
\tag{3.1}
$$

The dynamical system of differential equation (3.1) is generated from the model. Summing all equations from the system (3.1), the feasible region of the model is obtained as follows:

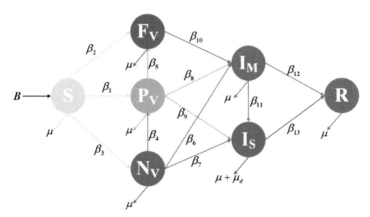

Figure 3.3   Schematic diagram of the model.

where $FR = \{(S, N_V, P_V, F_V, I_M, I_S, R) \in R_+^7 : S > 0, N_V, P_V, F_V, I_M, I_S, R \geq 0\}$.

The model (3.1) is modified by applying the Caputo fractional order derivative.

Definition 3.1. A y function's derivative in Caputo fractional order in the interval $[0, T]$ is defined by ${}^C D_{0+}^\alpha y(t) = \frac{1}{\Gamma(n-\alpha)} \int_0^t (t-s)^{n-\alpha-1} y^{(n)}(s) \, ds$, where C shows Caputo derivative, D^α denotes Caputo fractional-order derivative of order $n = [\alpha] + 1$ and $[\alpha]$ represents the integer part of α.

Definition 3.2. As defined by the Laplace transform of the Caputo derivative,

$$L\{D^\alpha y(t)\} = s^\alpha y(s) - \sum_{k=0}^{n-1} s^{\alpha-k-1} y^{(k)}(0), n - 1 < \alpha < n, n \in N.$$

The Ministry of Health and Family Welfare's website was consulted for the partial and complete vaccination data that are used to calculate the parametric values in Table 3.1 on 9 August 2021, where state-wise vaccination

Table 3.1   A description of the model's input parameters.

Parameter	Description
B	Birth rate
β_1/β_2	The rate of partial and complete vaccination rate in susceptible population class in the respective region
β_3	The rate by which susceptible population remain totally non-vaccinated
β_4	The frequency with which non-vaccinated people receive their initial dose or a half dose of the COVID-19 vaccine
β_5	The rate at which partial vaccinated individuals gets complete vaccination
β_6/β_7	The rate at which non-vaccinated individuals gets mild/severe infection
β_8/β_9	The rate at which partial vaccinated individuals gets mild/severe infection
β_{10}	The rate at which fully vaccinated individuals gets mild infection
β_{11}	The rate at which mild infected individuals moves to class of severely infected individuals
β_{12}/β_{13}	The rate at which mild/severe infected individuals gets recovered
μ_d	Death rate attributable to COVID-19 infection
μ	Natural death rate

data are also reported. Other parametric values are calculated from the data worldometer.com and other websites as mentioned below.

https://indianexpress.com/article/india/92-of-fully-vaccinated-hcws-who-got-covid-had-mild-infections-7364167/

https://www.thehindu.com/data/daily-covid-19-vaccination-rate-remains-consistently-poor-in-may/article34719605.ece

3.2.1 Equilibrium Points

(i) $E_0\left(\frac{B}{\mu},0,0,0,0,0,0\right)$

(ii) $E_1(S^1, N_V^1, P_V^1, F_V^1, I_M^1, I_S^1, R^1)$, where $S^1 = \frac{B}{\beta_2+\mu}, N_V^1 = 0, P_V^1 = 0$,

$$F_V^1 = \frac{\beta_2 S^1 + \beta_5 P_V^1}{\beta_{10} + \mu}, I_M^1 = \frac{\beta_{10}F_V^1 + \beta_8 P_V^1 + \beta_6 N_V^1}{\beta_{11} + \beta_{12} + \mu},$$

$$I_S^1 = \frac{\beta_7 N_V^1 + \beta_9 P_V^1 + \beta_{11}I_M^1}{\beta_{13} + \mu_d + \mu}, R^1 = \frac{\beta_{12}I_M^1 + \beta_{13}I_S^1}{\mu}.$$

(iii) $E^*(S^*, N_V^*, P_V^*, F_V^*, I_M^*, I_S^*, R^*)$, where

$$S^* = \frac{\beta_4 + \beta_6 + \beta_7 + \mu}{\beta_3}, N_V^* = \frac{B - (\beta_2 + \mu)S^*}{(\beta_1 + \beta_3)S^*}, P_V^* = \frac{\beta_1 S^* N_V^* + \beta_4 N_V^*}{\beta_5 + \beta_8 + \beta_9 + \mu},$$

$$F_V^* = \frac{\beta_2 S^* + \beta_5 P_V^*}{\beta_{10} + \mu}, I_M^* = \frac{\beta_{10}F_V^* + \beta_8 P_V^* + \beta_6 N_V^*}{\beta_{11} + \beta_{12} + \mu},$$

$$I_S^* = \frac{\beta_7 N_V^* + \beta_9 P_V^* + \beta_{11}I_M^*}{\beta_{13} + \mu_d + \mu}, R^* = \frac{\beta_{12}I_M^* + \beta_{13}I_S^*}{\mu}.$$

According to Caputo (1967), applying Caputo derivative to the system (3.1) [2]

$$^C D^\alpha S = B - \beta_1 SN_V - \beta_2 S - \beta_3 SN_V - \mu S,$$

$$^C D^\alpha N_V = \beta_3 SN_V - \beta_4 N_V - \beta_6 N_V - \beta_7 N_V - \mu N_V,$$

$$^C D^\alpha P_V = \beta_1 SN_V + \beta_4 N_V - (\beta_5 + \beta_8 + \beta_9)P_V - \mu P_V,$$

$$^C D^\alpha F_V = \beta_2 S + \beta_5 P_V - \beta_{10}F_V - \mu F_V, \tag{3.2}$$

$$^C D^\alpha I_M = \beta_{10}F_V + \beta_8 P_V + \beta_6 N_V - \beta_{11}I_M - \beta_{12}I_M - \mu I_M,$$

$$^C D^\alpha I_S = \beta_7 N_V + \beta_9 P_V + \beta_{11}I_M - \beta_{13}I_S - \mu_d I_S - \mu I_S,$$

$$^C D^\alpha R = \beta_{12}I_M + \beta_{13}I_S - \mu R.$$

Here, C shows Caputo derivative with order α having initial conditions $S(0) = S_0$, $N_V(0) = N_{V_0}$, $P_V(0) = P_{V_0}$, $F_V(0) = F_{V_0}$, $I_M(0) = I_{M_0}$, $I_S(0) = I_{S_0}$ and $R(0) = R_0$.

$$S(i+1) = S(i) + \frac{r^\alpha}{\Gamma(\alpha+1)}(B - \beta_1 SN_V - \beta_2 S - \beta_3 SN_V - \mu S),$$

$$N_V(i+1) = N_V(i) + \frac{r^\alpha}{\Gamma(\alpha+1)},$$
$$(\beta_3 S N_V - \beta_4 N_V - \beta_6 N_V - \beta_7 N_V - \mu N_V),$$

$$P_V(i+1) = P_V(i) + \frac{r^\alpha}{\Gamma(\alpha+1)}$$
$$(\beta_1 S N_V + \beta_4 N_V - (\beta_5 + \beta_8 + \beta_9) P_V - \mu P_V),$$

$$F_V(i+1) = F_V(i) + \frac{r^\alpha}{\Gamma(\alpha+1)} (\beta_2 S + \beta_5 P_V - \beta_{10} F_V - \mu F_V),$$

$$I_M(i+1) = I_M(i) + \frac{r^\alpha}{\Gamma(\alpha+1)}$$
$$(\beta_{10} F_V + \beta_8 P_V + \beta_6 N_V - \beta_{11} I_M - \beta_{12} I_M - \mu I_M),$$

$$I_S(i+1) = I_S(i) + \frac{r^\alpha}{\Gamma(\alpha+1)}$$
$$(\beta_7 N_V + \beta_9 P_V + \beta_{11} I_M - \beta_{13} I_S - \mu_d I_S - \mu I_S),$$

$$R(i+1) = R(i) + \frac{r^\alpha}{\Gamma(\alpha+1)} (\beta_{12} I_M + \beta_{13} I_S - \mu R). \tag{3.3}$$

The fundamental reproduction number for an integer order system is derived using the next-generation matrix approach.

$$F = \begin{bmatrix} \beta_3 S N_V \\ \beta_1 S N_V \\ 0 \\ 0 \\ 0 \\ 0 \\ 0 \end{bmatrix} \text{ and}$$

$$V = \begin{bmatrix} \beta_4 N_V + \beta_6 N_V + \beta_7 N_V + \mu N_V \\ -\beta_4 N_V + (\beta_5 + \beta_8 + \beta_9) P_V + \mu P_V \\ -\beta_2 S - \beta_5 P_V + \beta_{10} F_V + \mu F_V \\ -\beta_{10} F_V - \beta_8 P_V - \beta_6 N_V + \beta_{11} I_M + \beta_{12} I_M + \mu I_M \\ -\beta_7 N_V - \beta_9 P_V - \beta_{11} I_M + \beta_{13} I_S + \mu_d I_S + \mu I_S \\ -\beta_{12} I_M - \beta_{13} I_S + \mu R \\ -B + \beta_1 S N_V + \beta_2 S + \beta_3 S N_V + \mu S \end{bmatrix}.$$

After calculating Jacobian matrix of F and V are denoted by matrix $f = \left[\frac{\partial F_i(E_0)}{\partial X_j}\right]$ and $v = \left[\frac{\partial V_i(E_0)}{\partial X_j}\right]$, respectively, here v is non-singular matrix.

Therefore, the basic reproduction number at equilibrium point E_0 is $R_0 = \frac{B}{\beta_2+\mu}\left(\frac{\beta_3}{\beta_4+\beta_6+\beta_7+\mu} + \frac{\beta_1}{\beta_5+\beta_8+\beta_9+\mu}\right)$ [4, 5].

3.3 Stability

The Jacobian matrix of system (3.1) is

$$
J = \begin{bmatrix}
J_{11} & -\beta_3 S & -\beta_1 S & 0 & 0 & 0 & 0 \\
\beta_3 N_V & J_{22} & 0 & 0 & 0 & 0 & 0 \\
\beta_1 N_V & \beta_1 S + \beta_4 & J_{33} & 0 & 0 & 0 & 0 \\
\beta_2 S & 0 & \beta_5 & -\beta_{10} & 0 & 0 & 0 \\
0 & \beta_6 & \beta_8 & \beta_{10} & J_{55} & 0 & 0 \\
0 & \beta_7 & \beta_9 & 0 & \beta_{11} & J_{66} & 0 \\
0 & 0 & 0 & 0 & \beta_{12} & \beta_{13} & -\mu
\end{bmatrix},
$$

where $J_{11} = -\beta_1 P_V - \beta_2 - \beta_3 N_V - \mu$, $J_{22} = \beta_3 S - \beta_4 - \beta_6 - \beta_7 - \mu$, $J_{33} = -\beta_5 - \beta_8 - \beta_9 - \mu$, $J_{55} = -\beta_{11} - \beta_{12} - \mu$, $J_{66} = -\beta_{13} - \mu_d - \mu$.
E_0 is locally stable equilibrium point with eigenvalues of Jacobian matrix as follows:

$$\omega_1 = -\mu, \omega_2 = -\beta_{10}, \omega_3 = -\beta_5 - \beta_8 - \beta_9 - \mu, \omega_4 = -\beta_2 - \mu,$$
$$\omega_5 = -\beta_{11} - \beta_{12} - \mu, \omega_6 = -\beta_{13} - \mu_d - \mu,$$
$$\omega_7 = \frac{B\beta_2 - \mu(\beta_4 + \beta_6 + \beta_7 + \mu)}{\mu}.$$

Note that above all eigenvalues are negative except ω_7, but the ω_7 is negative when $B\beta_2 < \mu(\beta_4 + \beta_6 + \beta_7 + \mu)$. Hence it is the necessary condition for an equilibrium point E_0 to be stable.
E_1 is locally stable having eigenvalues:

$$\omega_1 = -\mu, \omega_2 = -\beta_{10}, \quad \omega_3 = -\beta_5 - \beta_8 - \beta_9 - \mu, \quad \omega_4 = -\beta_2 - \mu,$$
$$\omega_5 = -\beta_{11} - \beta_{12} - \mu, \quad \omega_6 = -\beta_{13} - \mu_d - \mu,$$
$$\omega_7 = \frac{B\beta_2 - (\beta_2 + \mu)(\beta_4 + \beta_6 + \beta_7 + \beta_2)}{(\beta_2 + \mu)}. \tag{3.4}$$

In similar way, the necessary condition for an equilibrium point E_1 to be stable is $B\beta_2 < (\beta_2 + \mu)(\beta_4 + \beta_6 + \beta_7 + \beta_2)$ and hence the E^* is locally stable under this condition.
Moreover, $|\arg(E_0)| > \frac{\pi}{2}, |\arg(E^1)| > \frac{\pi}{2}$ and $|\arg(E^*)| > \frac{\pi}{2}$. Hence, equilibrium points E_0, E^1 and E^* are locally asymptotically stable equilibrium points.

3.4 Numerical Simulation

The simulation is used to show how vaccination affects COVID-19 transmission in India. The parametric value ($B = 0.0176$, $\beta_1 = 0.2895$, $\beta_2 = 0.0827$, $\beta_3 = 0.6278$, $\beta_4 = 0.4611$, $\beta_5 = 0.2856$, $\beta_6 = 0.1$, $\beta_7 = 0.02$, $\beta_8 = 0.041$, $\beta_9 = 0.01$, $\beta_{10} = 0.01$, $\beta_{11} = 0.0230$, $\beta_{12} = 0.9895$, $\beta_{13} = 0.8001$, $\mu = 0.0072$ and $\mu_d = 0.0135$) is based on information about the COVID-19 epidemic in India that began on 30 June 2021, and it was used in the numerical simulation.

When the entire population is fully vaccinated, the intensity of COVID-19 disease in different regions of India is depicted in Figure 3.4. It can be observed that the transmission rate is high in some of the northern regions of the country (Uttar Pradesh, Bihar) where population density is high. Since the data are taken for fully vaccinated conditions, the transmission rate is less than 1. This situation shows that the infection will die out in some period. Hence vaccinating more and more populations can break the spread of COVID-19.

Figure 3.5 shows the variations in the compartments with time. The figure shows that in India, the number of vaccinated individuals improves during August 2021. Also, as a result of vaccination, mild and severe infection cases

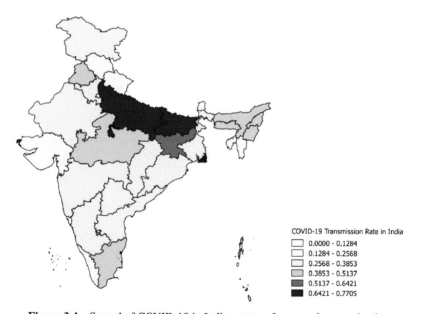

Figure 3.4   Spread of COVID-19 in Indian states after complete vaccination.

are shown reduced in the region. Moreover, increasing vaccination provides enough immunity against the infection; hence, high recovery rate is observed in the graph.

Graphs in figure 3.6 (a) and (b) show the bifurcation diagram on the class of non-vaccinated and partially vaccinated individuals concerning the frequency of transmission of COVID-19 in India. The graph is plotted for different values of the fractional order (α) of the differential equations generated from the model. The figure shows that as we decrease the value of α, a high variation in the partially vaccinated class is observed. Moreover, the sensitivity of alpha is increased for higher value of R_0.

In figure 3.7 (a) and 3.7 (b) shows variation in R_0 with respect to most sensitive parameters of the model. In figure 3.7 (a) impact of rate at which susceptible population is fully vaccinated β_2 on the threshold value of the infection R_0 is shown. Whereas figure 3.7 (b) shows variation in for the rate by which susceptible populations remain non vaccinated β_3, the rate by which non-vaccinated individuals got the first dose β_4, and the rate by which partially vaccinated individuals got the second dose β_5. It can be observed from the figures that β_2 is the most effective parameter to control the transmission of the infection. The rate β_2 can be improved by rising rates β_4 and β_5. Moreover, the rate β_3 is inversely affecting R_0 hence we need to stimulate individuals from the susceptible class to have the vaccine.

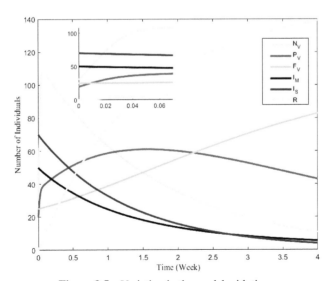

Figure 3.5 Variation in the model with time.

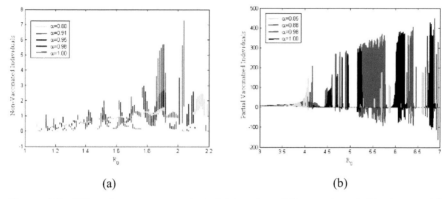

Figure 3.6 Bifurcation on non-vaccinated (a) and partially vaccinated (b) classes with respect to R_0.

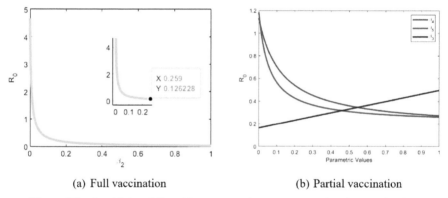

(a) Full vaccination   (b) Partial vaccination

Figure 3.7 Sensitivity of R_0 with respect to the parameters related to vaccination.

3.5 Conclusion

On 16 January 2021, India launched the largest COVID-19 vaccination programme in history. To study the impact of vaccination on the transmission of COVID-19 in India, a fractional order model is constructed in the chapter. In the compartmental model, the impact of partial and complete vaccination on the mild and severely infected classes is studied using the reported data for the Indian states. Section 3.3 provides confirmation that all equilibrium points are stable. QGIS software is used to forecast and plot the COVID-19 transmission rate in different parts of India following the immunization of the entire population. It is observed that the transmission rate is high in the region where population density is high. After analysing Figure 3.4, we can say that the infection will die out in some period if the whole population is fully vaccinated. Figure 3.6 shows the bifurcation on non-vaccinated and partially vaccinated classes with respect to the transmission rate (R_0). Moreover, the figure shows that the sensitivity of fractional order is increased at a higher value of R_0. The sensitivity analysis of R_0 demonstrates that the most effective criterion to prevent the spread of the infection is the rate at which the susceptible population gets fully immunised. Hence the study suggests that we need to motivate individuals from the susceptible class for vaccination.

Acknowledgements

The Department of Mathematics at Gujarat University acknowledges the technical assistance provided by DST-FIST file # MSI-097 to all the writers. A Junior Research Fellowship from the Council of Scientific & Industrial Research (file no.-09/070(0061)/2019-EMR-I) supports the work of the second author (AHS). The UGC awarded National Fellowship for Other Backward Classes to the third author (ENJ) (NFO-2018-19-OBC-GUJ-71790).

References

[1] Andreadakis, Z., Kumar, A., Román, R. G., Tollefsen, S., Saville, M., & Mayhew, S. (2020). The COVID-19 vaccine development landscape. Nature reviews. Drug discovery, 19(5), 305-306.

[2] Caputo, M. (1967). Linear models of dissipation whose Q is almost frequency independent—II. *Geophysical Journal International*, *13*(5), 529-539.

[3] Chatterjee, K., Chatterjee, K., Kumar, A., & Shankar, S. (2020). Healthcare impact of COVID-19 epidemic in India: A stochastic mathematical model. *Medical Journal Armed Forces India*, *76*(2), 147-155.

[4] Diekmann, O., Heesterbeek, J. A. P., & Metz, J. A. (1990). On the definition and the computation of the basic reproduction ratio R_0 in models for infectious diseases in heterogeneous populations. *Journal of Mathematical Biology*, *28*(4), 365-382.

[5] Driessche, P. & Watmough, J. (2002). Reproduction numbers and sub-threshold endemic equilibria for compartmental models of disease transmission, *Mathematical Biosciences*, *180*(1), 29-48.

[6] Fisayo, T., & Tsukagoshi, S. (2021). Three waves of the COVID-19 pandemic. *Postgraduate Medical Journal*, *97*(1147), 332-332.

[7] Foy, B. H., Wahl, B., Mehta, K., Shet, A., Menon, G. I., & Britto, C. (2021). Comparing COVID-19 vaccine allocation strategies in India: A mathematical modelling study. *International Journal of Infectious Diseases*, *103*, 431-438.

[8] Iftimie, S., López-Azcona, A. F., Vallverdú, I., Hernández-Flix, S., De Febrer, G., Parra, S., Hernández-Aguilera, A., Riu, F., Joven, J., Andreychuk, N. & Castro, A. (2021). First and second waves of coronavirus disease-19: A comparative study in hospitalized patients in Reus, Spain. *PloS one*, *16*(3), e0248029.

[9] Mandal, S., Arinaminpathy, N., Bhargava, B., & Panda, S. (2021). India's pragmatic vaccination strategy against COVID-19: A mathematical modelling based analysis. *medRxiv*.

[10] Matrajt, L., Eaton, J., Leung, T., & Brown, E. R. (2021). Vaccine optimization for COVID-19: Who to vaccinate first?. *Science Advances*, *7*(6), eabf1374.

[11] Praveen, S. V., Ittamalla, R., & Deepak, G. (2021). Analyzing the attitude of Indian citizens towards COVID-19 vaccine–A text analytics study. *Diabetes & Metabolic Syndrome: Clinical Research & Reviews*, *15*(2), 595-599.

[12] Zhou, S., Zhu, T., Wang, Y., & Xia, L. (2020). Imaging features and evolution on CT in 100 COVID-19 pneumonia patients in Wuhan, China. *European radiology*, *30*(10), 5446-5454.

4

A Computational Approach for Regulation of Biomedical Waste Expulsion in a Novel Coronavirus Pandemic

Oshin Rawlley[1], Yatendra Sahu[2], Rajeev Kumar Gupta[3], Amit Kumar Mishra[4], Ramakant Bhardwaj[5], and Satyendra Narayan[6]

[1]Vellore Institute of Technology, India
[2] Indian Institute of Information Technology and Management, India
[3] Pandit Deendayal Energy University, India
[4]Amity School of Engineering and Technology, Amity University, India
[5]Department of Mathematics, Amity University, India
[6]Department of Applied Computing, Sheridan Institute of Technology and Advanced Learning in Oakville, Canada
E-mail: oshinrawlley12@gmail.com; yatensahu@gmail.com; Rajeevmanit12276@gmail.com; akmishra1@gwa.amity.edu; rkbhardwaj100@gmail.com; narayan.satyendra@gmail.com

Abstract

The year 2020 for India turned out to be a nightmare for the world as the novel coronavirus penetrated every walk of our lives. In December 2019, the world got word of the pandemic caused by the novel coronavirus. On January 30, India reported its very first case in Kerala. The person had a travel history from Wuhan, China. Bygone times have also witnessed pandemics in every era. Right from the emergence of the 1720 plague, 1820 cholera, 1920 Spanish Flu, to COVID-2019, the epidemics have dilapidated the very foundation of the world in ways that have caused the very existential basis of humankind to be on crutches. The virus proliferated inevitably in nearby areas of Kerala, having a butterfly effect in continuation. In this situation, the health facilities were to be primed and proper management was required to

65

cater to the mass outbreak of the sick. The medical homes showed a surge in contaminated waste expelled from the hospitals. This biomedical waste creates a nuisance to nature and to people trying to manage the biomedical waste produced by hospitals. One aspect of living, that being waste management, has been sparked. This chapter hogs the limelight specifically on the bio-medical waste management that is being produced in the course of COVID-19. With increasing demands and usage of medical equipment, PPEs, etc., health care waste generation shows a surging trend. The sophisticated methods for waste disposal can be used to achieve minimal damage to nature, which is the intention of this chapter. Since the anti-drug for COVID is yet to come, proper management of waste dispensing should be held in priority as it may further invite unprecedented diseases that may cause greater social upheaval. Also, we acknowledge certain positive effects of the coronavirus on nature, which we term 'Lessons Learnt from Nature'.

Keywords: COVID, Environment, Water Pollution, Air Pollution, Medical Wastage.

4.1 Introduction

The world is in the grip of a pandemic that has claimed many lives so far. A spate of deaths has occurred, and still the incessant toll on health is on the rise. Human intervention has intruded into nature's well-being for decades [1, 2]. The overexploitation of flora and fauna has sabotaged our life-giving basics and given birth to new viruses which are unparalleled [3]. It should be no wonder that coronavirus has also originated from the uncontrolled and unethical exploiting of animals which may carry novel zoonotic diseases [4]. COVID-19 was declared a pandemic by the WHO on 11 March 2020. This outbreak led to people shuttling in hospitals with the increased use of medical instruments, masks, PPE kits, etc., for proper detection and care of themselves. Observing this, a precarious state was anticipated to characterize biomedical waste management, which is culminating in soaring risks to the environment and social health. Biomedical waste (BMW) is a type of pollutant in the form of outdated or expired drugs, human body parts, plastics (head/face masks, gloves, plastic shields, plastic body wraps, etc.), cotton, used syringes, cardboard, rubber and metals [5]. Figure 4.1 shows the history of the pandemics.

It is our responsibility to be aware of waste disposal techniques and to have a system in place that is resilient enough to meet the needs of the

Figure 4.1   History of pandemics.

outbreak crisis. Tonnes of biomedical pollutants are generated on an everyday basis without being able to manage to dispose of them. This is leading to another problem of land degradation, water pollution and the proliferation of diseases through oblivious respiration and contact. Since the last two decades, India has witnessed expeditious maturation in industries and technology, evident from the mass export of consignments of hydroxyl chloroquine. But there are rising concerns over waste generated while treating novel coronavirus patients. The sanitation workers, doctors and health workers who wear protection kits and one-time-use masks discard them every day. Proper management of waste is yet to be revised again and again to cope with the current situation. With a population of 1.2 billion, India has only 118 government-approved testing laboratories, 1.1363 beds per thousand patients (China has 4.2) and more than one million tests done [6]. On the contrary, India has recorded the highest recovery rate of 80% from 41.39% earlier.

But the challenge is rising slowly: handling large quantities of waste generated by the infected pupils. In Wuhan, where the novel coronavirus first emerged, the authorities had to build new hospitals to cater to the sudden influx of patients. They also had to establish a new medical waste plant for the treatment of waste. They deployed 46 facilities for mobile waste treatment. The waste generation statistics revealed about six times more production of medical waste than the hospitals generated when at the culmination point. China's hospitals produced 240 metric tonnes per day [7].

This improper waste management leads to serious diseases, respiratory ailments and degrading water bodies, and open burning leads to air pollution, like when crop burning takes place in Haryana and the whole state, along with Delhi, gets affected. The significance of BMW is to spread awareness among the pupils about keeping society and nature clean. It is our social responsibility also to have an effective waste management program. More sophisticated models are required for the monitoring of hospital waste. Technologies like non-burn waste for an eco-friendly practice to be adopted that is beneficial for the environment.

4.2 Literature Review

The management of bio-medical waste in times of a pandemic is gaining increased attention all over the world. Biomedical wastes (M&H) Rules 1998-E(P) Act 1986, as amended in 2000 and 2003. The BMW rules were proposed to be within the legal ambit and to be provided with guidelines and directives [8].

A literature review of the existing practices and models of waste management with a prime focus on better management of pollutants is acknowledged. The chapter reviews the practical problems encountered in segregation, transportation and disposal. Segregating at the very source has proven. It is essential to have a background of already existing practices for modelling new changes according to the situation in COVID. This review endeavours to accomplish the same.

4.2.1 BMW disposal

Normally, the process of BMW disposal from 'cradle to grave' includes characterization, quantification, segregation, storage, transportation, treatment and disposal [7]. The characterization process consists of segregation done at the ward or unit level. For the correct segregation process, colour coding for bags is used for the correct identification of waste. The identified waste is treated accordingly.

Infectious waste is only 13–20% of the total waste. Every type of waste has its own disposal method. Like yellow-coded bags, they carry infectious

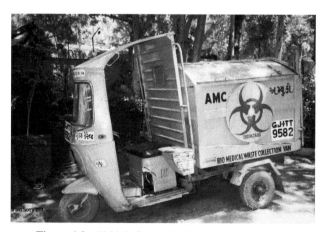

Figure 4.2 Vehicle for medical waste transportation.

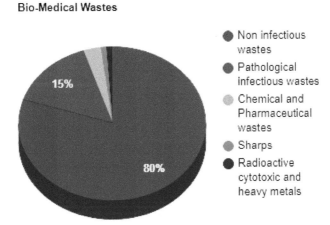

Bio-Medical Wastes

- Non infectious wastes
- Pathological infectious wastes
- Chemical and Pharmaceutical wastes
- Sharps
- Radioactive cytotoxic and heavy metals

Figure 4.3 Waste distribution of BMW.

waste in India. It contains contaminated cotton, tissues, body parts, cultures, etc., and finally it goes for incineration. While contaminated needles, plastics and rubber fall into another category, they have different routes. There are dedicated auto/vans to carry such waste, as shown in Figure 4.2. Also, you can see the waste distribution in Figure 4.3.

BMW for waste management is depicted below:

4.2.1.1 Mechanical processes
The physical form of waste is changed in a form so that handling becomes easy or another process is used in conjunction. The two processes are as follows:

- compaction and
- shredding.

4.2.1.2 Thermal processes
It uses heat to destroy medical waste. Most microorganisms are destroyed at 50°C and most living organisms are killed at 100°C systems which operate at 150°C, which is insufficient to destroy the materials. The high-heat system operates at more than 600°C to 2000°C. The three thermal processes are listed as follows:

a) **Thermal Techniques for Low Heat (90°C to 177°C):** Low-heat thermal techniques include the following techniques:

Autoclave – Steam sterilization is a low-heat thermal process in which waste is put in direct contact with steam for a long enough time to clean it.

Hydroclave – Indirect heating is done by providing steam in the outer jacket of a double-walled container while the waste inside the wall is turned on by a mechanism. The moisture content of the waste turns into steam. This causes waste to be fragmented. Sterilization takes 15 min at 132 and 30 min at 121°C.

Microwave – Unlike others, it heats waste internally. The electromagnetic radiation spectrum lies between the frequencies of 300 and 300,000 MHz.

Incineration – It is the controlled burning of waste at high temperatures to turn it into inert mineral residue and gases

Some other methods are primary and secondary combustion chambers, vacuum steam compaction, shredding and steam mixing, electro-thermal deactivation, dry heat treatment, etc. Figure 4.4 depicts the various thermal techniques used to destroy wastage.

Various thermal methods are used for the management of hazardous wastes. Incinerators usually burn the wastes at a high temperature, which has a dry oxidation process. A hydroclave is a cylindrical, full-jacketed vessel with a shredder attached. It rips apart the waste at a high temperature.

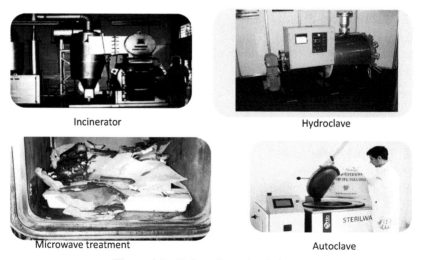

Incinerator

Hydroclave

Microwave treatment

Autoclave

Figure 4.4 Various thermal techniques.

Microwaving is a technique that uses wet heat generated by microwave energy, similar to incinerator hydroclaves, which also use high-heat systems. Autoclaving uses heat and pressure at the same time to sterilize medical items.

b) **Thermal Techniques for Medium Heat (177 °C–370 °C):** This technique includes the following:

Reverse Polymerization: Reverse polymerization, or RP, is a way to destroy organic material by using microwave energy in an atmosphere with a lot of nitrogen.
Thermal Depolymerization: Thermal depolymerization (TDP) is a method of depolymerization that transforms complex organic materials into light crude oil using hydrous pyrolysis.

c) **Thermal Techniques for High Heat (above 540 °C):** It includes methods like the following:

- pyrolysis–oxidation,
- plasma pyrolysis,
- laser-based pyrolysis,
- induction-based pyrolysis,
- superheated steam reforming and
- advanced thermal oxidation.

Due to incineration at a temperature lower than the standard ones, the standard ones are not achieved in many cases, which lead to the emission of toxic gases like dioxins and furans. The formula for evaluating the efficiency of an incinerator is evaluated as

$$CE = (\%CO_2 * 100)/\%CO_2 + \%CO,$$

where the temperature of primary chamber is $800 °C \pm 50 °C$ and secondary chamber gas residence time; minimum is 01 s at $1050 °C \pm 50 °C$ with 3% oxygen.

Biomedical waste incineration may be the best option for incineration of biomedical waste. It has high combustion efficiency and lower emissions than all other incinerators. It is suitable for watery, low-heat content waste like biomedical waste. It is suitable for small modular units and is easy to start-up. The types of incinerators have pretty much remained the same, but the emission limits have changed. Earlier emission limits are shown in Table 4.1.Figure 4.5 shown the emission limit for the different chemical components.

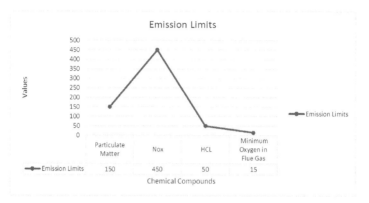

Figure 4.5   Emission limits before.

Table 4.1   Incineration emissions range.

Combustion efficiency	$<99\%$
HCl emission	49–196 mg/Nm^3
Particulate emission	132–8600 mg/Nm^3
NO$_x$ emission	Usually within the limits

Monitoring of incineration units was done which resulted in the following observations.

For the results obtained above, there could be numerous reasons listed below:

- improper handling of the unit,
- lack of waste segregation,
- batch operation of the unit,
- excess capacity of the unit,
- lack of mixing (turbulence) and
- absence of air pollution control equipment.

Remedial measures:

There are certain remedial measures which can be undertaken:

- proper segregation of wastes,
- continuous operation as far as possible,
- central facilities rather than small individual units,
- avoid excess capacity,
- air pollution control equipment and
- operate primary chamber in the starved air mode and the secondary in the excess air mode

4.2.1.3 Chemical processes

It is synonymous with chemical disinfection. Various chemicals are used, like alcohols, iodine, phenolic compounds and chlorine compounds.

Water is needed to bring the chemicals and microorganisms together as necessary to achieve inactivation.

4.2.1.4 Irradiation processes

It is synonymous with EM or ionizing radiation. The process utilizes Cobalt-60 and an electron beam gun.

4.2.1.5 Biological processes

Biological enzymes are used for treating BMW. It is claimed that biological reactions will not only decontaminate the waste but will also cause the destruction of all organic constituents, so that only plastics, glass and other inerts will remain in the residue.

The authors, like Wang et al. [19], studied the impacts of air quality due to the complete lockdowns in China in January and February 2020. The difficulties of scaling up pharmaceutical supplies on a global scale for mass production of potential vaccines and drugs were also mentioned. The sustainability of proper production and supply is discussed by Sarkis et al., where they explored the opportunities of doing so after the COVID-19 pandemic [9].

4.3 Latest Methodology

To ensure proper disposal of BMW, the Central Pollution Board of India has declared certain special guidelines. The pollution watchdog, the CPCB, has released steps for treatment and disposal of biomedical waste. The health care facilities, i.e. HCFs, must do strict segregation of BMW, the waste generated from immunization, treatment things like syringes, cotton swabs, needles, cytotoxic drugs, etc. [2]. Figure 4.6 shows the incineration process after segregation.Table 4.2 shown the comparative analysis of different methods used for the red bags disposal.

Yellow-, blue- and red-coloured bags are chosen for clarity of segregation and, according to the colour codes, different treatments are applied to them. Table 4.3 depicts the various treatments the colour-coded bags call for.

The Figure 4.7 depicts the production and treatment trends of the medical wastage during 2007-2020. According to the 2011 draft, all HCFs must have BMW operational facilities to facilitate required treatment and disposal of pollutants [3]. Adding to this, no HCF should be setting up the treatment plant on-site if CBMWF is reachable within a 75 km range. We can see the

Figure 4.6   Incineration after segregation.

Table 4.2   Methods used for red bag disposal.

Red bin	PPEs	Local disinfection method to be used
Red Bin-01	Goggles/Face shield	Immerse in a 0.5% sodium hypochlorite solution (freshly prepared) for 10 minutes, dry and wipe with a 70% alcohol swab.
Red Bin-02	N-95 Masks, Cover-alls	Store in double bags (red) and hand over to authorized sanitation and housekeeping staff twice daily.
Red Bin-03	Disposable PPEs	To be handed over to the authorized waste-collecting staff of M/s Biotic Waste Solutions Pvt. Ltd.

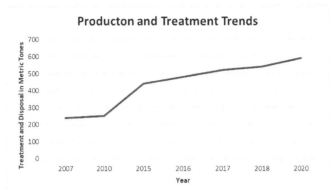

Figure 4.7 Production and treatment trend.

Table 4.3 Colour coding and type of containers for BMW.

Colour coding	Type of container	Waste category	Treatment/disposal
Yellow	Plastic bags	1,2,3,6	Incineration/deep burial
Red	Disinfected container/plastic bags	3,6,7	Autoclaving, microwaving and chemical treatment
Blue/white (translucent)	Plastic bags/ puncture proof containers	4 & 7	Autoclaving, microwaving and chemical treatment, destruction/shredding
Black	Plastic bags	5,9,10	Disposal in the secure landfill

surging trend of production and treatment of bio-medical waste from earlier years to COVID-19, 20 below.

4.4 Emission New Rules

- The incinerator's stricter emissions range means revised standards of 50 mg/Nm^3 from the earlier acceptable ones of 150 mg/Nm^3 at 12% CO_2 correction.
- The standards of retention time in the secondary chamber are updated to 2 seconds from the earlier one second. Earlier, due to non-achievability of standard temperatures, releases of furans and dioxins were reported, which have been checked by maintaining the above-mentioned conditions.

- To comply with the rules of social distancing, a new surveillance technique is being used by Natural Resources Wales to deal with reprobate waste operators showing a callous approach to their duty, leading to a lag in the cleanliness in the COVID crisis. Drone technology has been implemented to have aerial surveillance of the operations involved in managing waste. This technology has turned down the greasy hands and forced them to carry out the tasks under strict vigilance to avoid any malaise [1].
- Also, satellite cameras have been flown for tracking purposes to aid in grabbing waste criminals.
- Drone technology helps in the identification of illegal waste disposal punks. The SMART Waste project has undertaken different satellite imagery trials for high-resolution imageries for distinct vigilance.
- Now the hospital waste, such as personal protective equipment, disposable food and disinfecting wipes, is to be disposed of in solid waste properly bagged in designated colour-coded bags.
- The patient's room waste should be immediately attended to, and waste bags must be replaced frequently.
- Alcohol/soap-based disinfectants should be used in regulated frequencies.
- Gloves, PPE kits, masks, face shields, plastic body covers, head covers and sanitizer bottles have a high calorific value. These can be disinfected by autoclaving or shredding and then perhaps recycled for better sustainability.
- If these measures are not taken seriously, they may further increase the risk of community spread of COVID.

Self-medication has been fatal in many cases of COVID because unobserved and minor things are avoided by patients in ignorance, which when severe becomes difficult to treat at a later stage [10, 11]. Because the cells may have multiplied by that time, which is unable to be subdued using unidentified anti-drugs. This may sometimes overwhelm the health services, recording an increase in the death toll [12, 13].

4.5 Effects on Nature

If we flip the coin, it has a positive side to the COVID crisis in nature. In broad strokes, in every era of 100 years, pandemics and epidemics have taken a mammoth toll on the health of people and claimed many lives [14, 15]. The precedents have enough evidence to acknowledge this fact. Nature had been throttled for many years in contemplation of industrial and technological development, envisaging a better world. Nature has been grappling with toxic pollutants in numerous ways, be it vehicular pollution, cutting trees in furtherance of erecting brick and cement, tainting water sources in the name of their own cultural shams, etc. Like, plastic-based materials are posing a problem for the environment [16]. Tracey Read, founder of a non-plastic organization called 'Seas without Plastic' in Hong Kong, stated that masks are made of polypropylene, a type of plastic which is very difficult to mix up in nature. There has been notable evidence reported by environmentalists now that discarded masks are found on the sea beaches in Mumbai (India), China, etc. But the pandemic has compulsively engaged humankind and the state of its very existence in such a way that Mother Earth has got a very good breather for herself [17].

Due to lockdown, air pollution levels have drastically come down. Delhi, one of India's most populous cities, has seen its air quality index (AQI) fall to 170 after reaching 240 and above (pre-lockdown), choking people's breaths and causing allergic reactions in eyes, as well as complaints of heavy heads. About 60% improvement has been shown due to reduced road activities and industrial emissions [18].

NO_2 and CO have also shrunk considerably during the lockdown. The picture below shows the satellite image after and before, respectively.The Figure 4.8 shown PM levels before and after the Covid for the Delhi region.

4.6 Conclusion

Waste management is one of the most vital sections of a country's economy that needs immediate governmental attention. Practices for proper disposal should be regulated periodically to ensure updated methods of managing the waste. Awareness of these methods among people will encourage them to be careful at the very source level for the good functioning of subsequent processes. Also, it will be a great contribution from humankind to nature's well-being. Due to the coronavirus, a huge amount of medical waste is generated per day, which is a type of pollutant in the form of outdated/expired drugs, human body parts, plastics (head/face masks, gloves, plastic shields, plastic body wraps etc.), cotton, used syringes, cardboard, rubber and metals. The coronavirus is the result of uncontrolled and unethical animal exploitation. Improper waste management leads to serious diseases, respiratory ailments and degraded water bodies, and open burning leads to air pollution. As a result, there is an urgent need for the efficient management of this medical waste. In Wuhan, where the novel coronavirus first emerged, the

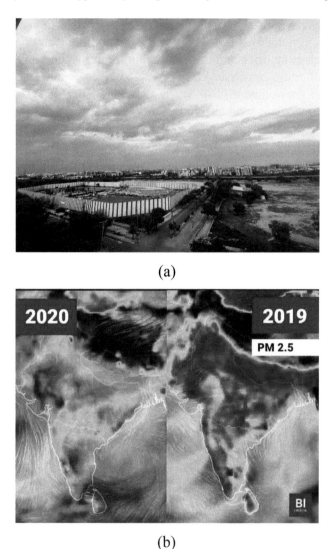

(a)

(b)

Figure 4.8   (a) Delhi regained its breath. With the onset of the pandemic Delhi saw a positive change in its nature. The animals started floating on the ground, river Yamuna was taking good breath also the pollution levels were gone down. This picture displays the same where the clear skies are flowing with drastic reduced pollution levels. (b) PM 2.5 levels in India's atmosphere compared between 30 March 2020 Italy also observed dips in the NO_2 levels seen in the satellite images. The NO_2 levels saw a great fall due to the reduced human intervention in the nature. The satellite images were able to capture these pictures showing a positive impact on the nature. Clearly indicating the fact that how we have destroyed the nature in various forms and how the pandemic showed the improved levels in air quality index.

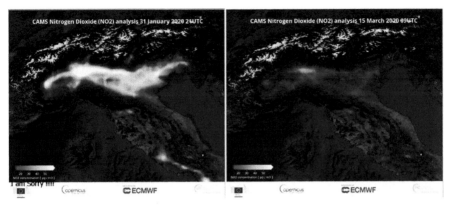

Figure 4.9 Surface concentrations of nitrogen dioxide over northern Italy, 31 January versus 15 March 2020. Copernicus Atmosphere Monitoring Service (CAMS); ECMWF. These are the dense surface concentrations of NO_2 which has shown a reduction trend after the pandemic. These pictures confirm a gradual decrease in the pollutants level for the Northern Italy.

authorities had to build new hospitals to cater to the sudden influx of patients. The Figure 4.9 shown the surface concentrations of Nitrogen Dioxide over northern Italy. They also had to establish a new medical waste plant for the treatment of waste. They deployed 46 facilities for mobile waste treatment. This chapter discussed various methods for disposing of medical waste.

References

[1] Fréderic Dutheil, Julien S Baker, and Valentin Navel. "COVID-19 as afactor influencing air pollution?" In:Environmental Pollution (Barking, Essex: 1987)263 (2020), p. 114466.

[2] Mohamed E El Zowalaty, Sean G Young, and Josef D Järhult. Environmental impact of the COVID-19 pandemic–a lesson for the future. 2020.

[3] Hasan Ero ğlu. "Effects of Covid-19 outbreak on environment and renewableenergy sector". In:Environment, Development and Sustainability(2020),pp. 1–9.

[4] Snehal Lokhandwala and Pratibha Gautam. "Indirect impact of COVID-19 on environment: A brief study in Indian context". In:Environmental research188 (2020), p. 109807.

[5] World Health Organization et al. Infection prevention and control during health care when COVID-19 is suspected: interim guidance, 19 March 2020.Tech. rep. World Health Organization, 2020.

[6] Stuart Oskamp. "Resource conservation and recycling: Behavior and pol-icy". In:Journal of Social Issues51.4 (1995), pp. 157–177

[7] Joana C Prata et al. "COVID-19 pandemic repercussions on the use andmanagement of plastics". In:Environmental Science & Technol-ogy54.13(2020), pp. 7760–7765.

[8] Jeffrey NT Squire. "Biomedical pollutants in the urban environment andimplications for public health: a case study". In:ISRN Public Health2013(2013).

[9] Manuel A Zambrano-Monserrate, Marıa Alejandra Ruano, and Luis Sanchez-Alcalde. "Indirect effects of COVID-19 on the environment". In:Science of the Total Environment(2020), p. 138813.

[10] Airborne Nitrogen Dioxide Plummets Over China. NASA's Earth Observing System (EOS), NASA Goddard Space Flight Center, National Aeronautics and Space Administration, USA. Accessed (20 March 2020)

[11] Myllyvirta L. Analysis: coronavirus has temporarily reduced China's CO2 emissions by a quarter. London, UK: Carbon Brief. (Accessed 23 March 2020)

[12] Amid COVID-19, biomedical waste turning more hazardous: The Financial Express: access on: 25/03/2020: Available at: https://thefin ancialexpress.com.bd/health/amid-covid-19-biomedical-waste-turning -morehazardous-1586504008.

[13] Untreated medical waste: A serious threat to public health: The Daily Star: access on: 28/10/2019 Available at: https://www.thedailystar.net/f rontpage/news/untreated-medical-waste-serious-threatpublic-health-18 19624.

[14] S. Jobling, M. Nolan, C. R. Tyler, G. Brighty, and J. P. Sumpter,"Widespread sexual disruption in wild fish," Environmental Science and Technology, vol.32, no.17, pp.2498–2506,1998.

[15] S. V. Manyele, "Effects of improper hospital-waste management on occupational health and safety," African Newsletter on Occupational Health and Safety, vol. 14, pp. 30–33, 2004.

[16] B. Mbongwe, B. T. Mmereki, and A. Magashula, "Healthcare waste management: current practices in selected healthcare facilities, Botswana," Waste Management, vol.28, no.1, pp. 226–233, 2008.

[17] P. Klangsin and A. K. Harding, "Medical waste treatment and disposal methods used by hospitals in Oregon, Washington, and Idaho," Journal of the Air and Waste Management Association, vol. 48, no. 6, pp. 516–526, 1998.

[18] A. Pruss, E. Giroult, and P. Rushbrook, Eds., SafeManagement of Wastes fromHealth-Care Activities, World Health Organization, Geneva, Switzerland, 1999.

[19] Wang et al., 2020 J. Wang, J. Shen, D. Ye, X. Yan, Y. Zhang, W. Yang, X. Li, J. Wang, L. Zhang, L. Pan Disinfection technology of hospital wastes and wastewater: suggestions for disinfection strategy during coronavirus disease 2019 (COVID-19) pandemic in China.

5

Solution for Fractional-order Pneumonia–COVID-19 Co-infection

Nita H. Shah, Nisha Sheoran, and Ekta Jayswal

Department of Mathematics, Gujarat University, India
E-mail: nitahshah@gmail.com, sheorannisha@gmail.com,
jayswal.ekta1993@gmail.com

Abstract

A Caputo fractional order mathematical model is proposed and analysed to study the dynamics of pneumonia–COVID-19 co-infection. The existence and uniqueness of solution of the model is established. Basic reproduction number is calculated for pneumonia and COVID-19 infection. The model possesses four equilibrium points, namely, disease free, pneumonia-free, COVID-19-free and endemic point where both the disease co-exists. Local stability conditions are established for disease-free, pneumonia-free and COVID-19-free equilibrium point. Also forward bifurcation is plotted where disease-free losses its stability and stable endemic equilibrium exists as $R_0 > 1$. In simulation, solution for each class is plotted and convergence is shown. The role of fractional order derivative is studied for different values $\alpha = 0.85, 0.9, 0.95$ and 1; moreover, the importance of memory is analysed. It is concluded that with the increase in memory the prevalence of pneumonia–COVID-19 co-infective are decreased. An increment in recovered population is also observed when memory effect is introduced by considering fractional order model.

Keywords: Pneumonia, COVID-19, Co-infection, Caputo Derivative, Stability, Forward Bifurcation

5.1 Introduction

Mathematical modelling enables us to develop mathematical models to analyse, investigate, predict and also to solve problems in different fields of sciences, engineering, etc. Compartmental modelling is a part of mathematical modelling which is a very useful techniques applied to study dynamics of infectious diseases. It assists society in controlling the spread of any ongoing pandemic by studying macroscopic behaviour of diseases in a population. One of the ongoing diseases is COVID-19 pandemic which aroused in December 2019, Wuhan in a small province of China [1]. COVID-19 virus spreads through respiratory droplets of infected person or by coming in contact with contaminated objects [2]. Individuals with some or other disease are at greater risk of developing COVID-19 than the healthy individuals. It was observed from the statistical data that individuals with COVID-19 having comorbidities, like diabetes and asthma, were transferred to hospitals or intensive care unit at large [3–5].

Pneumonia stands out to be one of the leading causes of death in China [6]. Pneumonia is an airborne disease that infects the lungs causing inflammation of one or both the lungs. The air sacs get filled with pus or fluid causing cough, fever and difficulty in breathing. Shi et al. [7] in his study analysed 81 patients who were confirmed of having COVID-19–pneumonia infection infected by SARS-CoV-2 virus. Pneumonia was the first infection which was identified in COVID-19 infected patients. Therefore, it demanded study the dynamics of pneumonia–COVID-19 co-infection.

Many researchers have developed compartmental models to study COVID-19 pandemic and pneumonia infection separately or with other co-infections. The authors like Ossaiugbo and Okposo [8] have modelled pneumonia infection dynamics, Tilahun et al. [9] studied pneumonia and typhoid co-infection and studied cost-effectiveness analysis and concluded that treatment for pneumonia and typhoid infection was the cheapest and He [10] developed SEIR model of COVID-19 and studied non-linear dynamics of the disease. Others include study by [11–13]. Some of the researchers have studied COVID-19 along with comorbidity which include study by Hezam et al. [14]; he framed COVID-19 cholera co-infection model and studied it for Yemen and concluded the importance of sufficient testing and maintaining social distance and quarantining of the infected individuals can flatten the curve of epidemic. Omame et al. [15] analysed deterministic COVID-19 and comorbidity co-infection model where he considered diabetes mellitus as the comorbidity infection. He investigated COVID-19 re-infection and also carried out optimal control analysis.

Mathematical models can be of two-type integer order model (deterministic model) and fractional order model. In recent years, fractional order modelling is becoming a topic of interest for many. Fractional order modelling deals with memory effect [16]. As memory effect has a great role to play in analysing the dynamics of infectious diseases, modelling through fractional order is more beneficial than integer order modelling providing more insight into the disease dynamics. Some of the fractional order models include model studied by [17–20].

To the best of the author's knowledge no pneumonia–COVID-19 co-infection is examined with a compartmental model. Encouraged by the aforementioned studies, we proposed mathematical model to study dynamics of pneumonia–COVID-19 co-infection by incorporating memory effect. The compartmental model is formulated in Section 2 along with the discussion of basic properties of the proposed model. The rest of the chapter is organized as follows: the equilibrium points and basic reproduction number are computed in Section 5.3. In Section 4, the stability analysis of the equilibrium points is carried out. Simulation is studied in Section 5 and finally conclusion is drawn in Section 5.6.

5.2 Formulation of Fractional Order Mathematical Model

5.2.1 Preliminaries of fractional order in Caputo sense

Definition 5.1 ([21]). The fractional order derivative in Caputo sense of order α of a function $f \in C^n$ is given by

$$_{\alpha}^{C}D_t^{\alpha} f(t) = \frac{1}{\Gamma(n-\alpha)} \int_{\alpha}^{t} \frac{f^n(\xi)}{(t-\xi)^{\alpha+n-1}} \, d\xi, \qquad n-1 < \alpha < n \in \yen. \quad (5.1)$$

A mathematical model is developed to gain insight into the spread of pneumonia–COVID-19 co-infection among different population classes using fractional derivative in Caputo sense since fractional derivative is known for its memory effect. The formulated model comprises of eight mutually exclusive classes: individuals who are prone to infection, Susceptible S; population exposed to COVID-19, E_{CO}-; exposed to pneumonia, E_{pn}; exposed to co-infection, i.e. both COVID-19 and pneumonia, E_{COpn};

COVID-19 infected, I_{CO}; pneumonia infected, I_{pn}; co-infected class, I_{COpn}; and recovered class, R.

The fractional order model is formulated considering following assumptions:

1. The new individuals to the susceptible class are recruited at the rate B.
2. Exposed to pneumonia, that is E_{pn}, is considered to be the class of individuals who carry pneumonia bacteria and are capable to transmit the infection.
3. Recovery class is assumed to be the class of recovered individuals either from COVID-19, pneumonia or both.
4. Natural death rate is taken to be μ from all the compartments. Also, death due to co-infection μ_{COpn} is taken into account.

Table 5.1   Parametric values with description used to model equations.

Notations	Description	Parametric values	References
B	Rate of birth	1	Assumed
β_1	Rate of exposure of susceptible to COVID-19	0.75	Assumed
β_2	Rate of exposure of susceptible to pneumonia	0.4	Assumed
β_3	Rate at which exposed gets infected with COVID-19	0.05	Calculated
β_4	Rate at which population exposed to COVID-19 gets exposed to co-infection	0.41	[22]
β_5	Rate at which population exposed to pneumonia gets exposed to co-infection	0.037	Assumed
β_6	Rate at which population exposed to pneumonia gets pneumonia infection	0.0109	[8]
β_7	Rate at which co-infection exposed population gets the infection	0.35	Assumed
β_8	Rate at which pneumonia infectives gets recovered	0.9	[23]
β_9	Recovery rate of co-infectives gets recovered	0.15	Assumed
β_{10}	Recovery rate of COVID-19 infectives	0.9	Calculated
μ_{COpn}	Death rate due to pneumonia–COVID-19 co-infection	0.04	[24]
μ	Natural mortality rate	0.002	Assumed

The parametric values and notations are described in Table 5.1 given below

The flow of populations from one class to another are connected by some rates as shown in Figure 5.1. Considering the above-mentioned assumptions following set of non-linear differential equations are formulated,

$$^{C}D^{\alpha}S = B - (\beta_1 I_{CO} + \beta_2 I_{pn} + \mu)S,$$
$$^{C}D^{\alpha}E_{CO} = \beta_1 S I_{CO} - (\beta_3 + \beta_4 I_{pn} + \mu)E_{CO},$$
$$^{C}D^{\alpha}E_{pn} = \beta_2 S I_{pn} - (\beta_5 I_{CO} + \beta_6 + \mu)E_{pn},$$
$$^{C}D^{\alpha}I_{CO} = \beta_3 E_{CO} - (\beta_{10} + \mu)I_{CO},$$
$$^{C}D^{\alpha}I_{CO} = \beta_3 E_{CO} - (\beta_{10} + \mu)I_{CO}, \tag{5.2}$$
$$^{C}D^{\alpha}E_{COpn} = \beta_4 E_{CO}I_{pn} + \beta_5 E_{pn}I_{CO} - (\beta_7 I_{COpn} + \mu)E_{COpn},$$
$$^{C}D^{\alpha}I_{pn} = \beta_6 E_{pn} - (\beta_8 + \mu)I_{pn},$$
$$^{C}D^{\alpha}I_{COpn} = \beta_7 E_{COpn}I_{COpn} - (\beta_9 + \mu + \mu_{COpn})I_{COpn},$$
$$^{C}D^{\alpha}R = \beta_8 I_{pn} + \beta_9 I_{COpn} + \beta_{10}I_{CO} - \mu R,$$

where $\alpha \in (0,1]$ is the order of derivative in Caputo sense with initial conditions $S(0) = S_0, E_{CO}(0) = E_{CO_0}, E_{pn}(0) = E_{pn_0}, I_{CO}(0) = I_{CO_0}, E_{COpn} = E_{COpn_0}, I_{pn}(0) = I_{pn_0}, I_{COpn}(0) = I_{COpn_0},$ and $R(0) = R_0.$ For $\alpha = 1$, the system behaves as an integer order model.

Taking Laplace transformation of eqn (5.2)

$$S = S_0 + L^{-1}\left[\frac{1}{r^{\alpha}}\{B - (\beta_1 I_{CO} + \beta_2 I_{pn} + \mu)S\}\right],$$

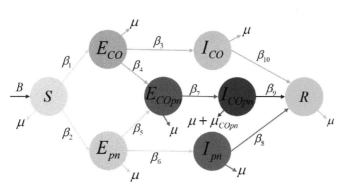

Figure 5.1 Transmission plot of population among different classes.

$$E_{CO} = E_{CO_0} + L^{-1}\left[\frac{1}{r^\alpha}\{\beta_1 SI_{CO} - (\beta_3 + \beta_4 I_{pn} + \mu)E_{CO}\}\right],$$

$$E_{pn} = E_{pn_0} + L^{-1}\left[\frac{1}{r^\alpha}\{\beta_2 SI_{pn} - (\beta_5 I_{CO} + \beta_6 + \mu)E_{pn}\}\right],$$

$$I_{CO} = I_{CO_0} + L^{-1}\left[\frac{1}{r^\alpha}\{\beta_3 E_{CO} - (\beta_{10} + \mu)I_{CO}\}\right],\qquad(5.3)$$

$$E_{COpn} = E_{COpn_0} + L^{-1}$$
$$\left[\frac{1}{r^\alpha}\{\beta_4 E_{CO}I_{pn} + \beta_5 E_{pn}I_{CO} - (\beta_7 I_{COpn} + \mu)E_{COpn}\}\right],$$

$$I_{pn} = I_{pn_0} + L^{-1}\left[\frac{1}{r^\alpha}\{\beta_6 E_{pn} - (\beta_8 + \mu)I_{pn}\}\right],$$

$$I_{COpn} = I_{COpn_0} + L^{-1}$$
$$\left[\frac{1}{r^\alpha}\{\beta_7 E_{COpn}I_{COpn} - (\beta_9 + \mu + \mu_{COpn})I_{COpn}\}\right],$$

$$R = R_0 + L^{-1}\left[\frac{1}{r^\alpha}\{\beta_8 I_{pn} + \beta_9 I_{COpn} + \beta_{10}I_{CO} - \mu R\}\right].$$

Applying discretization method discussed by Sayed [25], we get

$$S(i+1) = S(i) + \frac{s^\alpha}{\Gamma(\alpha+1)}$$
$$\{B - (\beta_1 I_{CO}(i) + \beta_2 I_{pn}(i) + \mu)S(i)\},$$

$$E_{CO}(i+1) = E_{CO}(i) + \frac{s^\alpha}{\Gamma(\alpha+1)}$$
$$\{\beta_1 S(i)I_{CO}(i) - (\beta_3 + \beta_4 I_{pn}(i) + \mu)E_{CO}(i)\},$$

$$E_{pn}(i+1) = E_{pn}(i) + \frac{s^\alpha}{\Gamma(\alpha+1)}$$
$$\{\beta_2 S(i)I_{pn}(i) - (\beta_5 I_{CO}(i) + \beta_6 + \mu)E_{pn}(i)\},$$

$$I_{CO}(i+1) = I_{CO}(i) + \frac{s^\alpha}{\Gamma(\alpha+1)}\{\beta_3 E_{CO}(i) - (\beta_{10} + \mu)I_{CO}(i)\},$$

$$E_{COpn}(i+1) = E_{COpn}(i) + \frac{s^\alpha}{\Gamma(\alpha+1)}$$
$$\{\beta_4 E_{CO}I_{pn} + \beta_5 E_{pn}I_{CO} - (\beta_7 I_{COpn} + \mu)E_{COpn}\},$$

$$(5.4)$$

$$I_{pn}(i+1) = I_{pn}(i) + \frac{s^{\alpha}}{\Gamma(\alpha+1)}\{\beta_6 E_{pn}(i) - (\beta_8 + \mu)I_{pn}(i)\},$$

$$I_{COpn}(i+1) = I_{COpn}(i) + \frac{s^{\alpha}}{\Gamma(\alpha+1)}$$
$$\{\beta_7 E_{COpn}(i)I_{COpn}(i) - (\beta_9 + \mu + \mu_{COpn})I_{COpn}(i)\},$$

$$R(i+1) = R(i) + \frac{s^{\alpha}}{\Gamma(\alpha+1)}$$
$$\{\beta_8 I_{pn}(i) + \beta_9 I_{COpn}(i) + \beta_{10}I_{CO}(i) - \mu R(i)\}.$$

5.2.2 Existence of positivity and boundedness of solutions

To prove the positivity and boundedness of solutions first we shall discuss some of the basic essentials to prove. Now, since all the eight compartments represent human population, it is necessary to show that they are positive for any time $t \geq 0$. The basics to establish the positivity in Caputo sense are as follows:

Lemma 5.2 ([26]). (Generalized mean value theorem) Suppose that $f(\omega) \in £[a,b]$ and $^C D^{\alpha} f(\omega) \in £[a,b]$ for $\alpha \in (0,1]$; then we have

$$f(\omega) = f(a) + \frac{1}{\Gamma(\alpha+1)}(^C D^{\alpha} f)(\xi)(\omega - a)^{\alpha},$$

with $a \leq \xi \leq \omega \forall \omega \in (a,b]$ and $\Gamma(.)$ represents gamma function.

Remark 5.3. Let $f(\omega) \in C[a,b]$ and $^C D^{\alpha} f(\omega) \in C[a,b]$ for $\alpha \in (0,1]$ then if

1. $^C D^{\alpha} f(\omega) \geq 0 \forall \omega \in (a,b)$ then $f(\omega)$ is increasing.
2. $^C D^{\alpha} f(\omega) \leq 0 \forall \omega \in (a,b)$ then $f(\omega)$ is decreasing.

Lemma 5.4 ([27]). Consider vector function $f : R^{\Lambda} + XR^{\Lambda}n \to R^{\Lambda}n.$ such that it satisfies following conditions:

1. Function $f(t, X(t))$ is Lebesgue measurable with respect to t on ¡$R^{\Lambda}+$.
2. It is continuous with respect to t on ¡$R^{\Lambda}n$.
3. $\frac{\partial f(t,X)}{\partial X}$ is continuous with respect to t on ¡$R^{\Lambda}n$.
4. $\|f(t,X)\| \leq \kappa + \delta \|X\|$ for $t \in¡$ $^+$ and $X \in¡R^{\Lambda}n$, where κ and δ are positive constants.

Then the initial value problem has a unique solution.

Theorem 5.5. The Biological feasible region for the system (5.2) is given by

$$
\Lambda = \left\{ \begin{array}{l} (S(t), E_{CO}(t), E_{pn}(t), I_{CO}(t), E_{COpn}(t), I_{pn}(t), I_{COpn}(t), R(t)) \in_{i+}^8, \\ S + E_{CO} + E_{pn} + I_{CO} + E_{COpn} + I_{pn} + I_{COpn} + R \le \frac{B}{\mu} \end{array} \right\},
$$

where existence, uniqueness and boundedness hold for the model.

Proof. Using Lemma 5.4 and the theorem given in Huo et al. [28], system (5.2) existence and uniqueness can be easily proved. Next, to prove Λ is a positively invariant we show that all the solutions of system (5.2) remain in Λ, for this we have

$$
{}^C D^\alpha S \big|_{S=0} = B \ge 0,
$$

$$
{}^C D^\alpha E_{CO} \big|_{E_{CO}=0} = \beta_1 S I_{CO} \ge 0,
$$

$$
{}^C D^\alpha E_{pn} \big|_{E_{pn}=0} = \beta_2 S I_{pn} \ge 0,
$$

$$
{}^C D^\alpha I_{CO} \big|_{I_{CO}=0} = \beta_3 E_{CO} \ge 0,
$$

$$
{}^C D^\alpha E_{COpn} \big|_{E_{COpn}=0} = \beta_4 E_{CO} I_{pn} + \beta_5 E_{pn} I_{CO} \ge 0, \qquad (5.5)
$$

$$
{}^C D^\alpha I_{pn} \big|_{I_{pn}=0} = \beta_6 E_{pn} \ge 0,
$$

$$
{}^C D^\alpha I_{COpn} \big|_{I_{COpn}=0} = 0,
$$

$$
{}^C D^\alpha R \big|_{R=0} = \beta_8 I_{pn} + \beta_9 I_{COpn} + \beta_{10} I_{CO} \ge 0,
$$

Thus, Λ is a positively invariant set by Remark 5.2. Next, to prove boundedness, we assume that $N(t) = S + E_{CO} + E_{pn} + I_{CO} + E_{COpn} + I_{pn} + I_{COpn} + R$.

Now from the system (5.2) we have

$$
{}^C D^\alpha (N(t)) \le B - \mu(N(t));
$$

then after solving, we obtain

$$
N(t) \le Bt^\alpha E_{\alpha,\alpha+1}(-\mu t^\alpha) + E_{\alpha,1}(-\mu t^\alpha) N(0),
$$

$$
N(t) \le \frac{B}{\mu} \frac{1}{\Gamma(1)} = \frac{B}{\mu}.
$$

Hence, the feasible region is positively invariant and is bounded by $\frac{B}{\mu}$. Therefore, our model is well posed epidemiologically and mathematically.

5.2.3 Equilibrium points and basic reproduction number

The model possesses four equilibrium points as follows:

1. Disease-free equilibria $E^0\left(\frac{B}{\mu}, 0, 0, 0, 0, 0, 0, 0\right)$.
2. Pneumonia-free equilibria

$$E^{0pn}\left(\frac{(\beta_3 + \mu)(\beta_{10} + \mu)}{\beta_1 \beta_3}, \frac{\mu(\beta_{10} + \mu)(R_{0co} - 1)}{\beta_1 \beta_3}, 0, \frac{\mu(R_{0co} - 1)}{\beta_1}, 0,\right.$$

$$\left. 0, 0, \frac{\beta_{10}(R_{0co} - 1)}{\beta_1}\right)$$

   exists if $R_{0co} > 1$.
3. COVID-19-free equilibria

$$E^{0CO}\left(\frac{(\beta_6 + \mu)(\beta_8 + \mu)}{\beta_2 \beta_6}, 0, \frac{\mu(\beta_8 + \mu)(R_{0pn} - 1)}{\beta_2 \beta_6}, 0, 0, \frac{\mu(R_{0pn} - 1)}{\beta_2},\right.$$

$$\left. 0, \frac{\beta_8(R_{0pn} - 1)}{\beta_2}\right)$$

   exists if $R_{0pn} > 1$.
4. Endemic point $E^*\left(S^*, E_{CO}^*, E_{pn}^*, I_{CO}^*, E_{COpn}^*, I_{pn}^*, I_{COpn}^*, R^*\right)$,

$$S^* = \frac{B}{(\beta_1 I_{CO}^* + \beta_2 I_{pn}^* + \mu)},$$

$$E_{CO}^* = \frac{B\beta_1 I_{CO}^*}{(\beta_3 + \beta_4 I_{pn}^* + \mu)(\beta_1 I_{CO}^* + \beta_2 I_{pn}^* + \mu)},$$

$$E_{pn}^* = \frac{B\beta_2 I_{pn}^*}{(\beta_5 I_{CO}^* + \beta_6 + \mu)(\beta_1 I_{CO}^* + \beta_2 I_{pn}^* + \mu)},$$

$$I_{CO}^* = \frac{B\beta_3 \beta_1 I_{CO}^*}{(\beta_{10} + \mu)(\beta_3 + \beta_4 I_{pn}^* + \mu)(\beta_1 I_{CO}^* + \beta_2 I_{pn}^* + \mu)},$$

$$E_{COpn}^* = \frac{\beta_9 + \mu + \mu_{COpn}}{\beta_7},$$

$$I_{pn}^* = \frac{1}{\beta_4}\left[\frac{\beta_1 \beta_3(\beta_8 + \mu)(\beta_5 I_{CO}^* + \beta_6 + \mu)}{\beta_2 \beta_6(\beta_{10} + \mu)} - (\beta_3 + \mu)\right],$$

$$I_{COpn}^* = \frac{\beta_4 E_{CO}^* I_{pn}^* + \beta_5 E_{pn}^* I_{CO}^*}{(\beta_9 + \mu + \mu_{COpn})} - \frac{\mu}{\beta_7},$$

$$R^* = \frac{\beta_8 I_{pn} + \beta_9 I_{COpn} + \beta_{10} I_{CO}}{\mu},$$

where I_{CO}^* is positive root of the following quadratic equation

$$f_1 I_{CO}^{*2} + f_2 I_{CO}^* + f_3 = 0.$$

Such that

$$f_1 = \beta_1 \beta_5 + \frac{\beta_1 \beta_3 \beta_5^2 (\beta_8 + \mu)}{\beta_4 \beta_6 (\beta_{10} + \mu)},$$

$$f_2 = \beta_1 (\beta_6 + \mu) + \beta_5 \mu + \frac{\beta_1 \beta_3 \beta_5 (\beta_6 + \mu)(\beta_8 + \mu)}{\beta_4 \beta_6 (\beta_{10} + \mu)} \left(2 - \frac{R_{0pn}}{R_{0co}} \right),$$

$$f_3 = \frac{\beta_1 \beta_3 \beta_5 (\beta_6 + \mu)(\beta_8 + \mu)}{\beta_4 \beta_6 (\beta_{10} + \mu)} \left(1 - \frac{R_{0pn}}{R_{0co}} \right) + \mu (\beta_6 + \mu)(1 - R_{0pn}),$$

Unique endemic point exits if $R_{0pn} > R_{0co} > 1$.

Next, the basic reproduction number is computed using the principle of next generation matrix method discussed in Diekmann et al. [29], which is given by $R_0 = \sqrt{R_{0pn} R_{0co}}$, where $R_{0pn} = \frac{B\beta_2 \beta_6}{\mu(\beta_6 + \mu)(\beta_8 + \mu)}$ and $R_{0co} = \frac{B\beta_1 \beta_3}{\mu(\beta_3 + \mu)(\beta_{10} + \mu)}$ are the basic reproduction number due to pneumonia and COVID-19, respectively.

5.3 Stability Analysis

In this section, local stability analysis is carried out for the equilibrium points as discussed in previous section. To prove the stability, Jacobian matrix of the system (5.2) is given by

$$J = \begin{bmatrix} -a_1 & 0 & 0 & -\beta_1 S & 0 & -\beta_2 S & 0 & 0 \\ \beta_1 I_{CO} & -a_2 & 0 & \beta_1 S & 0 & -\beta_4 E_{CO} & 0 & 0 \\ \beta_2 I_{pn} & 0 & -a_3 & -\beta_5 E_{pn} & 0 & \beta_2 S & 0 & 0 \\ 0 & \beta_3 & 0 & -(\beta_{10} + \mu) & 0 & 0 & 0 & 0 \\ 0 & \beta_4 I_{pn} & \beta_5 I_{CO} & \beta_5 E_{pn} & -(\beta_7 I_{COpn} + \mu) & \beta_4 E_{CO} & 0 & 0 \\ 0 & 0 & \beta_6 & 0 & 0 & -(\beta_8 + \mu) & 0 & 0 \\ 0 & 0 & 0 & 0 & \beta_7 I_{COpn} & 0 & -a_4 & 0 \\ 0 & 0 & 0 & \beta_{10} & 0 & \beta_8 & \beta_9 & -\mu \end{bmatrix},$$

(5.6)

where

$$a_1 = \beta_1 I_{CO} + \beta_2 I_{pn} + \mu, \, a_2 = \beta_3 + \beta_4 I_{pn} + \mu,$$
$$a_3 = \beta_5 I_{CO} + \beta_6 + \mu, \, a_4 = \beta_9 + \mu_{COpn} + \mu.$$

Theorem 5.6. The disease-free equilibrium point $E^0 \left(\frac{B}{\mu}, 0, 0, 0, 0, 0, 0, 0 \right)$ is locally asymptotically stable if $R_{0pn} < 1$ and $R_{0co} < 1$ otherwise unstable.

Proof. The Jacobian matrix J at disease-free equilibrium point is given by

$$
J_{E^0} = \begin{bmatrix}
-\mu & 0 & 0 & \frac{-\beta_1 B}{\mu} & 0 & \frac{-\beta_2 B}{\mu} & 0 & 0 \\
0 & -(\beta_3 + \mu) & 0 & \frac{\beta_1^* B}{\mu} & 0 & 0 & 0 & 0 \\
0 & 0 & -(\beta_6 + \mu) & 0 & 0 & \frac{\beta_2 B}{\mu} & 0 & 0 \\
0 & \beta_3 & 0 & -(\beta_{10} + \mu) & 0 & 0 & 0 & 0 \\
0 & 0 & 0 & 0 & -\mu & 0 & 0 & 0 \\
0 & 0 & \beta_6 & 0 & 0 & -(\beta_8 + \mu) & 0 & 0 \\
0 & 0 & 0 & 0 & 0 & 0 & -(\beta_9 + \mu_{COpn} + \mu) & 0 \\
0 & 0 & 0 & \beta_{10} & 0 & \beta_8 & \beta_9 & -\mu
\end{bmatrix}.
$$

The four eigenvalues of J_{E^0} are $-\mu, -\mu, -\mu, -(\beta_9 + \mu_{COpn} + \mu)$ and other eigenvalues can be calculated from the given polynomial equation

$$\lambda^4 + p_1 \lambda^3 + p_2 \lambda^2 + p_3 \lambda + p_4 = 0,$$

where

$$
\begin{aligned}
p_1 &= (\beta_3 + \beta_6 + \beta_8 + \beta_{10} + 4\mu), \\
p_2 &= (\beta_3 + \mu)(\beta_{10} + \mu)(1 - R_{0co}) + (\beta_6 + \mu)(\beta_8 + \mu)(1 - R_{0pn}) + \\
&\quad (\beta_3 + \beta_{10} + 2\mu)(\beta_6 + \beta_8 + 2\mu), \\
p_3 &= (\beta_3 + \mu)(\beta_{10} + \mu)(1 - R_{0co})(\beta_6 + \beta_8 + 2\mu) + (\beta_6 + \mu)(\beta_8 + \mu) \\
&\quad (1 - R_{0pn})(\beta_3 + \beta_{10} + 2\mu), \\
p_4 &= (1 - R_{0pn})(1 - R_{0co})(\beta_3 + \mu)(\beta_6 + \mu)(\beta_8 + \mu)(\beta_{10} + \mu).
\end{aligned}
$$

Using fractional Routh Hurwitz criterion [30], $p_1 > 0, p_2 > 0, p_3 > 0, p_4 > 0$ and $p_1 p_2 p_3 > p_3^2 + p_1^2 p_4$ then all the eigenvalues will have negative real part and satisfy $\arg(\lambda_i) > \frac{\pi}{2}$. Hence, disease-free equilibrium point is locally asymptotically stable if $R_{0pn} < 1$ and $R_{0co} < 1$ (for $p_2 > 0, p_3 > 0, p_4 > 0$, this is only possible if $R_{0pn} < 1$ and $R_{0co} < 1$).

Theorem 5.7. The pneumonia-free equilibrium point E^{0pn} is locally asymptotically stable if $R_{0co} > 1$ and $R_{0co} > R_{0pn}$.

Proof. The Jacobian at pneumonia-free equilibrium is computed whose eigenvalues are $-\mu, -\mu, -\mu, -(\beta_9 + \mu_{COpn} + \mu)$ and other eigenvalues are calculated from the following equation:

$$\lambda^5 + q_1 \lambda^4 + q_2 \lambda^3 + q_3 \lambda^2 + q_4 \lambda + q_5 = 0,$$

where

$$q_1 = \mu(R_{0co} - 1)\left(1 + \tfrac{\beta_5}{\beta_1}\right) + \beta_3 + \beta_6 + \beta_8 + \beta_{10} + 5\mu,$$

$$\begin{aligned}
q_2 &= \tfrac{1}{\beta_1}(\mu(R_{0co} - 1)^2\beta_5 + \beta_1\mu(R_{0co} - 1)(\beta_3 + \beta_6 + \beta_8 + \beta_{10} + 4\mu) \\
&\quad +\mu^2(R_{0co} - 1)\beta_5 + \mu(R_{0co} - 1)\beta_5(\beta_3 + \beta_8 + \beta_{10} + 3\mu)) + \\
&\quad \mu(\beta_3 + \beta_6 + \beta_8 + \beta_{10} + 4\mu) + (\beta_3 + \beta_{10} + 2\mu)(\beta_6 + \beta_8 + 2\mu) \\
&\quad + (\beta_6 + \mu)(\beta_8 + \mu)\left(1 - \tfrac{R_{0pn}}{R_{0co}}\right),
\end{aligned}$$

$$\begin{aligned}
q_3 &= \tfrac{1}{\beta_1}(\mu(R_{0co} - 1)(\beta_5(\beta_3 + \beta_8 + \beta_{10} + 3\mu)(R_{0co} - 1 + \mu) \\
&\quad +\beta_1((\beta_3 + \mu)(\beta_6 + \mu) + (\beta_6 + \mu)(\beta_{10} + \mu) + (\beta_{10} + \mu)(\beta_8 + \mu) \\
&\quad +(\beta_3 + \mu)(\beta_{10} + \mu)) + \beta_5(\beta_8 + \mu)(\beta_3 + \beta_{10} + 2\mu))) + \mu(R_{0co} - 1) \\
&\quad (\beta_6 + \mu)(\beta_8 + \mu)\left(1 - \tfrac{R_{0pn}}{R_{0co}}\right) + \left(1 - \tfrac{R_{0pn}}{R_{0co}}\right)((\beta_6 + \mu)(\beta_8 + \mu) \\
&\quad (\beta_3 + \beta_{10} + 3\mu)) + \mu(\beta_3 + \beta_{10} + 2\mu)(\beta_6 + \beta_8 + 2\mu),
\end{aligned}$$

$$\begin{aligned}
q_4 &= \tfrac{1}{\beta_1}(\mu(R_{0co} - 1)^2\beta_5((\beta_6 + \mu)(\beta_8 + \mu) + (\beta_8 + \mu)(\beta_{10} + \mu) \\
&\quad +(\beta_3 + \mu)(\beta_{10} + \mu)) + \mu^2\,(R_{0co} - 1)\beta_5(\beta_8 + \mu)(\beta_3 + \beta_{10} + 2\mu)) \\
&\quad +\mu(R_{0co} - 1)(\beta_3 + \mu)(\beta_{10} + \mu)(\beta_2 + \beta_8 + 2\mu) + \mu R_{0co}(\beta_2 + \mu) \\
&\quad (\beta_8 + \mu)(\beta_3 + \beta_{10} + 2\mu)\left(1 - \tfrac{R_{0pn}}{R_{0co}}\right),
\end{aligned}$$

$$\begin{aligned}
q_5 &= \tfrac{1}{\beta_1}(\mu(R_{0co} - 1)^2\beta_5(\beta_3 + \mu)(\beta_8 + \mu)(\beta_{10} + \mu)) + \mu(R_{0co} - 1) \\
&\quad (\beta_3 + \mu)(\beta_6 + \mu)(\beta_8 + \mu)(\beta_{10} + \mu)\left(1 - \tfrac{R_{0pn}}{R_{0co}}\right).
\end{aligned}$$

Using fractional Routh–Hurwitz criterion,
$q_1 > 0, q_2 > 0, q_3 > 0, q_4 > 0, q_5 > 0$ (this is possible if $R_{0co} > 1$ and $R_{0co} > R_{0pn}$),
$q_1q_2q_3 > q_3^2 + q_1^2q_4$ and $(q_1q_4 - q_5)(q_1q_2q_3 - q_1^2q_4 + q_3^2) > q_5(q_1q_2 - q_3)^2$; then real part of all the eigenvalues under these conditions is negative and can easily satisfy $\arg(\lambda_i) > \tfrac{\pi}{2}$. Hence, the pneumonia-free equilibrium point is locally asymptotically stable.

Theorem 5.8. The COVID-19-free equilibrium point E^{0CO} is locally asymptotically stable if
$R_{0pn} > 1$ and $R_{0pn} > R_{0co}$.

Proof. The eigenvalues of the Jacobian matrix at COVID-19-free equilibrium are given by $-\mu, -\mu, -\mu, -(\beta_9 + \mu_{COpn} + \mu)$ and other eigenvalues are computed from the following equation:

$$\lambda^5 + r_1\lambda^4 + r_2\lambda^3 + r_3\lambda^2 + r_4\lambda + r_5 = 0,$$

where

$$r_1 = \mu(R_{0pn} - 1)\left(1 + \frac{\beta_4}{\beta_2}\right) + \beta_3 + \beta_6 + \beta_8 + \beta_{10} + 5\mu,$$

$$r_2 = \frac{1}{\beta_2}(\mu(R_{0pn} - 1)^2\beta_4 + \beta_2\mu(R_{0pn} - 1)(\beta_3 + \beta_6 + \beta_8 + \beta_{10} + 4\mu)$$
$$+ \mu^2(R_{0pn} - 1)\beta_4 + \mu(R_{0pn} - 1)\beta_4(\beta_6 + \beta_8 + \beta_{10} + 3\mu))$$
$$+ \mu(\beta_3 + \beta_6 + \beta_8 + \beta_{10} + 4\mu) + (\beta_3 + \beta_{10} + 2\mu)(\beta_6 + \beta_8 + 2\mu)$$
$$+ (\beta_3 + \mu)(\beta_{10} + \mu)\left(1 - \frac{R_{0co}}{R_{0pn}}\right),$$

$$r_3 = \frac{1}{\beta_2}(\mu(R_{0pn} - 1)(\beta_4(\beta_6 + \beta_8 + \beta_{10} + 3\mu)(R_{0pn} - 1 + \mu)$$
$$+ \beta_2((\beta_3 + \mu)(\beta_6 + \mu) + (\beta_6 + \mu)(\beta_{10} + \mu) + (\beta_3 + \mu)(\beta_8 + \mu)$$
$$+ (\beta_{10} + \mu)(\beta_8 + \mu) + (\beta_6 + \mu)(\beta_8 + \mu)) + \beta_4(\beta_{10} + \mu)$$
$$(\beta_6 + \beta_8 + 2\mu))) + \mu(R_{0pn} - 1)(\beta_3 + \mu)(\beta_{10} + \mu)\left(1 - \frac{R_{0co}}{R_{0pn}}\right)$$
$$+ \left(1 - \frac{R_{0co}}{R_{0pn}}\right)((\beta_3 + \mu)(\beta_{10} + \mu)(\beta_6 + \beta_8 + 3\mu)) + \mu(\beta_3 + \beta_{10} + 2\mu)$$
$$(\beta_6 + \beta_8 + 2\mu),$$

$$r_4 = \frac{1}{\beta_2}(\mu(R_{0pn} - 1)^2\beta_4((\beta_6 + \mu)(\beta_8 + \mu) + (\beta_6 + \mu)(\beta_{10} + \mu)$$
$$+ (\beta_8 + \mu)(\beta_{10} + \mu)) + \mu^2(R_{0pn} - 1)\beta_4(\beta_{10} + \mu)(\beta_6 + \beta_8 + 2\mu))$$
$$+ \mu(R_{0pn} - 1)(\beta_6 + \mu)(\beta_8 + \mu)(\beta_3 + \beta_{10} + 2\mu) + \mu R_{0pn}(\beta_3 + \mu)$$
$$(\beta_{10} + \mu)(\beta_6 + \beta_8 + 2\mu)\left(1 - \frac{R_{0co}}{R_{0pn}}\right),$$

$$r_5 = \frac{1}{\beta_2}(\mu(R_{0pn} - 1)^2\beta_4(\beta_6 + \mu)(\beta_8 + \mu)(\beta_{10} + \mu)) + \mu(R_{0pn} - 1)$$

$$(\beta_3 + \mu)(\beta_6 + \mu)(\beta_8 + \mu)(\beta_{10} + \mu)\left(1 - \frac{R_{0co}}{R_{0pn}}\right).$$

Using fractional Routh Hurwitz criterion, $r_1 > 0, r_2 > 0, r_3 > 0, r_4 > 0, r_5 > 0$(possible only if $R_{0pn} > 1$ and $R_{0pn} > R_{0co}$), $r_1 r_2 r_3 > r_3^2 + r_1^2 r_4$ and $(r_1 r_4 - r_5)(r_1 r_2 r_3 - r_1^2 r_4 + r_3^2) > r_5(r_1 r_2 - r_3)^2$; then real part of all the eigenvalues under these conditions is negative and satisfy $\arg(\lambda_i) > \frac{\pi}{2}$. Hence the theorem.

5.4 Numerical Simulation

This section covers numerical solution for the model and the results obtained are analysed. The simulation is carried out using data of Table 5.1. The COVID-19 infected rate and its recovery rate are calculated from Ref. [31] in Table 5.1.

Fractional order model for pneumonia–COVID-19 co-infection is solved using Laplace iterative scheme. The approximate solution of each class of the system (5.2) is been plotted for different values of $\alpha \in (0, 1]$ shown in Figures 5.2– 5.9. It is clearly seen that solution obtained by iterative scheme of Laplace transformation gives us the promising solution predicting the behaviour of each population density within the region discussed in Section 5.2. Initially we observe that susceptible population increases once and then decreases since the susceptible population is exposed to COVID-19 and pneumonia. As shown in Figure 5.2 The population exposed to COVID-19 increases as α increases which is observed in Figure 5.3. Similarly, population exposed to pneumonia increases with the increase in α from 0.85 to 1 depicted in Figure 5.4. Also, COVID-19 infectives are increasing as shown in Figure 5.5. The population exposed to co-infection decreases with the increase in α since we can observe from Figure 5.8 that co-infected

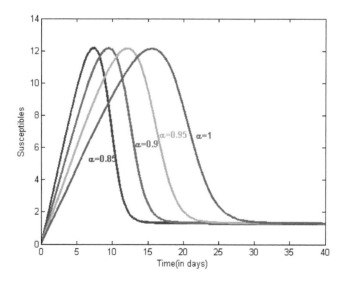

Figure 5.2   Solution of susceptible population.

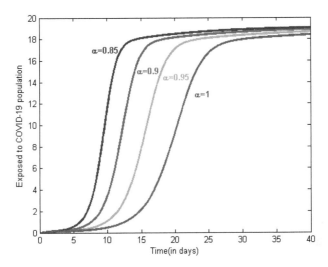

Figure 5.3 Solution of individuals exposed to COVID-19.

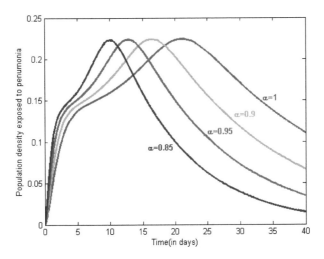

Figure 5.4 Solution of population exposed to pneumonia.

population is also decreasing as α decreases. Also, from Figure 5.7, we conclude that pneumonia infectives are decreasing with the decrease in α. As the pneumonia infectives are getting recovered thus increasing recovery with the decrease in α.

For the integer order system, it's been observed that the co-infected population is more at $\alpha = 1$ as compared to $\alpha = 0.85$, which states that growing memory results in reductions of the infection rate as shown in Figure 5.8. Increasing memory effect decreases infection as it holds the past information which proves to be helpful in current situation.

Hence, memory has a huge role to play also the system stabilizes faster with the increase in memory from $\alpha = 1$ to $\alpha = 0.85$ which can be very well observed in Figures 5.2– 5.9.

Also, the impact of integer order model is studied even though its impact is less than fractional order model.

Directional plots are also studied in order to observe the movements of some population through the system. In Figure 5.10, it is observed that susceptible population and population exposed to pneumonia are getting exposed to COVID-19. Similarly, Figure 5.11 illustrates that the flow of susceptible population exposed to pneumonia is getting infected with pneumonia. Figure 5.12 indicates that the population exposed to co-infection of pneumonia and COVID-19 is getting infected and after some time moves to recovery class.

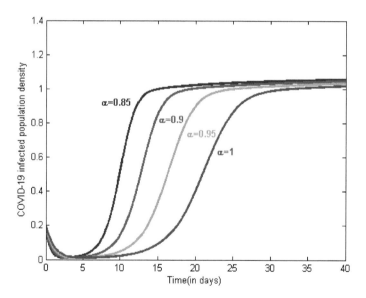

Figure 5.5   Solution of population infected by COVID-19.

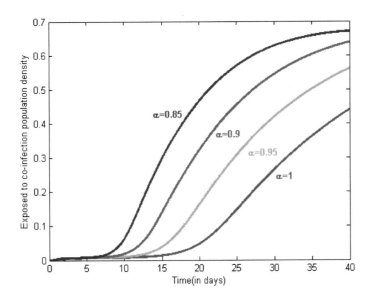

Figure 5.6 Solution of population exposed to co-infection.

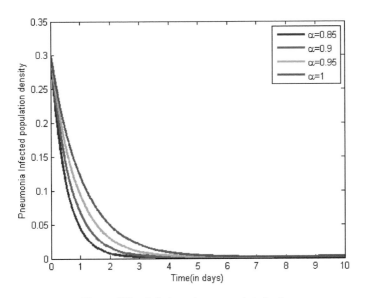

Figure 5.7 Solution of pneumonia infectives.

Figure 5.13 illustrates that proposed pneumonia–COVID-19 co-infection model undergoes forward bifurcation. The existence of stable disease-free equilibrium point is observed when there are no COVID-19 infectives, and as basic reproduction number cross the value one, there exist stable endemic equilibria and unstable disease-free equilibria. It indicates that elimination of COVID-19 infectives from society is possible if $R_0 < 1$.

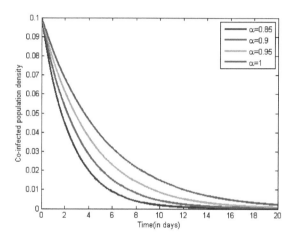

Figure 5.8   Solution of co-infected class.

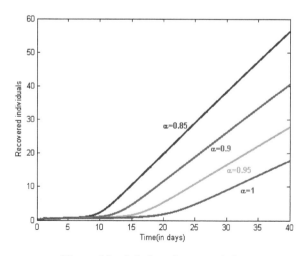

Figure 5.9   Solution of recovered class.

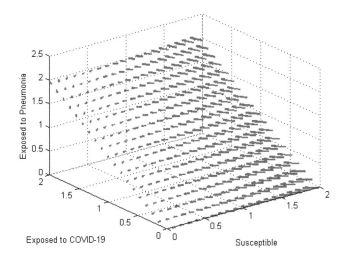

Figure 5.10 Directional plot among the susceptible class, COVID-19 exposed class and pneumonia exposed class.

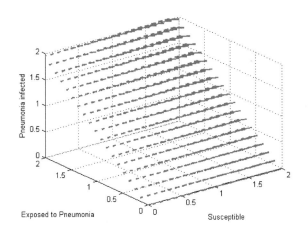

Figure 5.11 Directional flow of population among the susceptible class, class exposed to pneumonia and pneumonia infected class.

Figure 5.14 shows the flow of infected population among COVID-19, pneumonia and co-infected classes. It is observed that COVID-19 infected and pneumonia infected population decreases with time t. Co-infected population increases for two weeks after that it tends to decrease. In Figure 5.15, recovery of each infected class is plotted which indicates that initially recovery of co-infectives is slow but after four weeks recovery increases. Also, recovery of pneumonia infected patients is faster than

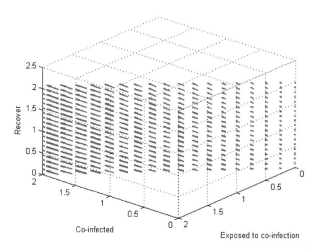

Figure 5.12   Directional plot of classes exposed to co-infection, co-infected and recover.

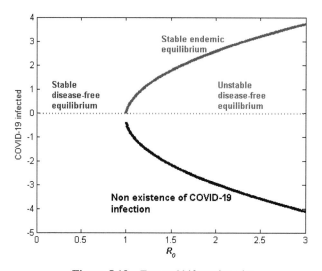

Figure 5.13   Forward bifurcation plot.

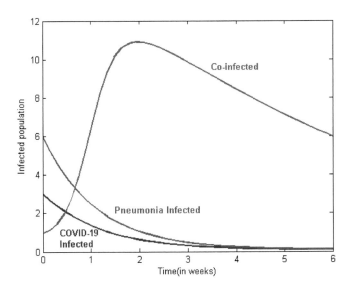

Figure 5.14 Behaviour of infected class considered in the model, i.e. COVID-19 infectives, pneumonia infected and co-infected.

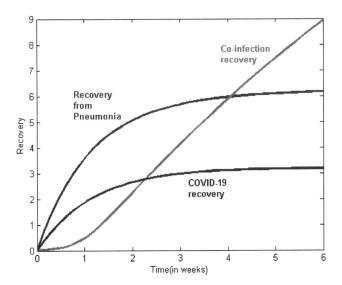

Figure 5.15 Recovery plotted for infected class.

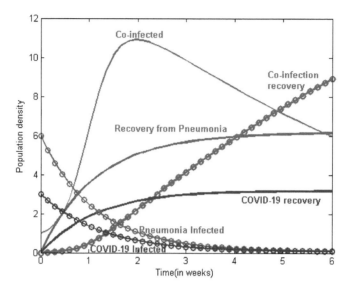

Figure 5.16   Plot of recovery and infectives of population of COVID-19, pneumonia and its co-infection.

recovery of COVID-19-infected patients. It also illustrates that the recovery curve for pneumonia and COVID-19 infectives flattens as time goes on. In Figure 5.16, the curve of infected class and recovery are plotted together which helps compare the recovery and number of infectives at a given time t. In the co-infected case, low recovery rate is observed when co-infection is at its peak phase while after two weeks, the number of co-infection cases drops with improvement in the recovery rate.

5.5 Conclusion

In the present chapter, the fractional order model of pneumonia–COVID-19 co-infection is studied. The model equilibrium points are computed which are disease-free, pneumonia-free, COVID-19 free and endemic equilibrium. Also, the existence and boundedness of the equilibrium points are verified. The expression of reproduction number R_{0pn} for pneumonia-only and R_{0co} for COVID-19 are calculated. The basic reproduction number for co-infection disease is given by $R_0 = \sqrt{R_{0pn}R_{0co}}$ which was calculated as 2.736. The local stability analysis is carried out for disease-free equilibrium point which is stable when $R_{0pn} < 1$ and $R_{0co} < 1$. Similarly, local stability conditions are established for pneumonia-free and COVID-19-free point. It is observed that the formulated model undergoes forward bifurcation which concludes that if $R_0 < 1$, COVID-19 infectives can be eradicated from the society. Therefore, it is possible to eliminate the co-infection of pneumonia–COVID-19 from the population. The model is also numerically simulated to study the effect of Caputo derivative for different α on the various population. The result showed that the rate of convergence is faster for fractional order rather than integer order. Also, with the increase in memory effect we observed decrease in co-infected population and increase in recovery of individuals. This concludes the importance of fractional order α. The simulation also showed which class of infection had the higher rate of recovery. Pneumonia patients were recovered at faster rate. The co-infected population was highest at second week and after that it started decreasing and recovery after some time crossed the infectives. From this, it is concluded that intensive care is to be provided to co-infectives for initial two weeks to slow down the spread of co-infection. Also, from the expression of R_0 we can observe that on reducing the exposure of susceptible population to COVID-19 and pneumonia one can reduce R_0 efficiently. The exposure to the disease can be prevented by using precautionary measures, like wearing mask, vaccination, social distancing, washing hands regularly, etc.

Acknowledgements

For technical support the authors thank DST-FIST file # MSI-097 provided to department of mathematics of Gujarat University. The second author (NS) thanks Gujarat government for giving SHODH Scholarship. The UGC awarded National Fellowship for Other Backward Classes to the third author (ENJ) (NFO-2018-19-OBC-GUJ-71790).

References

[1] C. Wang, P. W. Horby, F. G. Hayden & G. F. Gao, 'A novel coronavirus outbreak of global health concern', The lancet, pp. 470-473, Feb., 2020.

[2] D. Hui, et al., 'The continuing 2019-nCoV epidemic threat of novel coronaviruses to global health—The latest 2019 novel coronavirus outbreak in Wuhan, China', Int. J. Infect. dis., pp. 264-266, Feb., 2020.

[3] R. Gupta, A. Hussain, A. Misra, 'Diabetes and COVID-19: evidence, current status and unanswered research questions', Eur. J. Clin. Nutr., pp. 864-870, Jun., 2020.

[4] S. C. Lee, et al., 'Impact of comorbid asthma on severity of coronavirus disease (COVID-19)', Scientific reports, pp.1-9, Dec., 2020.

[5] A. Paramasivam, J. V. Priyadharsini, S. Raghunandhakumar, P. Elumalai, 'A novel COVID-19 and its effects on cardiovascular disease. Hypertension Research', pp. 729-730, Jul., 2020.

[6] X. Guan, et. al., 'Pneumonia incidence and mortality in Mainland China: systematic review of Chinese and English literature', PloS One, pp. 1985-2008, Jul., 2010.

[7] H. Shi, et al., 'Radiological findings from 81 patients with COVID-19 pneumonia in Wuhan, China: a descriptive study', The Lancet infectious diseases, pp. 425-434, Apr., 2020.

[8] M. I. Ossaiugbo, N. I. Okposo, 'Mathematical modeling and analysis of pneumonia infection dynamics', Sci. World J., pp. 73-80, Aug., 2021.

[9] G. T. Tilahun, O. D. Makinde, D. Malonza, 'Co-dynamics of pneumonia and typhoid fever diseases with cost effective optimal control analysis', Appl. Math. Comput., pp. 438-459. Jan., 2018.

[10] S. He, Y. Peng, K. Sun, 'SEIR modeling of the COVID-19 and its dynamics', Nonlinear dyn., pp. 1667-1680, Aug., 2020.

[11] S. I. Araz, 'Analysis of a Covid-19 model: optimal control, stability and simulations', Alex. Eng. J., pp. 647-658, Feb., 2021.

[12] O. J. Peter, et al., 'A new mathematical model of COVID-19 using real data from Pakistan', Results Phys., May, 2021.

[13] I. A. Baba, 'Mathematical model to assess the imposition of lockdown during COVID-19 pandemic', Results Phys., May, 2021.

[14] I. M. Hezam, A. Foul, A., A. Alrasheedi, 'A dynamic optimal control model for COVID-19 and cholera co-infection in Yemen', Adv. Differ. Equ., pp. 1-30. Dec., 2021.

[15] A. Omame, 'Analysis of COVID-19 and comorbidity co-infection model with optimal control', Optim. Control Appl. Methods, pp. 1-23, Nov, 2021.

[16] L. C. de Barros, 'The memory effect on fractional calculus: an application in the spread of COVID-19', Comput. Appl. Math., pp. 1-21, Apr., 2021.

[17] A. Tanvi, R. Aggarwal, R., Y. A. Raj, 'A fractional order HIV-TB co-infection model in the presence of exogenous reinfection and recurrent TB', Nonlinear Dyn., pp. 4701-4725, Jun., 2021.

[18] P. Agarwal, R. Singh, 'Modelling of transmission dynamics of Nipah virus (Niv): a fractional order approach', Phys. A: Stat. Mech. Appl., Jun., 2020.

[19] A. S. Alshomrani, M. Z. Ullah, D. Baleanu, 'Caputo SIR model for COVID-19 under optimized fractional order', Adv. Differ. Equ., pp.1-17, Dec., 2021.

[20] N. H. Tuan, H. Mohammadi, S. Rezapour, 'A mathematical model for COVID-19 transmission by using the Caputo fractional derivative', Chaos Solitons Fractals, Nov., 2020.

[21] I. Podlubny, 'Fractional Differential Equations: An Introduction to Fractional Derivatives, Fractional Differential Equations, to Methods of Their Solution and Some of Their Applications', Academic Press, 1998.

[22] D. Wang, et al., 'Clinical characteristics of 138 hospitalized patients with 2019 novel coronavirus–infected pneumonia in Wuhan, China', JAMA, pp. 1061-1069. Feb., 2020.

[23] G. T. Tilahun, 'Modeling co-dynamics of pneumonia and meningitis diseases', Adv. Differ. Equ., pp. 1-18, Dec., 2019.

[24] R. H. Du, et al., 'Predictors of mortality for patients with COVID-19 pneumonia caused by SARS-CoV-2: a prospective cohort study' Eur. Respir. J., May, 2020.

[25] A. M. A. El-Sayed, R. P. Agarwal, S. M. Salman, 'Fractional-order Chua's system: discretization, bifurcation and chaos', Adv. Differ. Equ., Dec., 2013.

[26] Z. M. Odibat, N. T. Shawagfeh, 'Generalized Taylor's formula', Appl. Math. Comput., pp. 286-293, Dec., 2007.

[27] W. Lin, 'Global existence theory and chaos control of fractional differential equations', J. Math. Anal. Appl., pp. 709-726, Aug., 2007.

[28] J. Huo, H. Zhao, L. Zhu, 'The effect of vaccines on backward bifurcation in a fractional order HIV model', Nonlinear Anal. Real World Appl, pp. 289-305. Dec., 2015.

[29] O. Diekmann, J. A. P. Heesterbeek, J. A. Metz, 'On the definition and the computation of the basic reproduction ratio R0 in models for infectious diseases in heterogeneous populations', J. Math. Biol., pp. 365-382. Jun., 1990.

[30] E. Ahmed, A. M. A. El-Sayed, H. A. El-Saka, 'On some Routh–Hurwitz conditions for fractional order differential equations and their applications in Lorenz, Rössler, Chua and Chen systems', Physics Letters A, pp. 1-4., Oct., 2006.

[31] https://www.newindianexpress.com/world/2020/mar/24/fresh-coronavirus-cases-rise-to-78-in-china-overall-count-in-country-crosses-81000-212088.html.

6

Optimal Control to Curtail the Spread of COVID-19 through Social Gatherings: A Mathematical Model

Nita H. Shah, Purvi M. Pandya, and Ankush H. Suthar

Department of Mathematics, Gujarat University, India
E-mail: nitahshah@gmail.com, pandya091@gmail.com,
ankush.suthar1070@gmail.com

Abstract

An outbreak from which now the whole world is fighting, a coronavirus disease-19 (COVID-19) was first identified in December 2019 in Wuhan, China. The current situation has become pandemic in nature in which social distancing and isolation are effective to date. Social gatherings or religious gatherings act as fuel to the spread of this virus. Due to such gatherings, individuals may get exposed who are either infected or quarantined. They are then hospitalized if the condition worsens and may recover over a period of time. So, a systematic pathway is designed to scrutinize the spread pattern of individuals due to the transmission of the virus through such gatherings. Moreover, a non-linear differential equation is formulated by the proposed mathematical model which is further used to calculate the threshold to study the impact of a viral pandemic due to social gatherings. To curtail the spread, optimal controls are applied which will help limit the disastrous outcomes to a certain level. Also, to substantiate the data numerical simulation are carried out and the transmission patterns for respective compartments are studied.

Keywords: Coronavirus, Pandemic, Social Gatherings, Mathematical Model, Threshold, Optimal Control, Simulation

Mathematics Subject Classification: 37Nxx

6.1 Introduction

World today is facing a public health crisis with global concern due to novel coronavirus disease-19 (nCoV-19) also termed as coronavirus disease-19 (COVID-19), which accounts for millions of deaths giving a reason for extensive morbidity and mortality around the world. The symptoms of coronavirus infection can range from mild which includes common cold, cough and fever, whereas lethal illness relates to severe respiratory tract infections, such as breathing difficulties. During this outbreak, to prevent the spread of the disease, strict restrictions are imposed on routine activities as part of social distancing norms. The pandemic is challenging in many ways which are in terms of education, business, psychosocial issues, ICU infrastructures, economic damages, etc. In this crisis, social gatherings work as rocket fuel to the spread of this virus. Social gatherings, such as parties, weddings and religious activities, increase the exposure rate of this virus spreading the infection extensively. A strict lockdown is imposed to control such gatherings. According to the studies, self-quarantine, social distancing, isolation and procuring immediate treatment while noticing symptoms of the virus are considered to be effective measure to stay away from the virus or get quick recovery.

Several experimental studies along with some statistical studies are done to understand the virus and prevent the spread. Many clinical trials are conducted to find the cure at the earliest. A review of COVID-19 was done which includes its history, origin, mode of spreading and prevention [12]. In addition, a brief review to summarize the studies and clinical trials till late February 2020 containing every information about COVID-19 from symptoms to treatments to guide frontline medical staff in clinical management [10]. Moreover, a case report on the potential application of lung ultrasound to identify the infected at the bedside and its findings helps health workers reduce their exposure to infected patients [3]. Additionally, guidance or different strategies for surgeries during this pandemic was discussed addressing all the related practices [2]. An article revealing myths versus truths regarding this virus was published to spread awareness among people [1]. A stochastic transmission model to measure the effectiveness of isolation of cases along with contact tracing was developed [9]. Furthermore, a stratified multistage random sampling technique was used to study the spread of infection among people under home quarantine in Shenzhen [18].

More studies related to mathematical modelling play a major role to analyse and understand the behaviour of this viral pandemic. A database study from 31 December 2019 to 14 April 2020 of the COVID-19 outbreak was done where a deterministic *SEIR* epidemic disease model was formulated [4]. Many other models based on *SEIR* dynamics were introduced using which the spread dynamics, sensitivity analysis and different simulations to validate the data used were analysed [6, 8, 19]. Moreover, the *SEIR* model was framed to examine the dynamics of this spread for India [13] and Italy as well [15]. Some models other than *SEIR* have considered other compartments for individuals who are symptomatic, asymptomatic, quarantined and hospitalized using which basic reproduction number and other simulations are scrutinized [16]. For more accuracy in results, the *SEIR* model for bats and hosts is considered and equations are formulated using fractional derivatives for the total population of people [11]. Furthermore, using a compartmental model consisting of exposed, infected, critical, hospitalized and dead individuals, control strategies to curtail transmission of the viral pandemic were analysed graphically [17].

In Section 6.2, the mathematical model to study the viral pandemic due to social gatherings is designed and formulated along with the computation of the threshold. Furthermore, in the next section, optimal control is designed and calculated for the model followed by numerical simulation in Section 6.4. Section 6.5 concludes the findings.

6.2 Mathematical Model

The rise of infection caused by coronavirus prevailing in society due to social gatherings cum religious gatherings is scrutinized here. As shown in Figure 6.1, six compartments are considered, namely social gatherings (S_G) and exposed individuals (E) due to such gatherings. They are either quarantined (Q) or infected (I) and need to be hospitalized (H) out of which some may recover (R) and quarantined as a precaution. Some incidences of quarantined individuals getting infected and vice versa are also recorded. The notations along with its parametric values used here are described in Table 6.1, which indicates the rate of flow of individuals among all compartments. Here, using the designed model shown in Figure 6.1, the system of non-linear differential equations is framed as follows:

$$\frac{dS_G}{dt} = B - \alpha_1 S_G E - \mu S_G,$$

$$\frac{dE}{dt} = \alpha_1 S_G E - \alpha_2 E - \alpha_3 E - \mu E,$$

Table 6.1 Notation with parametric values.

Notation	Description	Parametric values	Source
B	Natural birth rate	1	Assumed
α_1	Rate of individuals who are exposed due to social gathering	0.88	Computed
α_2	Rate where exposed individuals are being quarantined	0.8	Computed
α_3	Rate from exposed individuals to infected individuals	0.25	Computed
α_4	Rate of quarantined individuals being infected	0.6	Computed
α_5	Rate where infected individuals get quarantined	0.07	Computed
α_6	Rate at which quarantined individuals get hospitalized	0.1	Computed
α_7	Rate of infected individuals being hospitalized	0.9	Computed
α_8	Rate where hospitalized individuals get recovered	0.27	Computed
α_9	Rate of recovered individuals being quarantined	0.8	Computed
μ	Natural death rate	0.1	Assumed
μ_d	Death rate due to infection	0.05	Computed

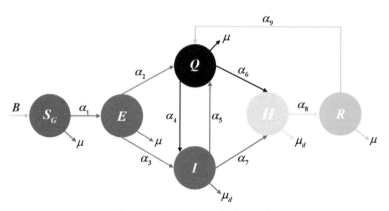

Figure 6.1 Mathematical model.

$$\frac{dQ}{dt} = \alpha_2 E - \alpha_4 Q + \alpha_5 I - \alpha_6 Q + \alpha_9 R - \mu Q, \qquad (6.1)$$

$$\frac{dI}{dt} = \alpha_3 E + \alpha_4 Q - \alpha_5 I - \alpha_7 I - \mu_d I,$$

$$\frac{dH}{dt} = \alpha_6 Q + \alpha_7 I - \alpha_8 H - \mu_d H,$$

$$\frac{dR}{dt} = \alpha_8 H - \alpha_9 R - \mu R,$$

where $S_G + E + Q + I + H + R \leq N$. Also, $S_G > 0$; $E, Q, I, H, R \geq 0$
Therefore, the feasible region for the model is considered as

$$\Lambda = \left\{ (S_G, E, Q, I, H, R) \in R^6 : S_G + E + Q + I + H + R \leq \frac{B}{\mu} \right\}.$$

Now, the disease-free equilibrium point of the model is $Y_0 = \left(\frac{B}{\mu}, 0, 0, 0, 0, 0 \right)$ and the endemic point is $Y^* = (S_G{}^*, E^*, Q^*, I^*, H^*, R^*)$, where $S_G{}^* = \frac{\alpha_2 + \alpha_3 + \mu}{\alpha_1}$, $E^* = \frac{B\alpha_1 - (\alpha_2 + \alpha_3)\mu - \mu^2}{\alpha_1(\alpha_2 + \alpha_3 + \mu)}$,

$$Q^* = \frac{\begin{pmatrix} (\alpha_8 + \mu_d)(\alpha_3\alpha_5 + \alpha_2\mu_d + \alpha_2(\alpha_5 + \alpha_7))\mu \\ +(\alpha_2\mu_d^2 + (\alpha_2\alpha_8 + \alpha_3\alpha_5 + \alpha_2(\alpha_5 + \alpha_7))\mu_d \\ +\alpha_8(\alpha_2 + \alpha_3)(\alpha_5 + \alpha_7))\alpha_9 \\ (B\alpha_1 - (\alpha_2 + \alpha_3)\mu - \mu^2) \end{pmatrix}}{\alpha_1(\alpha_2 + \alpha_3 + \mu)\begin{pmatrix} (\alpha_8 + \mu_d)(\alpha_5 + \alpha_7 + \mu_d)\mu^2 \\ + (\alpha_8 + \mu_d)\begin{pmatrix} (\alpha_4 + \alpha_6 + \alpha_9)\mu_d \\ +(\alpha_5 + \alpha_7)\alpha_9 \\ +(\alpha_4 + \alpha_6)\alpha_7 + \alpha_5\alpha_6 \end{pmatrix}\mu \\ +\mu_d\alpha_9(\alpha_4\alpha_8 + \alpha_5\alpha_6 \\ + (\alpha_4 + \alpha_6)(\alpha_7 + \mu_d)) \end{pmatrix}},$$

$$I^* = \frac{\begin{pmatrix} \alpha_3\mu^2(\alpha_8 + \mu_d) + (\alpha_8 + \mu_d)(\alpha_2\alpha_4 + \alpha_3 \\ (\alpha_4 + \alpha_6 + \alpha_9))\mu + (((\alpha_4 + \alpha_6)\alpha_3 + \alpha_2\alpha_4)\mu_d \\ +\alpha_4\alpha_8(\alpha_2 + \alpha_3))\alpha_9 \end{pmatrix}}{\alpha_1(\alpha_2 + \alpha_3 + \mu)\begin{pmatrix} (\alpha_8 + \mu_d)(\alpha_5 + \alpha_7 + \mu_d)\mu^2 \\ + (\alpha_8 + \mu_d)\begin{pmatrix} (\alpha_4 + \alpha_6 + \alpha_9)\mu_d \\ +(\alpha_5 + \alpha_7)\alpha_9 \\ +(\alpha_5 + \alpha_7)\alpha_6 + \alpha_4\alpha_7 \end{pmatrix}\mu \\ +\mu_d\alpha_9(\alpha_4\alpha_8 + \alpha_4\alpha_7 + (\alpha_4 + \alpha_6)\mu_d \\ +(\alpha_5 + \alpha_7)\alpha_6) \end{pmatrix}},$$

$$H^* = \cfrac{\begin{array}{c}(\alpha_3\alpha_7\mu + \alpha_2\alpha_6\mu_d + (\alpha_2 + \alpha_3)\left((\alpha_4 + \alpha_6)\right.\\\left.\alpha_7 + \alpha_5\alpha_6)\right)\left(B\alpha_1 - (\alpha_2 + \alpha_3)\mu - \mu^2\right)(\alpha_9 + \mu)\end{array}}{\alpha_1(\alpha_2 + \alpha_3 + \mu)\left(\begin{array}{c}(\alpha_8 + \mu_d)(\alpha_5 + \alpha_7 + \mu_d)\mu^2\\((\alpha_4 + \alpha_6 + \alpha_9)(\alpha_7 + \mu_d)\\+\alpha_5(\alpha_6 + \alpha_9))\mu + \mu_d\alpha_9(\alpha_4\alpha_8\\+\alpha_5\alpha_6 + (\alpha_4 + \alpha_6)(\alpha_7 + \mu_d))\end{array}\right)},$$

and

$$R^* = \cfrac{\begin{array}{c}(\alpha_3\alpha_7\mu + \alpha_2\alpha_6\mu_d + (\alpha_2 + \alpha_3)\left((\alpha_4 + \alpha_6)\alpha_7 + \alpha_5\alpha_6\right))\\\left(B\alpha_1 - (\alpha_2 + \alpha_3)\mu - \mu^2\right)\alpha_8\end{array}}{\alpha_1(\alpha_2+\alpha_3+\mu)\left(\begin{array}{c}(\alpha_8 + \mu_d)(\alpha_5 + \alpha_7 + \mu_d)\mu^2 + (\alpha_8 + \mu_d)\\((\alpha_4 + \alpha_6 + \alpha_9)(\alpha_7 + \mu_d) + \alpha_5(\alpha_6 + \alpha_9))\mu\\+\mu_d\alpha_9(\alpha_4\alpha_8 + \alpha_5\alpha_6 + (\alpha_4 + \alpha_6)(\alpha_7 + \mu_d))\end{array}\right)}.$$

Now, using the next-generation matrix method the threshold (R_0) is evaluated [5].

Let $X' = (S_G, E, Q, I, H, R)'$ and $X' = \frac{dX}{dt} = F(X) - V(X)$ where $F(X)$ shows the rate of new individuals entering the compartment and $V(X)$ as the rate of spread of disease which are evaluated as

$$F(X) = \begin{bmatrix} \alpha_1 S_G E \\ 0 \\ 0 \\ 0 \\ 0 \\ 0 \end{bmatrix}$$

and

$$V(X) = \begin{bmatrix} \alpha_2 E + \alpha_3 E + \mu E \\ -\alpha_2 E + \alpha_4 Q - \alpha_5 I + \alpha_6 Q - \alpha_9 R + \mu Q \\ -\alpha_3 E - \alpha_4 Q + \alpha_5 I + \alpha_7 I + \mu_d I \\ -\alpha_6 Q - \alpha_7 I + \alpha_8 H + \mu_d H \\ -\alpha_8 H + \alpha_9 R + \mu R \\ -B + \alpha_1 S_G E + \mu S_G \end{bmatrix}.$$

Now, $DF(X_0) = \begin{bmatrix} f & 0 \\ 0 & 0 \end{bmatrix}$ and $DV(X_0) = \begin{bmatrix} v & 0 \\ J_1 & J_2 \end{bmatrix}$, where f and v are 6×6 matrices defined as $f = \left[\frac{\partial F_i(X_0)}{\partial X_j}\right]$ and $v = \left[\frac{\partial V_i(X_0)}{\partial X_j}\right]$.

Here, we get $f = \begin{bmatrix} \alpha_1 S_G & 0 & 0 & 0 & 0 & \alpha_1 E \\ 0 & 0 & 0 & 0 & 0 & 0 \\ 0 & 0 & 0 & 0 & 0 & 0 \\ 0 & 0 & 0 & 0 & 0 & 0 \\ 0 & 0 & 0 & 0 & 0 & 0 \\ 0 & 0 & 0 & 0 & 0 & 0 \end{bmatrix}$

and

$$v = \begin{bmatrix} \alpha_2 + \alpha_3 + \mu & 0 & 0 & 0 & 0 & 0 \\ -\alpha_2 & \alpha_4 + \alpha_6 + \mu & -\alpha_5 & 0 & -\alpha_9 & 0 \\ -\alpha_3 & -\alpha_4 & \alpha_5 + \alpha_7 + \mu_d & 0 & 0 & 0 \\ 0 & -\alpha_6 & -\alpha_7 & \alpha_8 + \mu_d & 0 & 0 \\ 0 & 0 & 0 & -\alpha_8 & \alpha_9 + \mu & 0 \\ \alpha_1 S_G & 0 & 0 & 0 & 0 & \alpha_1 E + \mu \end{bmatrix}.$$

Here, v is the non-singular matrix.
For this model, the spectral radius of the matrix fv^{-1} gives the threshold R_0 at Y_0.

$$\therefore R_0 = \frac{\alpha_1 B}{\mu(\alpha_2 + \alpha_3 + \mu)}. \tag{6.2}$$

6.3 Optimal Control

The main purpose to apply optimal control is to curtail the spread of the infection due to coronavirus and curtail the hospitalization rate and increase the recovery rate. Here, three bounded Lebesgue integrable controls, namely u_1, u_2 and u_3, are proposed as shown in Figure 6.2.

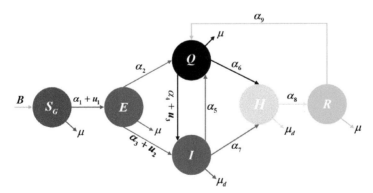

Figure 6.2   Model with control variables.

After applying optimal control, system (6.1) will be as follows:

$$\frac{dS_G}{dt} = B - \alpha_1 S_G E - u_1 S_G - \mu S_G,$$

$$\frac{dE}{dt} = \alpha_1 S_G E + u_1 S_G - \alpha_2 E - \alpha_3 E - u_2 E - \mu E,$$

$$\frac{dQ}{dt} = \alpha_2 E - \alpha_4 Q - u_3 Q + \alpha_5 I - \alpha_6 Q + \alpha_9 R - \mu Q, \qquad (6.3)$$

$$\frac{dI}{dt} = \alpha_3 E + u_2 E + \alpha_4 Q + u_3 Q - \alpha_5 I - \alpha_7 I - \mu_d I,$$

$$\frac{dH}{dt} = \alpha_6 Q + \alpha_7 I - \alpha_8 H - \mu_d H,$$

$$\frac{dR}{dt} = \alpha_8 H - \alpha_9 R - \mu R.$$

The objective function is given by

$$J(u_i, \Lambda) = \int_0^T (A_1 S_G{}^2 + A_2 E^2 + A_3 Q^2 + A_4 I^2 + A_5 H^2 + A_6 R^2$$

$$+ w_1 u_1{}^2 + w_2 u_2{}^2 + w_3 u_3{}^2) \, dt, \qquad (6.4)$$

where Λ denotes set of all compartmental variables, $A_i, i = 1 \text{ to } 6$ are non-negative weight constants for S_G, E, Q, I, H, R, respectively, and $w_i, i = 1 \text{ to } 3$ are weight constants for control variables $u_i, i = 1 \text{ to } 3$, respectively.

Now, the calculated control variables are from $t = 0$ to $t = T$ such that

$$J(u_i(t)) = \min\{J(u_i{}^*, \Lambda)/u_i \in \phi\}, \, i = 1, 2, 3 \quad,$$

where ϕ is a smooth function on $[0, 1]$. The optimal controls are found by accumulating all the integrands of eqn (6.4) using the lower bounds and upper bounds, respectively [7].

Now, the Pontryagin's principle [14] is used to optimize the function in eqn (6.4) where a Lagrangian function is constructed consisting of adjoint variables $\lambda_i, i = 1 \text{ to } 6$ which is as follows:

$$\begin{aligned}
L(\Lambda, A_i) =\ & A_1 S_G{}^2 + A_2 E^2 + A_3 Q^2 + A_4 I^2 + A_5 H^2 + A_6 R^2 + w_1 u_1{}^2 \\
& + w_2 u_2{}^2 + w_3 u_3{}^2 + \lambda_1 (B - \alpha_1 S_G E - u_1 S_G - \mu S_G) \\
& + \lambda_2 (\alpha_1 S_G E + u_1 S_G - \alpha_2 E - \alpha_3 E - u_2 E - \mu E) \\
& + \lambda_3 (\alpha_2 E - \alpha_4 Q - u_3 Q + \alpha_5 I - \alpha_6 Q + \alpha_9 R - \mu Q) \\
& + \lambda_4 (\alpha_3 E + u_2 E + \alpha_4 Q + u_3 Q - \alpha_5 I - \alpha_7 I - \mu_d I) \\
& + \lambda_5 (\alpha_6 Q + \alpha_7 I - \alpha_8 H - \mu_d H) + \lambda_6 (\alpha_8 H - \alpha_9 R - \mu R)
\end{aligned}$$

Now, the partial derivative of the Lagrangian function w.r.t. each variable of the compartment are as follows:

$$\dot{\lambda_1} = -\frac{\partial L}{\partial S_G} = -2A_1 S_G + (\lambda_1 - \lambda_2)(\alpha_1 E + u_1) + \lambda_1 \mu,$$

$$\dot{\lambda_2} = -\frac{\partial L}{\partial E} = -2A_2 E + (\lambda_1 - \lambda_2)\alpha_1 S_G + (\lambda_2 - \lambda_3)\alpha_2 + (\lambda_2 - \lambda_4)$$

$$(\alpha_3 + u_2) + \lambda_2 \mu,$$

$$\dot{\lambda_3} = -\frac{\partial L}{\partial Q} = -2A_3 Q + (\lambda_3 - \lambda_4)(\alpha_4 + u_3) + (\lambda_3 - \lambda_5)\alpha_6 + \lambda_3 \mu,$$

$$\dot{\lambda_4} = -\frac{\partial L}{\partial I} = -2A_4 I + (\lambda_4 - \lambda_3)\alpha_5 + (\lambda_4 - \lambda_5)\alpha_7 + \lambda_4 \mu_d,$$

$$\dot{\lambda_5} = -\frac{\partial L}{\partial H} = -2A_5 H + (\lambda_5 - \lambda_6)\alpha_8 + \lambda_5 \mu_d,$$

$$\dot{\lambda_6} = -\frac{\partial L}{\partial R} = -2A_6 R + (\lambda_6 - \lambda_3)\alpha_9 + \lambda_6 \mu.$$

The necessary conditions for a Lagrangian function L to be optimal for control are

$$\dot{u_1} = -\frac{\partial L}{\partial u_1} = -2w_1 u_1 + (\lambda_1 - \lambda_2)S_G = 0$$

$$\dot{u_2} = -\frac{\partial L}{\partial u_2} = -2w_2 u_2 + (\lambda_2 - \lambda_4)E = 0$$

and $\dot{u_3} = -\frac{\partial L}{\partial u_3} = -2w_3 u_3 + (\lambda_3 - \lambda_4)Q = 0.$

Hence, $u_1 = \frac{(\lambda_1 - \lambda_2)S_G}{2w_1}$, $u_2 = \frac{(\lambda_2 - \lambda_4)E}{2w_2}$ and $u_3 = \frac{(\lambda_3 - \lambda_4)Q}{2w_3}$.

This gives optimal control conditions as

$$u_1{}^* = max\left(a_1, min\left(b_1, \frac{(\lambda_1 - \lambda_2)S_G}{2w_1}\right)\right),$$

$$u_2{}^* = max\left(a_2, min\left(b_2, \frac{(\lambda_2 - \lambda_4)E}{2w_2}\right)\right) \text{ and }$$

$$u_3{}^* = max\left(a_3, min\left(b_3, \frac{(\lambda_3 - \lambda_4)Q}{2w_3}\right)\right).$$

In the next section, using parametric values given in Table 6.1, the analytical results for calculated optimal control and other numerical simulations are carried out.

6.4 Numerical Simulation

Figure 6.3 shows the transmission pattern of individuals who suffer due to social gatherings. In Figure 6.3 (a), it is observed that 52.71% of exposed individuals get quarantined in approximately 0.36 weeks. Also, 48.16% of exposed individuals get infected in approximately 0.62 weeks and immediately get hospitalized. Furthermore, 48.58% of infected individuals may recover in approximately 1.46 weeks who are further kept under observation out of which 33.79% may be said to be fully cured in approximately 3.36 weeks. From Figure 6.3 (b), it is noted that a decrease in social gatherings will result in a decrease in exposure to the virus and an increase in recovery rate.

In Figure 6.4, the direction flow of individuals towards recovery is observed. As shown in Figure 6.4 (a), individuals exposed to the virus may get recovered if they take proper precautions and care; otherwise it may lead them away from recovery. Also, in the case of infected individuals, they

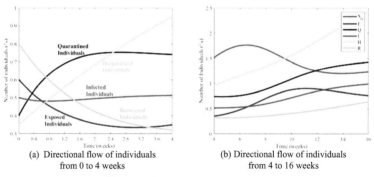

(a) Directional flow of individuals from 0 to 4 weeks   (b) Directional flow of individuals from 4 to 16 weeks

Figure 6.3   Transmission pattern.

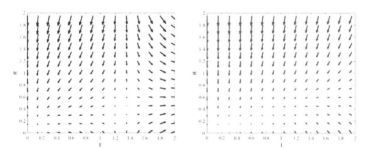

Figure 6.4   Directional flows of state variables. (a) Behaviour of exposed individuals towards recovery and (b) Behaviour of infected individuals towards recovery.

will recover if they get proper treatment at the proper time as monitored in Figure 6.4 (b).

Using the dynamical system (6.1), the trajectory around the endemic equilibrium point is shown for respective compartments in Figure 6.5. Here, if the direction of the cures moves towards an equilibrium point, then the condition is said to be stabilized. So, in cases shown in Figure 6.5 (a) and

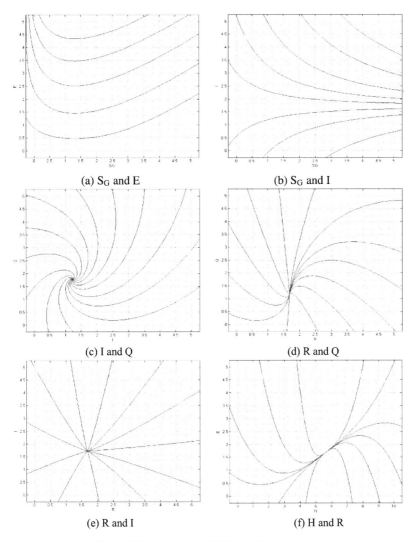

(a) S_G and E

(b) S_G and I

(c) I and Q

(d) R and Q

(e) R and I

(f) H and R

Figure 6.5 Trajectory field and solution curve.

(b), the rate of exposed and infected individuals keeps increasing due to social gatherings, which makes the system unstable. However, if exposed or infected individuals are quarantined for a time being or get hospitalized to get proper treatment, then the system is asymptotically stable as observed in Figure 6.5 (c)–(f).

Figure 6.6 (a)–(c) represents the variations or oscillations due to the viral pandemic through social gatherings. Performed analysis, the increased oscillations, and the periodic behaviour show that the problem will be pandemic in nature if not controlled at its initial stage itself. It is also observed that increased infected individuals result in increased quarantine and hospitalization rates. More social gatherings will result in more spread of infection.

As shown in Figure 6.7, due to 27% of social gatherings, 11% of individuals get exposed to coronavirus. Overall, 18% of individuals are quarantined; still, 13% get infected. Due to this exposure and infection due to the virus,

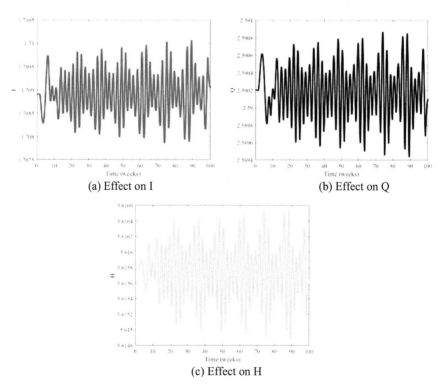

(a) Effect on I

(b) Effect on Q

(c) Effect on H

Figure 6.6   Effect on compartment through oscillations.

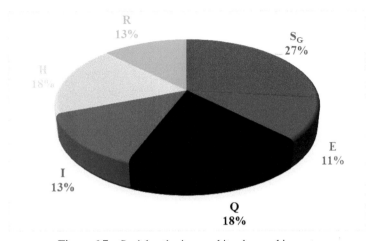

Figure 6.7 Social gatherings and its observed impact.

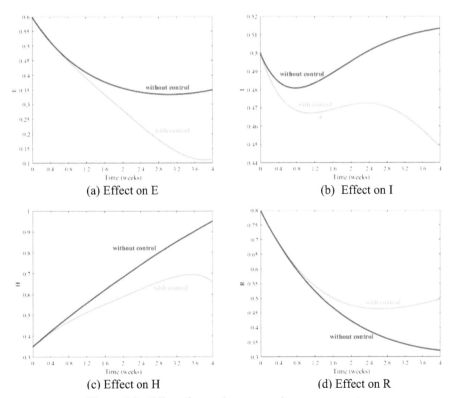

(a) Effect on E

(b) Effect on I

(c) Effect on H

(d) Effect on R

Figure 6.8 Effect of controls on respective compartments.

18% of individuals get hospitalized from which 13% may recover if treated properly.

After applying optimal control, a 67.48% decrease in exposed individuals is observed (Figure 6.8 (a)), which results in a 12.62% decrease in infected individuals (Figure 6.8 (b)). Moreover, a 31.10% decrease in hospitalization rate is observed (Figure 6.8 (c)) and a 35.29% increase in recovery rate is noted (Figure 6.8 (d)).

6.5 Conclusion

A systematic pathway is designed to understand and analyse the nature of coronavirus spread due to social gatherings. To study this spread, a system of non-linear differential equations is constructed. Moreover, by using the next-generation matrix method the threshold R_0 is formulated which connotes that through social gatherings and individual can infect approximately seven other individuals. In this pandemic situation, controlling the spread of this virus is a must to live a healthy and safe life. So, optimal control theory is directed to curtail the spread and show the effect of controls on respective compartments. Furthermore, numerical simulation is conducted to substantiate the data, which includes the transmission pattern of individuals among all compartments and the directional flow or intensity showing the behaviour of exposed and infected individuals towards recovery. Also, the trajectory field implies that social gathering will increase exposure to coronavirus resulting in increased infection which makes the system unstable. In addition, the system will be stable if individuals stay quarantined or if infected, procure timely and proper treatment. Simple modifications in lifestyle and taking precautionary measures, such as washing hands after a certain period of time, wearing a mask while staying out and social distancing, can help reduce the chances of infection in this pandemic situation. "Distance is rescued" till no cure or vaccines are found.

Acknowledgements

The authors thank DST-FIST file # MSI-097 for technical support to the Department of Mathematics, Gujarat University. The second author (Purvi M. Pandya) thanks the Education Department, Gujarat for scholarship under SHODH (student AisheCode: 201901380135). The third author (Ankush H. Suthar) is funded by a Junior Research Fellowship from the Council of Scientific and Industrial Research (file no.:09/070(0061)/2019-EMR-I).

References

[1] Amgain, K., Neupane, S., Panthi, L., and Thapaliya, P. (2020). Myths versus Truths regarding the Novel Coronavirus Disease (COVID-2019) Outbreak. *Journal of Karnali Academy of Health Sciences, 3*(1), 1-6.

[2] Brücher, B. L., Nigri, G., Tinelli, A., Lapeña, J. F. F., Espin-Basany, E., Macri, P., Matevossian, E., Ralon, S., Perkins, R., Lück, R., Kube, R., Costa, J. MC., Mintz, Y., Tez, M., Allert, S., Sökmen, S., Spychala, A., Zilberstein, B., Marusch, F., Kermansaravi, M., Kycler, W., Vicente, D., Scherer, M. A., Rivkind, A., Elias, N., Wallner, G., Roviello, F., Santos, L. L., Araujo, R. J.C., Szold, A., Oleas, R., Rupnik, M. S., Salber, J., Jamall, I. S., and Engel, A. (2020). COVID-19: Pandemic surgery guidance. *4open, 3*(1), 1-19.

[3] Buonsenso, D., Piano, A., Raffaelli, F., Bonadia, N., Donati, K. D. G., and Franceschi, F. (2020). Point-of-Care Lung Ultrasound findings in novel coronavirus disease-19 pneumonia: a case report and potential applications during COVID-19 outbreak. *European Review for Medical and Pharmacological Sciences, 24*, 2776-2780.

[4] Carcione, J. M., Santos, J. E., Bagaini, C., and Ba, J. (2020). A simulation of a COVID-19 epidemic based on a deterministic *SEIR* model. *arXiv preprint arXiv:2004.03575.*

[5] Diekmann, O., Heesterbeek, J. A. P., and Roberts, M. G. (2009) The construction of next-generation matrices for compartmental epidemic models. *Journal of the Royal Society Interface, 7*(47), 873-885.

[6] Fang, Y., Nie, Y., and Penny, M. (2020). Transmission dynamics of the COVID-19 outbreak and effectiveness of government interventions: A data-driven analysis. *Journal of medical virology*, 1-15.

[7] Fleming, W.H. and Rishel, R.W. (2012) *Deterministic and Stochastic Optimal Control.* Vol. 1, Springer Science & Business Media, New York.

[8] Grant, A. (2020). Dynamics of COVID-19 epidemics: *SEIR* models underestimate peak infection rates and overestimate epidemic duration. *medRxiv.*

[9] Hellewell, J., Abbott, S., Gimma, A., Bosse, N. I., Jarvis, C. I., Russell, T. W., Munday, J. D., Kucharski, A. J., Edmunds, W J., Funk, S., ... and Eggo, R. M. (2020). Feasibility of controlling COVID-19 outbreaks by isolation of cases and contacts. *The Lancet Global Health, 8*, e488-e496.

[10] Jiang, F., Deng, L., Zhang, L., Cai, Y., Cheung, C. W., and Xia, Z. (2020). Review of the clinical characteristics of coronavirus disease 2019 (COVID-19). *Journal of General Internal Medicine*, 1-5.

[11] Khan, M. A., and Atangana, A. (2020). Modeling the dynamics of novel coronavirus (2019-nCov) with fractional derivative. *Alexandria Engineering Journal*, 1-11.

[12] Kumar, D., Malviya, R., and Sharma, P. K. (2020). Corona Virus: A Review of COVID-19. *Eurasian Journal of Medicine and Oncology*, *4*(1), 8-25.

[13] Pandey, G., Chaudhary, P., Gupta, R., and Pal, S. (2020). *SEIR* and Regression Model-based COVID-19 outbreak predictions in India. *arXiv preprint arXiv:2004.00958*.

[14] Pontriagin, L.S., Boltyanskii, V.G., Gamkrelidze, R.V. and Mishchenko, E.F. (1986) *The Mathematical Theory of Optimal Process*. vol 4, Gordon and Breach Science Publishers, New York.

[15] Rovetta, A., and Bhagavathula, A. S. (2020). Modelling the epidemiological trend and behavior of COVID-19 in Italy. *medRxiv*.

[16] Senapati, A., Rana, S., Das, T., and Chattopadhyay, J. (2020). Impact of intervention on the spread of COVID-19 in India: A model-based study. *arXiv preprint arXiv:2004.04950*.

[17] Shah, N. H., Suthar, A. H., and Jayswal, E. N. (2020). Control Strategies to Curtail Transmission of COVID-19. *medRxiv*.

[18] Wang, J., Liao, Y., Wang, X., Li, Y., Jiang, D., He, J., Zhang, S., and Xia, J. (2020). Incidence of novel coronavirus (2019-nCoV) infection among people under home quarantine in Shenzhen, China. *Travel Medicine and Infectious Disease*, 101660.

[19] Xianzhi Yuan, G., Di, L., Gu, Y., Qian, G., and Qian, X. (2020). The Framework for the Prediction of the Critical Turning Period for Outbreak of COVID-19 Spread in China based on the *SEIR* Model. *arXiv*, arXiv-2004.

7

Effectiveness of a Booster Dose of COVID-19 Vaccine

Nita H. Shah[1] and Moksha H. Satia[2]

[1]Department of Mathematics, Gujarat University, India
[2]Department of Mathematics, Aditya Silver Oak Institute of Technology,
Silver Oak University, India
E-mail: mokshasatia.05@gmail.com

Abstract

The drive for vaccinations is playing crucial role to manage the ongoing pandemic. These vaccinations have various strategies. Especially in India, two doses of each vaccine are given. Every dose holds the reinforcement into the human body. These doses subsequently cannot hold the immunity. In this term, the booster dose of vaccination is desired. In this paper, the effectiveness of these two doses and one booster dose is calculated on advancement of COVID-19 through a system of non-linear differential equations. This system has divided population into two parts, that is exposed individuals and unexposed individual. The improvement of the doses on these individuals is calculated through the threshold. The stability of the model along with the simulation is considered to brace the ongoing situation.

Keywords: COVID-19; Booster Dose; Dynamical System; Threshold; Stability Analysis

MSC Code: 37N30, 37N35

7.1 Introduction

The pandemic COVID-19 has changed the entire scenario of the world. It commenced in the year 2019 in Wuhan, China. The disease spread rapidly when humans come in the contact of each other. The entry of the virus in the body directly attacks to the immune system and decreases the possibility of the fighting with the virus hence risk the life. Therefore, it is essential to cease the pandemic. In order to stop the spread of virus the world needs to build a strong immune system in people. The only way to achieve this milestone is to get vaccinated. COVID-19 vaccination is a technology that protects against the virus. But at this stage, the duration of the protection is not known. To this end, clinical research of the COVID-19 vaccine has been developed and clinical research is being conducted to establish the characteristics of the COVID-19 vaccine. These characteristics give an idea about the efficiency, effectiveness and safety. Many vaccine candidates around the world have tried to develop the COVID-19 vaccine. Most of the vaccinations have two doses. Each dose is crucial to defeat the pandemic. Each dose develops the body's immune system for a specific period of time. Of the world's population, 65.4% of individuals have completed at least one dose of the vaccine. There are around thirty-three vaccines approved by the national government for public use by at least one official. India has approved nine COVID-19 vaccines, namely, Corbenvax, Covaxin, Covishield, Johnson & Johnson, Moderna, Novavax, Sputnik V, Sputnik Light and Zydus Cadila, out of which five are produced locally; but only two are widely used. Out of the total Indians, 72.9% have taken at least one dose of any of these vaccines, while 62.9% of the population has been fully vaccinated. But the statistics shows that these doses do not provide protection for a long time. As a consequence, booster dose is needed. Booster dose is an additional dose of vaccine that provides protection against COVID-19 when the effectiveness of earlier dose is reducing. In India, 2% of population has done his jab by taking this dose. This dose helps individuals fight with COVID-19 and reduce the risk of hospitalization or fall into any kind of critical condition.

In order to fight with this pandemic the only present remedy is its vaccine. Many scientists have tried to develop vaccine, but researchers are also not staying behind to support scientist by their ends with their findings. Le et al. have studied about the COVID-19 vaccine development on landscape in 2020 [17]. After the couple of years in 2021, J. Croda and O. Ranzani have studied about booster doses for inactivated COVID-19 vaccines [8]. They were able to answer the essential questions that 'when is the vaccination needed?' and 'who can take it?' On the same note, Jr. Moreira et al. have focused on safety and efficacy of a third dose COVID-19 Vaccine in 2022 [7]. The safety of COVID-19 vaccine for booster dose among adults is monitored by A. Hause et al. in the year 2022 [3]. Lately in November 2021, new variant of COVID-19 named Omicron has been discovered so that many researchers have done their work for combination of COVID-19 vaccine and Omicron variant. In 2022, N. Andrews et al. have discovered effectiveness of COVID-19 vaccine against the Omicron variant [10] and in the same year. S. Chenchula et al. have reviewed the efficacy of COVID-19 booster dose vaccination against the Omicron variant [16]. Indian researchers have also preformed research for vaccination. In India, Shah et al. have compared the impact of BCG vaccination on COVID-19 transmission, severity and mortality in 2021 [11]. COVID-19 vaccine preferences are expressed in 2022 by P. Bansal et al [14]. Also, in Gujarat, factors for adoption of COVID-19 vaccine are come in lime light by V. Tolia in 2022 [18]. Mathematical modelling that area which supports the results using mathematics and its fundamental. S. Mandal et al. have analysed India's pragmatic vaccination strategy against COVID-19 through a mathematical modelling in 2021 [15]. B. Foy et al. have compared COVID-19 vaccine allocation strategies for India using a mathematical modelling in 2021 [4], whereas C. MacIntyre et al. have compared vaccination strategies and herd immunity when individuals are limited and fully vaccinated for Australia in 2021 [6] .

In this chapter, the effectiveness of COVID-19 booster dose is studied. In Section 2, the mathematical model and its characteristics are discussed. The stability of the model is calculated in Section 3. Section 4 of numerical analysis supports the results which have been derived for the model.

7.2 Mathematical Model

During the pandemic, the spread of virus was at its peak. Due to this, scientists have to work day and night to find the right vaccination. A year ago, many of the scientists have found various vaccines. In India, several vaccinations are approved, from which two have multiple doses. The research has shown that two doses are not enough to control the spread and that is why booster dose is needed. The scientists have claimed that booster dose help increase the immunity of human body. To support this, the system of non-linear differential equation is developed with eleven compartments along with some variables (displayed in Table 7.1). Compartments are described as susceptible individuals (S) (individuals who are at risk of COVID-19), exposed individuals (E_C) (individuals who have come in direct contact with the individual who is ill with COVID-19), unexposed individuals (U_C) (individuals who are yet safe for COVID-19 infection), individuals having dose I of vaccination (D_1) (individuals who have done his first jab), individuals

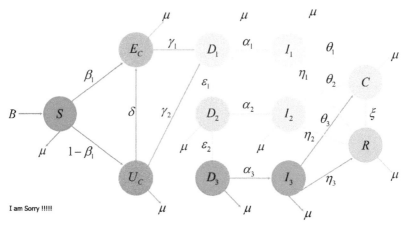

Figure 7.1   Flow diagram of individual.

having dose II of vaccination (D_2) (individuals who have done his second jab), individuals having booster dose of vaccination (D_3) (individuals who have done his third jab), infected individuals after dose I (I_1) (individuals who have caught COVID-19 infection after his first jab), infected individuals after dose II (I_2) (individuals who have caught COVID-19 infection after his second jab), infected individuals after booster dose (I_3) (individuals who have caught COVID-19 infection after his third jab), individuals in critical condition (C) (individuals who are infected with COVID-19 and needs to be hospitalized) and recovered individuals (R) (individuals who are recovered with COVID-19 infection).

The mathematical model for transmission of individuals to find effectiveness of booster dose can be seen from Figure 7.1. Using eleven compartments and various transmission rates, dynamical system is developed as follows:

$$\frac{dS}{dt} = B - \beta_1 S E_C - (1 - \beta_1) S U_C - \mu_s,$$

$$\frac{dE_C}{dt} = \beta_1 S E_C + \delta U_C - \gamma_1 E_C - \mu E_C,$$

$$\frac{dU_C}{dt} = (1 - \beta_1) S U_C - \delta U_C - \gamma_2 U_C - \mu U_C,$$

$$\frac{dD_1}{dt} = \gamma_1 E_C + \gamma_2 U_C - \varepsilon_1 D_1 - \alpha_1 D_1 - \mu D_1,$$

$$\frac{dD_2}{dt} = \varepsilon_1 D_1 - \varepsilon_2 D_2 - \alpha_2 D_2 - \mu D_2,$$

$$\frac{dD_3}{dt} = \varepsilon_2 D_2 - \alpha_3 D_3 - \mu D_3,$$

$$\frac{dI_1}{dt} = \alpha_1 D_1 - \theta_1 I_1 - \eta_1 I_1 - \mu I_1,$$

$$\frac{dI_2}{dt} = \alpha_2 D_2 - \theta_2 I_2 - \eta_2 I_2 - \mu I_2,$$

$$\frac{dI_3}{dt} = \alpha_3 D_3 - \theta_3 I_3 - \eta_3 I_3 - \mu I_3,$$

$$\frac{dC}{dt} = \theta_1 I_1 + \theta_2 I_2 + \theta_3 I_3 - \xi C - \mu C,$$

$$\frac{dR}{dt} = \eta_1 I_1 + \eta_2 I_2 + \eta_3 I_3 + \xi C - \mu R, \tag{7.1}$$

where $S + E_C + U_C + D_1 + D_2 + D_3 + I_1 + I_2 + I_3 + C + R = N$.
This represents non-negative R_+^{11} manifold with initial condition: $S \geq 0, E_C > 0, U_C > 0, D_1 > 0, D_2 > 0, D_3 > 0, I_1 > 0, I_2 > 0, I_3 > 0, C > 0, R > 0$.

Adding all the equations of the system, we have

$$\frac{d}{dt}(S + E_C + U_C + D_1 + D_2 + D_3 + I_1 + I_2 + I_3 + C + R)$$
$$= B - \mu(S + E_C + U_C + D_1 + D_2 + D_3 + I_1 + I_2 + I_3 + C + R) \geq 0,$$

which implies that $\lim_{t \to \infty} sup(S + E_C + U_C + D_1 + D_2 + D_3 + I_1 + I_2 + I_3 + C + R) \leq \frac{B}{\mu}$.

Therefore, the feasible region of the model is

$$\Lambda = \left\{ (S, E_C, U_C, D_1, D_2, D_3, I_1, I_2, I_3, C, R) \in \right.$$

$$\left. R^{11} : S + E_C + U_C + D_1 + D_2 + D_3 + I_1 + I_2 + I_3 + C + R \leq \frac{B}{\mu} \right\}.$$

The compact set Λ is a positively invariant set and attracts all positive orbits in R_+^{11}, and then the solutions are bounded.

7.2.1 Equilibrium point

In this section, equilibrium points of the system are discussed. They are the solution of the system. Solving the system, we have three equilibrium points.

i Virus-free equilibrium point

$$E_0\left(\frac{B}{\mu},0,0,0,0,0,0,0,0,0,0\right).$$

ii Unexposed individual's free equilibrium point
$E_1\left(S^1,E_C^1,0,D_1^1,D_2^1,D_3^1,I_1^1,I_2^1,I_3^1,C^1,R^1\right)$, where

$$S^1=\frac{\gamma_1+\mu}{\beta_1},E_C^1=\frac{B\beta_1-\mu(\gamma_1+\mu)}{\beta_1(\gamma_1+\mu)},$$

$$D_1^1=\frac{\gamma_1\left(B\beta_1-\mu(\gamma_1+\mu)\right)}{\beta_1(\alpha_1+\varepsilon_1+\mu)(\gamma_1+\mu)},$$

$$D_2^1=\frac{\gamma_1\varepsilon_1\left(B\beta_1-\mu(\gamma_1+\mu)\right)}{\beta_1(\alpha_1+\varepsilon_1+\mu)(\alpha_2+\varepsilon_2+\mu)(\gamma_1+\mu)},$$

$$D_3^1=\frac{\gamma_1\varepsilon_1\varepsilon_2\left(B\beta_1-\mu(\gamma_1+\mu)\right)}{\beta_1(\alpha_1+\varepsilon_1+\mu)(\alpha_2+\varepsilon_2+\mu)(\alpha_3+\mu)(\gamma_1+\mu)},$$

$$I_1^1=\frac{\alpha_1\gamma_1\left(B\beta_1-\mu(\gamma_1+\mu)\right)}{\beta_1(\alpha_1+\varepsilon_1+\mu)(\eta_1+\theta_1+\mu)(\gamma_1+\mu)},$$

$$I_2^1=\frac{\alpha_1\gamma_1\varepsilon_1\left(B\beta_1-\mu(\gamma_1+\mu)\right)}{\beta_1(\alpha_1+\varepsilon_1+\mu)(\alpha_2+\varepsilon_2+\mu)(\eta_2+\theta_2+\mu)(\gamma_1+\mu)},$$

$$I_3^1=\frac{\alpha_1\gamma_1\varepsilon_1\varepsilon_2\left(B\beta_1-\mu(\gamma_1+\mu)\right)}{\beta_1(\alpha_1+\varepsilon_1+\mu)(\alpha_2+\varepsilon_2+\mu)(\alpha_3+\mu)(\eta_3+\theta_3+\mu)(\gamma_1+\mu)},$$

$$C^1=\frac{\begin{array}{l}\gamma_1\left[B\beta_1-\mu(\gamma_1+\mu)\right]\left[\alpha_1\theta_1\left(\alpha_2+\varepsilon_2+\mu\right)\right.\\(\alpha_3+\mu)(\eta_2+\theta_2+\mu)(\eta_3+\theta_3+\mu)\\+\alpha_2\theta_2\varepsilon_2(\alpha_3+\mu)(\eta_1+\theta_1+\mu)(\eta_3+\theta_3+\mu)\\+\alpha_3\theta_3\varepsilon_1\varepsilon_2(\eta_1+\theta_1+\mu)(\eta_2+\theta_2+\mu)\end{array}}{\begin{array}{l}\beta_1(\alpha_1+\varepsilon_1+\mu)(\alpha_2+\varepsilon_2+\mu)(\alpha_3+\mu)(\eta_1+\theta_1+\mu)\\(\eta_2+\theta_2+\mu)(\eta_3+\theta_3+\mu)(\gamma_1+\mu)\end{array}},$$

$$R^1=\frac{\begin{array}{l}\gamma_1\left[B\beta_1-\mu(\gamma_1+\mu)\right]\left[\alpha_1\left(\alpha_2+\xi+\mu\right)\right.\\(\alpha_3+\mu)(\eta_2+\theta_2+\mu)(\eta_3+\theta_3+\mu)\\(\eta_1(\xi+\mu)+\theta_1\xi)+\alpha_2\varepsilon_1(\alpha_3+\mu)(\eta_1+\theta_1+\mu)\\(\eta_3+\theta_3+\mu)(\eta_2(\xi+\mu)+\theta_2\xi)\\+\alpha_3\varepsilon_1\varepsilon_2(\eta_1+\theta_1+\mu)(\eta_2+\theta_2+\mu)(\eta_3(\xi+\mu)+\theta_2\xi)\end{array}}{\begin{array}{l}\beta_1(\alpha_1+\varepsilon_1+\mu)(\alpha_2+\varepsilon_2+\mu)(\alpha_3+\mu)(\eta_1+\theta_1+\mu)\\(\eta_2+\theta_2+\mu)(\eta_3+\theta_3+\mu)(\gamma1+\mu)\end{array}}$$

$\rightarrow$ The equilibrium point E^1 exists only when $\beta_1 > \frac{\mu}{B}(\gamma_1 + \mu)$.

iii Endemic equilibrium point $E^*(S^*, E_C^*, U_C^*, D_1^*, D_2^*, D_3^*, I_1^*, I_2^*,$
$I_3^*, C^*, R^*)$,

where

$$S^* = \frac{\delta + \gamma_2 + \mu}{(1 - \beta_1)},$$

$$E_C^* = \frac{\delta\left[B(1 - \beta_1) - (\delta + \gamma_2 + \mu)\mu\right]}{(\delta + \gamma_2 + \mu)\left[(1 - \beta_1)(\gamma_1 + \mu) - \beta_1(\gamma_2 + \mu)\right]},$$

$$U_C^* = \frac{\begin{array}{c}[B(1 - \beta_1) - (\delta + \gamma_2 + \mu)\mu]\\ {}[(1 - \beta_1)(\gamma_1 + \mu) - \beta_1(\delta + \gamma_2 + \mu)]\end{array}}{(\delta + \gamma_2 + \mu)\left[(1 - \beta_1)(\gamma_2 + \mu) - \beta_1(\gamma_2 + \mu)\right]},$$

$$D_1^* = \frac{\begin{array}{c}[(1 - \beta_1)(\delta\gamma_1 + \gamma_2(\gamma_1 + \mu)) - \beta_1\gamma_2(\delta + \gamma_2 + \mu)]\\ {}[B(1 - \beta_1) - (\delta + \gamma_2 + \mu)\mu]\end{array}}{\begin{array}{c}(1 - \beta_1)(\alpha_1 + \varepsilon_1 + \mu)(\delta + \gamma_2 + \mu)\\ {}[(1 - \beta_1)(\gamma_1 + \mu) - \beta_1(\gamma_2 + \mu)]\end{array}},$$

$$D_2^* = \frac{\begin{array}{c}\varepsilon_1\left[(1 - \beta_1)(\delta\gamma_1 + \gamma_2(\gamma_1 + \mu)) - \beta_1\gamma_2(\delta + \gamma_2 + \mu)\right]\\ {}[B(1 - \beta_1) - (\delta + \gamma_2 + \mu)\mu]\end{array}}{\begin{array}{c}(1 - \beta_1)(\alpha_1 + \varepsilon_1 + \mu)(\alpha_2 + \varepsilon_2 + \mu)(\delta + \gamma_2 + \mu)\\ {}[(1 - \beta_1)(\gamma_1 + \mu) - \beta_1(\gamma_2 + \mu)]\end{array}},$$

$$D_3^* = \frac{\begin{array}{c}\varepsilon_1\varepsilon_2\left[(1 - \beta_1)(\delta\gamma_1 + \gamma_2(\gamma_1 + \mu)) - \beta_1\gamma_2(\delta + \gamma_2 + \mu)\right]\\ {}[B(1 - \beta_1) - (\delta + \gamma_2 + \mu)\mu]\end{array}}{\begin{array}{c}(1 - \beta_1)(\alpha_1 + \varepsilon_1 + \mu)(\alpha_2 + \varepsilon_2 + \mu)(\alpha_3 + \mu)(\delta + \gamma_2 + \mu)\\ {}[(1 - \beta_1)(\gamma_1 + \mu) - \beta_1(\gamma_2 + \mu)]\end{array}},$$

$$I_1^* = \frac{\begin{array}{c}\alpha_1\left[(1 - \beta_1)(\delta\gamma_1 + \gamma_2(\gamma_1 + \mu)) - \beta_1\gamma_2(\delta + \gamma_2 + \mu)\right]\\ {}[B(1 - \beta_1) - (\delta + \gamma_2 + \mu)\mu]\end{array}}{\begin{array}{c}(1 - \beta_1)(\alpha_1 + \varepsilon_1 + \mu)(\eta_1 + \theta_1 + \mu)(\delta + \gamma_2 + \mu)\\ {}[(1 - \beta_1)(\gamma_1 + \mu) - \beta_1(\gamma_2 + \mu)]\end{array}},$$

$$I_2^* = \frac{\begin{array}{c}\alpha_2\varepsilon_1\left[(1 - \beta_1)(\delta\gamma_1 + \gamma_2(\gamma_1 + \mu)) - \beta_1\gamma_2(\delta + \gamma_2 + \mu)\right]\\ {}[B(1 - \beta_1) - (\delta + \gamma 2 + \mu)\mu]\end{array}}{\begin{array}{c}(1 - \beta_1)(\alpha_1 + \varepsilon_1 + \mu)(\alpha_2 + \varepsilon_2 + \mu)(\eta_2 + \theta_2 + \mu)\\ (\delta + \gamma_2 + \mu)\left[(1 - \beta_1)(\gamma_1 + \mu) + \beta_1(\gamma_2 + \mu)\right]\end{array}},$$

$$I_3^* = \frac{\alpha_2 \varepsilon_1 \varepsilon_2 \left[(1 - \beta_1)(\delta \gamma_1 + \gamma_2(\gamma_1 + \mu)) - \beta_1 \gamma_2(\delta + \gamma_2 + \mu) \right] \left[B(1 - \beta_1) - (\delta + \gamma_2 + \mu)\mu \right]}{\begin{array}{c} (1 - \beta_1)(\alpha_1 + \varepsilon_1 + \mu)(\alpha_2 + \varepsilon_2 + \mu)(\alpha_3 + \mu) \\ (\eta_3 + \theta_3 + \mu)(\delta + \gamma_2 + \mu) \times \left[(1 - \beta_1)(\gamma_1 + \mu) - \beta_1(\gamma_2 + \mu) \right] \end{array}}$$

$$C^* = \frac{\begin{array}{c} \left[(1 - \beta_1)(\delta \gamma_1 + \gamma_2(\gamma_1 + \mu)) - \beta_1 \gamma_2(\delta + \gamma_2 + \mu) \right] \\ \left[B(1 - \beta_1) - (\delta + \gamma_2 + \mu)\mu \right] \\ \times \left[a_1 \theta_1 (\alpha_2 + \varepsilon_2 + \mu)(\alpha_3 + \mu)(\eta_2 + \theta_2 + \mu)(\eta_3 + \theta_2 + \mu) \right. \\ + \alpha_2 \theta_2 \varepsilon_1 (\alpha_3 + \mu)(\eta_1 + \theta_1 + \mu)(\eta_3 + \theta_3 + \mu) + \alpha_3 \theta_3 \varepsilon_1 \varepsilon_2 \\ \left. (\eta_1 + \theta_1 + \mu)(\eta_2 + \theta_2 + \mu) \right] \end{array}}{\begin{array}{c} (1 - \beta_1)(\eta_1 + \theta_1 + \mu)(\eta_2 + \theta_2 + \mu)(\delta + \gamma_2 + \mu)(\xi + \mu) \\ \left[(1 - \beta_1)(\gamma_1 + \mu) - \beta_1(\gamma_2 + \mu) \right], \end{array}}$$

$$R^* = \frac{\begin{array}{c} \left[(1 - \beta_1)(\delta \gamma_1 + \gamma_2(\gamma_1 + \mu)) - \beta_1 \gamma_2(\delta + \gamma_2 + \mu) \right] \\ \left[B(1 - \beta_1) - (\delta + \gamma_2 + \mu)\mu \right] \\ \times \left[\alpha_1 (\alpha_2 + \varepsilon_2 + \mu)(\alpha_3 + \mu)(\eta_2 + \theta_2 + \mu)(\eta_3 + \theta_2 + \mu) \right. \\ (\eta_1(\xi + \mu) + \xi \theta_1) \\ + \alpha_2 \varepsilon_1 (\alpha_3 + \mu)(\eta_1 + \theta_1 + \mu)(\eta_3 + \theta_3 + \mu)(\eta_2(\xi + \mu) + \xi \theta_2) \\ \left. + \alpha_3 \varepsilon_1 \varepsilon_2 (\eta_1 + \theta_1 + \mu)(\eta_2 + \theta_2 + \mu)(\eta_3(\xi + \mu) + \xi \theta_3) \right] \end{array}}{\begin{array}{c} (1 - \beta_1)(\alpha_1 + \varepsilon_1 + \mu)(\alpha_2 + \varepsilon_2 + \mu)(\alpha_3 + \mu) \\ (\eta_1 + \theta_1 + \mu)(\eta_2 + \theta_2 + \mu) \\ (\eta_3 + \theta_3 + \mu)(\delta + \gamma_2 + \mu)(\xi + \mu) \\ \left[(1 - \beta_1)(\gamma_1 + \mu) - \beta_1(\gamma_2 + \mu) \right] \end{array}}.$$

The endemic equilibrium point E^* exists when following conditions are satisfied:

1. $\beta_1 < 1$,
2. $B(1 - \beta_1) > \mu(\delta + \gamma_2 + \mu)$,
3. $(1 - \beta_1)(\gamma_1 + \mu) > \beta_1(\gamma_2 + \mu)$,
4. $(1 - \beta_1)(\gamma_1 + \mu) > \beta_1(\delta + \gamma_2 + \mu)$,
5. $(1 - \beta_1)(\delta \gamma_1 + \gamma_2(\gamma_1 + \mu)) > \beta_1 \gamma_2(\delta + \gamma_2 + \mu)$.

Condition (1) is obvious as β_1 is the transmission rate. From conditions (3), (4) and (5), we can say

$$(1 - \beta_1) > \beta_1 max \left\{ \frac{\gamma_2 + \mu}{\gamma_1 + \mu}, \frac{\delta + \gamma_2 + \mu}{\gamma_1 + \mu}, \frac{\delta + \gamma_2 + \mu}{\delta \frac{\gamma_1}{\gamma_2} + \gamma_1 + \mu} \right\}$$

$$= \beta_1 \left(\frac{\delta + \gamma_2 + \mu}{\gamma_1 + \mu} \right). \tag{*}$$

Hence, using condition (2) and (*), we have $1 - \beta_1 > max\left\{ \frac{\mu}{B}, \frac{\beta_1}{\gamma_1 + \mu} \right\}$ $(\delta + \gamma_2 + \mu)$.

→ If $1 - \beta_1 > max\left\{ \frac{\mu}{B}, \frac{\beta_1}{\gamma_1 + \mu} \right\} (\delta + \gamma_2 + \mu)$ satisfied then only endemic equilibrium point exists.

These equilibrium points occur when each equation of the system is zero. Hence, the condition of equilibrium points gives the existence of system. And if the conditions are satisfied it gives the positive result, then the system gives the singular solution which the time independent solutions which does not change with time and gives the proper awareness about the transmission of COVID-19.

7.2.2 Basic reproduction number

The basic reproduction number gives the threshold of disease spread. It gives the impression that how fast disease spreads and whether it becomes epidemic or die out in nearer future. The basic reproduction number is the ratio of newly infected individuals affected by secondary infectious individual. To find this, the next generation matrix method (Diekmann et al 2009) [13] is adopted. The Jacobian matrix f and v are calculated as

$$f = \begin{bmatrix} \beta_1 S & 0 & 0 & 0 & 0 & 0 & 0 & 0 & 0 & \beta_1 E_C \\ 0 & (1-\beta_1)S & 0 & 0 & 0 & 0 & 0 & 0 & 0 & (1-\beta_1)U_C \\ 0 & 0 & 0 & 0 & 0 & 0 & 0 & 0 & 0 & 0 \\ 0 & 0 & 0 & 0 & 0 & 0 & 0 & 0 & 0 & 0 \\ 0 & 0 & 0 & 0 & 0 & 0 & 0 & 0 & 0 & 0 \\ 0 & 0 & 0 & 0 & 0 & 0 & 0 & 0 & 0 & 0 \\ 0 & 0 & 0 & 0 & 0 & 0 & 0 & 0 & 0 & 0 \\ 0 & 0 & 0 & 0 & 0 & 0 & 0 & 0 & 0 & 0 \\ 0 & 0 & 0 & 0 & 0 & 0 & 0 & 0 & 0 & 0 \\ 0 & 0 & 0 & 0 & 0 & 0 & 0 & 0 & 0 & 0 \\ 0 & 0 & 0 & 0 & 0 & 0 & 0 & 0 & 0 & 0 \end{bmatrix} \quad \text{and}$$

$$v = \begin{bmatrix}
-\gamma_1-\mu & \delta & 0 & 0 & 0 & 0 & 0 & 0 & 0 & 0 & 0 \\
0 & -\delta-\gamma_2-\mu & 0 & 0 & 0 & 0 & 0 & 0 & 0 & 0 & 0 \\
\gamma_1 & \gamma_2 & -\varepsilon_1-\alpha_1-\mu & 0 & 0 & 0 & 0 & 0 & 0 & 0 & 0 \\
0 & 0 & \varepsilon_1 & -\varepsilon_2-\alpha_2-\mu & 0 & 0 & 0 & 0 & 0 & 0 & 0 \\
0 & 0 & 0 & \varepsilon_2 & -\alpha_3-\mu & 0 & 0 & 0 & 0 & 0 & 0 \\
0 & 0 & \alpha_1 & 0 & 0 & -\theta_1-\eta_1-\mu & 0 & 0 & 0 & 0 & 0 \\
0 & 0 & 0 & \alpha_2 & 0 & 0 & -\theta_2-\eta_2-\mu & 0 & 0 & 0 & 0 \\
0 & 0 & 0 & 0 & \alpha_3 & 0 & 0 & -\theta_3-\eta_3-\mu & 0 & 0 & 0 \\
0 & 0 & 0 & 0 & 0 & \theta_1 & \theta_2 & \theta_3 & -\xi-\mu & 0 & 0 \\
0 & 0 & 0 & 0 & 0 & \eta_1 & \eta_2 & \eta_3 & \xi & -\mu & 0 \\
-\beta_1 S & -(1-\beta_1)S & 0 & 0 & 0 & 0 & 0 & 0 & 0 & 0 & -\beta_1 E_C-(1-\beta_1)U_C
\end{bmatrix}.$$

Here, v is non-singular matrix. Additionally, fv^{-1} is the next generation matrix whose largest eigenvalue is known as the basic reproduction number.

i The basic reproduction number for virus-free equilibrium point E_0 is expressed as R_0 and as given below:

$$R_0 = \frac{B\beta_1}{\mu(\gamma_1+\mu)}.$$

ii The basic reproduction number for unexposed individuals' free equilibrium point E^1 is denoted as R_1 and given as

$$R_1 = \frac{\mu(\gamma_1+\mu)}{B\beta_1}.$$

iii The basic reproduction number R_2 for endemic equilibrium point E^* is given as

$$R_2 = \frac{\beta_1 \left(\delta + \gamma_2 + \mu\right) \left[\begin{matrix} B\left(1 - \beta_1\right)\left\{\left(1 - \beta_1\right)\left(\gamma_1 + \mu\right) \\ +\beta_1\left(\delta + \gamma_2 + \mu\right)\right\} + \beta_1 d\mu\left(\delta + \gamma_2 + \mu\right) \end{matrix}\right]}{B\left(\gamma_1 + \mu\right)\left(1 - \beta_1\right)^2 \left[\left(1 - \beta_1\right)\left(\gamma_1 + \mu\right) + \beta_1\left(\gamma_2 + \mu\right)\right].}$$

→ Substituting the numeric values mentioned in Table 7.1, we have $R_1 = 2.9160$, which suggest that unexposed individual helps manage the pandemic and they can never be zero. Otherwise, situation will not be under controlled. Hence, such individuals should go for vaccination at the earliest. Therefore, now onwards, we will discuss our model for only two equilibrium points, i.e. virus-free equilibrium point and endemic equilibrium point.

7.3 Stability Analysis

In this section, the stability of the developed system is derived. The local and global asymptotical stability about the virus-free equilibrium point E_0 and endemic equilibrium point E^* is performed.

7.3.1 Local stability

Here, local stability about the remaining two equilibrium points E_0 and E^* is discussed using eigenvalues. The Jacobian J of the system is

$$J = \begin{bmatrix} -\beta_1 E_C - (1 - \beta_1)U_C - \mu & -\beta_1 S & -(1 - \beta_1)S & 0 \\ \beta_1 E_C & \beta_1 S - \gamma_1 - \mu & \delta & 0 \\ (1 - \beta_1)U_C & 0 & (1 - \beta_1)S - \delta - \gamma_2 - \mu & 0 \\ 0 & \gamma_1 & \gamma_2 & -\alpha_1 - \varepsilon_1 - \mu \\ 0 & 0 & 0 & \varepsilon_1 \\ 0 & 0 & 0 & 0 \\ 0 & 0 & 0 & \alpha_1 \\ 0 & 0 & 0 & 0 \\ 0 & 0 & 0 & 0 \\ 0 & 0 & 0 & 0 \end{bmatrix}$$

$$
\begin{bmatrix}
0 & 0 & 0 & 0 & 0 & 0 & 0 \\
0 & 0 & 0 & 0 & 0 & 0 & 0 \\
0 & 0 & 0 & 0 & 0 & 0 & 0 \\
-\alpha_2 - \varepsilon_2 - \mu & 0 & 0 & 0 & 0 & 0 & 0 \\
\varepsilon_2 & -\alpha_3 - \mu & 0 & 0 & 0 & 0 & 0 \\
0 & 0 & -\eta_1 - \theta_1 - \mu & 0 & 0 & 0 & 0 \\
\alpha_2 & 0 & 0 & -\eta_2 - \theta_2 - \mu & 0 & 0 & 0 \\
0 & \alpha_3 & 0 & 0 & -\eta_3 - \theta_3 - \mu & 0 & 0 \\
0 & 0 & \theta_1 & \theta_2 & \theta_3 & -\xi - \mu & 0 \\
0 & 0 & \eta_1 & \eta_2 & \eta_3 & \xi & -\mu
\end{bmatrix}.
$$

Finding Jacobian for the virus-free equilibrium point $E_0(\frac{B}{\mu}, 0, 0, 0, 0, 0, 0, 0, 0, 0, 0)$, we have J_0 as

$$
J_0 =
\begin{bmatrix}
-\mu & -\frac{B\beta_1}{\mu} & -\frac{(1-\beta_1)B}{\mu} & 0 & 0 \\
0 & \frac{B\beta_1}{\mu} - \gamma_1 - \mu & \delta & 0 & 0 \\
0 & 0 & \frac{(1-\beta_1)B}{\mu} - \delta - \gamma_2 - \mu & 0 & 0 \\
0 & \gamma_1 & \gamma_2 & -\alpha_1 - \varepsilon_1 - \mu & 0 \\
0 & 0 & 0 & \varepsilon_1 & -\alpha_2 - \varepsilon_2 - \mu \\
0 & 0 & 0 & 0 & \varepsilon_2 \\
0 & 0 & 0 & \alpha_1 & 0 \\
0 & 0 & 0 & 0 & \alpha_2 \\
0 & 0 & 0 & 0 & 0 \\
0 & 0 & 0 & 0 & 0 \\
0 & 0 & 0 & 0 & 0
\end{bmatrix}
$$

$$
\begin{bmatrix}
0 & 0 & 0 & 0 & 0 & 0 \\
0 & 0 & 0 & 0 & 0 & 0 \\
0 & 0 & 0 & 0 & 0 & 0 \\
0 & 0 & 0 & 0 & 0 & 0 \\
0 & 0 & 0 & 0 & 0 & 0 \\
-\alpha_3 - \mu & 0 & 0 & 0 & 0 & 0 \\
0 & -\eta_1 - \theta_1 - \mu & 0 & 0 & 0 & 0 \\
0 & 0 & -\eta_2 - \theta_2 - \mu & 0 & 0 & 0 \\
\alpha_3 & 0 & 0 & -\eta_3 - \theta_3 - \mu & 0 & 0 \\
0 & \theta_1 & \theta_2 & \theta_3 & -\xi - \mu & 0 \\
0 & \eta_1 & \eta_2 & \eta_3 & \xi & -\mu
\end{bmatrix}.
$$

The above Jacobian J_0 has eleven eigenvalues:
$\lambda_1 = -\left(\frac{B\beta_1 + \delta\mu + \gamma_2\mu + \mu^2 - B}{\mu}\right)$, $\lambda_2 = -\left(\frac{-B\beta_1 + \gamma_1\mu + \mu^2}{\mu}\right)$, $\lambda_3 = -\mu$, $\lambda_4 = -\mu$, $\lambda_5 = -(\alpha_1 + \varepsilon_1 + \mu)$, $\lambda_6 = -(\alpha_2 + \varepsilon_2 + \mu)$, $\lambda_7 = -(\alpha_3 + \mu)$, $\lambda_8 = -(\eta_1 + \theta_1 + \mu)$, $\lambda_9 = -(\eta_2 + \theta_2 + \mu)$, $\lambda_{10} = -(\eta_3 + \theta_3 + \mu)$, $\lambda_{11} = -(\xi + \mu)$.

It can be easily seen that $\lambda_3, \lambda_4, \lambda_5, \lambda_6, \lambda_7, \lambda_8, \lambda_9, \lambda_{10}$ and λ_{11} are negative.

Now, $\lambda_1 = -\left(\frac{B\beta_1 + \delta\mu + \gamma_2\mu + \mu^2 - B}{\mu}\right) = \frac{B(1-\beta_1) - \mu(\delta + \gamma_2 + \mu)}{\mu} = \frac{B(1-\beta_1)}{\mu} -$
$(\delta + \gamma_2 + \mu)$.

So, if $\frac{B(1-\beta_1)}{\mu} < (\delta + \gamma_2 + \mu)$ then only λ_1 is negative.

Also, λ_2 can be written as

$$\lambda_2 = \frac{B\beta_1}{\mu(\gamma_1 + \mu)} - 1 = R_0 - 1.$$

So, if $R_0 < 1$ then only λ_2 is negative.

Theorem 7.1. If $R_0 < 1$ and $\frac{B(1-\beta_1)}{\mu} < (\delta + \gamma_2 + \mu)$ then E_0 is locally asymptotically stable [12].

Next, finding Jacobian for the endemic equilibrium point $E^*(S^*, E_C^*, U_C^*, D_1^*, D_2^*, D_3^*, I_1^*, I_2^*, I_3^*, C^*, R^*)$, we get J^* as

$$J^* = \begin{bmatrix} -\beta_1 E_C^* - (1-\beta_1) U_C^* - \mu & -\beta_1 S^* & -(1-\beta_1) S^* & 0 \\ \beta_1 E_C^* & \beta_1 S^* - \gamma_1 - \mu & \delta & 0 \\ (1-\beta_1) U_C^* & 0 & (1-\beta_1) S^* - \delta - \gamma_2 - \mu & 0 \\ 0 & \gamma_1 & \gamma_2 & -\alpha_1 - \varepsilon_1 - \mu \\ 0 & 0 & 0 & \varepsilon_1 \\ 0 & 0 & 0 & 0 \\ 0 & 0 & 0 & \alpha_1 \\ 0 & 0 & 0 & 0 \\ 0 & 0 & 0 & 0 \\ 0 & 0 & 0 & 0 \\ 0 & 0 & 0 & 0 \end{bmatrix}$$

$$\begin{bmatrix} 0 & 0 & 0 & 0 & 0 & 0 & 0 \\ 0 & 0 & 0 & 0 & 0 & 0 & 0 \\ 0 & 0 & 0 & 0 & 0 & 0 & 0 \\ 0 & 0 & 0 & 0 & 0 & 0 & 0 \\ -\alpha_2 - \varepsilon_2 - \mu & 0 & 0 & 0 & 0 & 0 & 0 \\ \varepsilon_2 & -\alpha_3 - \mu & 0 & 0 & 0 & 0 & 0 \\ 0 & 0 & -\eta_1 - \theta_1 - \mu & 0 & 0 & 0 & 0 \\ \alpha_2 & 0 & 0 & -\eta_2 - \theta_2 - \mu & 0 & 0 & 0 \\ 0 & \alpha_3 & 0 & 0 & -\eta_3 - \theta_3 - \mu & 0 & 0 \\ 0 & 0 & \theta_1 & \theta_2 & \theta_3 & -\xi - \mu & 0 \\ 0 & 0 & \eta_1 & \eta_2 & \eta_3 & \xi & -\mu \end{bmatrix}.$$

The Jacobian J^* has eleven distinct eigenvalues among which eight are

$$\lambda_1 = -(\alpha_1 + \varepsilon_1 + \mu), \lambda_2 = -(\alpha_2 + \varepsilon_2 + \mu),$$

$$\lambda_3 = -(\alpha_3 + \mu), \lambda_4 = -(\eta_1 + \theta_1 + \mu),$$

$$\lambda_5 = -\left(\eta_2 + \theta_2 + \mu\right), \lambda_6 = -\left(\eta_3 + \theta_3 + \mu\right), \lambda_7 = -\left(\xi + \mu\right), \lambda_8 = -\mu.$$

Hence, the characteristic polynomial of Jacobian matrix J^* is

$$\left(\alpha_1 + \varepsilon_1 + \mu + \lambda\right)\left(\alpha_2 + \varepsilon_2 + \mu + \lambda\right)\left(\alpha_3 + \mu + \lambda\right)\left(\eta_1 + \theta_1 + \mu + \lambda\right)$$

$$\left(\eta_2 + \theta_2 + \mu + \lambda\right)\left(\eta_3 + \theta_3 + \mu + \lambda\right)\left(\xi + \mu + \lambda\right)\left(\lambda + \mu\right)Q\left(\lambda\right),$$

where $Q\left(\lambda\right)$ is polynomial of resultant 3×3 matrix J',

$$J' = \begin{bmatrix} -\beta_1 E_C - \left(1 - \beta_1\right)U_C - \mu & -\beta_1 S & -\left(1 - \beta_1\right)S \\ \beta_1 E_C & \beta_1 S - \gamma_1 - \mu & \delta \\ \left(1 - \beta_1\right)U_C & 0 & \left(1 - \beta_1\right)S - \delta - \gamma_2 - \mu \end{bmatrix}.$$

The resultant matrix J' has -1.0150, -0.2705 and 0.0002 as eigenvalues, among which first two are negative but the last one is non-negative. Hence, the endemic equilibrium point is not asymptotically stable, but it can be stable if γ_2 is increased by 1%, that is at least 1% more unexposed individuals should also start taking vaccination.

Theorem 7.2. E^* is not locally asymptotically stable at the time.

7.3.2 Global stability

In this, we will carry out the global stability about virus-free equilibrium point (E_0) and endemic equilibrium point (E^*). For this first we construct Lyapunov function for each equilibrium points and solve it using the theory of LaSalle [9]. For virus-free equilibrium point $E_0\left(\frac{B}{\mu}, 0, 0, 0, 0, 0, 0, 0, 0, 0, 0\right)$, consider the Lyapunov function

$$L\left(t\right) = E_C\left(t\right) + U_C\left(t\right).$$

Then

$$L'\left(t\right) = E_C{}'\left(t\right) + U_C{}'\left(t\right) = \beta_1 S E_C + \delta U_C - \gamma_1 E_C - \mu E_C + \left(1 - \beta_1\right)$$

$$SU_C - \delta U_C - \gamma_2 U_C - \mu U_C = B - \mu S - \gamma_1 E_C - \mu E_C - \gamma_2 U_C - \mu U_C.$$

Since $S \leq \frac{B}{\mu}$

$$L'\left(t\right) \leq B - \mu\left(\frac{B}{\mu}\right) - \gamma_1 E_C - \gamma_2 U_C - \mu\left(E_C + U_C\right)$$

$$= -\gamma_1 E_C - \gamma_2 U_C - \mu\left(E_C + U_C\right),$$

that is $L'(t) \leq 0$ and $L'(t) = 0$ is $E_C = U_C = 0$.

Therefore, E_0 is the only solution of system.

Hence, every solution of system reaches to E_0 as $t \to \infty$ with initial conditions.

Theorem 7.3. E_0 is globally asymptotically stable.

For endemic equilibrium point $E^*(S^*, E_C^*, U_C^*, D_1^*, D_2^*, D_3^*, I_1^*, I_2^*, I_3^*, C^*, R^*)$, consider the Lyapunov function

$$L = k_1 (S - S^* lnS) + k_2 (E_C - E_C^* lnE_C) + k_3 (U_C - U_C^* lnU_C)$$

$$+ k_4 (D_1 - D_1^* lnD_1) + k_5 (D_2 - D_2^* lnD_2) + k_6 (D_3 - D_3^* lnD_3)$$

$$+ k_7 (I_1 - I_1^* lnI_1) + k_8 (I_2 - I_2^* lnI_2) + k_9 (I_3 - I_3^* lnI_3)$$

$$+ k_{10} (C - C^* lnC) + k_{11} (R - R^* lnR),$$

where $k_i; i = 1, 2, \ldots, 11$ are constants.

Differentiating L w.r.t. t, we get

$$\frac{dL}{dt} = k_1 \left(1 - \frac{S^*}{S}\right) \frac{dS}{dt} + k_2 \left(1 - \frac{E_C^*}{E_C}\right) \frac{dE_C}{dt} + k_3 \left(1 - \frac{U_C^*}{U_C}\right) \frac{dU_C}{dt}$$

$$+ k_4 \left(1 - \frac{D_1^*}{D_1}\right) \frac{dD_1}{dt} + k_5 \left(1 - \frac{D_2^*}{D_2}\right) \frac{dD_2}{dt} + k_6 \left(1 - \frac{D_3^*}{D_3}\right) \frac{dD_3}{dt}$$

$$+ k_7 \left(1 - \frac{I_1^*}{I_1}\right) \frac{dI_1}{dt} + k_8 \left(1 - \frac{I_2^*}{I_2}\right) \frac{dI_2}{dt} + k_9 \left(1 - \frac{I_3^*}{I_3}\right) \frac{dI_3}{dt}$$

$$+ k_{10} \left(1 - \frac{C^*}{C}\right) \frac{dC}{dt} + k_{11} \left(1 - \frac{R^*}{R}\right) \frac{dR}{dt}$$

$$= -k_1 \left(1 - \frac{S^*}{S}\right)^2 \mu S + k_1 \left(1 - \frac{S^*}{S}\right) \left(\frac{S^* E_C^*}{S E_C} - 1\right) \beta_1 S E_C$$

$$+ k_1 \left(1 - \frac{S^*}{S}\right) \left(\frac{S^* U_C^*}{S U_C} - 1\right) \beta_1 S U_C - k_2 \left(1 - \frac{E_C^*}{E_C}\right)^2 \mu E_C$$

$$- k_3 \left(1 - \frac{U_C^*}{U_C}\right)^2 \mu U_C - k_4 \left(1 - \frac{D_1^*}{D_1}\right)^2 \mu D_1 - k_5 \left(1 - \frac{D_2^*}{D_2}\right)^2 \mu D_2$$

$$- k_6 \left(1 - \frac{D_3^*}{D_3}\right)^2 \mu D_3 - k_7 \left(1 - \frac{I_1^*}{I_1}\right)^2 \mu I_1 - k_8 \left(1 - \frac{I_2^*}{I_2}\right)^2 \mu$$

$$I_2 - k_9 \left(1 - \frac{I_3^*}{I_3}\right)^2 \mu I_3 - k_{10} \left(1 - \frac{C^*}{C}\right)^2 \mu C - k_{11} \left(1 - \frac{R^*}{R}\right)^2 \mu R$$

$$= -k_1 \left(1 - \frac{S^*}{S}\right)^2 \mu S - k_2 \left(1 - \frac{E_C^*}{E_C}\right)^2 \mu E_C - k_3 \left(1 - \frac{U_C^*}{U_C}\right)^2 \mu U_C$$

$$-k_4 \left(1 - \frac{D_1^*}{D_1}\right)^2 \mu D_1 - k_5 \left(1 - \frac{D_2^*}{D_2}\right)^2 \mu D_2 - k_6 \left(1 - \frac{D_3^*}{D_3}\right)^2 \mu D_3$$

$$-k_7 \left(1 - \frac{I_1^*}{I_1}\right)^2 \mu I_1 - k_8 \left(1 - \frac{I_2^*}{I_2}\right)^2 \mu I_2 - k_9 \left(1 - \frac{I_3^*}{I_3}\right)^2 \mu I_3$$

$$-k_{10} \left(1 - \frac{C^*}{C}\right)^2 \mu C - k_{11} \left(1 - \frac{R^*}{R}\right)^2 \mu R + Z(E),$$

where $Z(E) = k_1 \left(1 - \frac{S^*}{S}\right) \left(\frac{S^*}{S} \frac{E_C^*}{E_C} - 1\right) \beta_1 S E_C + k_1 \left(1 - \frac{S^*}{S}\right)$ $\left(\frac{S^*}{S} \frac{U_C^*}{U_C} - 1\right) \beta_1 S U_C.$

Here, $Z(E)$ is negative using the approaches derived as $\left(1 - \frac{S^*}{S}\right) \left(\frac{S^*}{S} \frac{E_C^*}{E_C} - 1\right) \leq 0$ and $\left(1 - \frac{S^*}{S}\right) \left(\frac{S^*}{S} \frac{U_C^*}{U_C} - 1\right) \leq 0$ from [5], [2] and [1].

Thus, $Z(E) \leq 0$ for all $E \geq 0$.

As $\frac{dL}{dt} \leq 0$ in E and $E = E^*$; the largest invariant set in E such that $\frac{dL}{dt} \leq 0$ is the singleton E^*, which is our endemic equilibrium point.

Theorem 7.4. The endemic equilibrium point E^* is globally stable.

7.4 Numerical Simulation

In this section, the simulation for the given system is designed using the parametric values given in Table 7.1.

Figure 7.2 displays the behaviour of the individuals during the pandemic. It shows susceptible individuals are directed to the exposed and unexposed individual. Also, this suggests that individuals can take the first dose of the vaccine within 90 days after getting exposed to COVID-19. Moreover, if an individual is not exposed, i.e. unexposed, then one should take the vaccine dose I immediately within a month as one starts getting exposed within 1.7 months. The figure also indicates that it takes less time to recover from the first infection than from the second.

Table 7.1   Notation and its description.

Notation	Description	Parametric value
B	The growth rate	0.4
β_1	The rate at which susceptible individuals are exposed to COVID-19	0.024
δ	The rate at which susceptible individuals are exposed to COVID-19	0.0147
γ_1	The rate at which exposed individuals go for dose I of vaccination	0.18
γ_2	The rate at which unexposed individuals go for dose I of vaccination	0.27
ε_1	The rate at which individual go for dose II after completion of dose I	0.15
ε_2	The rate at which individual go for booster dose after completion of both the doses	0.079
α_1	Infection rate after dose I of vaccination	0.155
α_2	Infection rate after dose II of vaccination	0.125
α_3	Infection rate after booster dose of vaccination	0.088
θ_1	The rate of infected individuals become critically ill after dose I of vaccination	0.10
θ_2	The rate of infected individuals become critically ill after dose II of vaccination	0.055
θ_3	The rate of infected individuals become critically ill after booster dose of vaccination	0.017
η_1	Recovery rate of infected individual after dose I of vaccination	0.82
η_2	Recovery rate of infected individual after dose II of vaccination	0.85
η_3	Recovery rate of infected individual after booster dose of vaccination	0.90
ξ	Recovery rate of critically ill individuals	0.977
μ	COVID-19 deceased individuals	0.10

Figure 7.3 suggests that individuals taking dose I are at their peak at the end of the first month. Then it decreases for the next seven months. Ongoing awareness for vaccination increases it later.

Figure 7.4 shows the rate of individuals taking dose II. The government has decided that dose II can be taken only after the completion of 84 days of dose I. So, initially, the number of individuals going for dose II is small but after about a year everyone will be vaccinated with dose I and this will result in an increase in dose II vaccine.

Figure 7.5 proposes that after approval of booster dose, people start taking it gradually for six months. But then as the effect of COVID-19 diminishes,

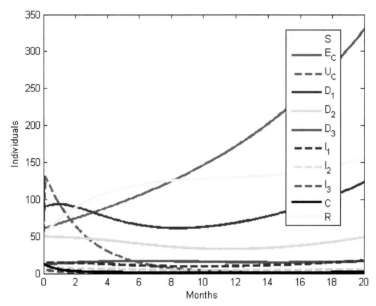

Figure 7.2   Transmission of individual.

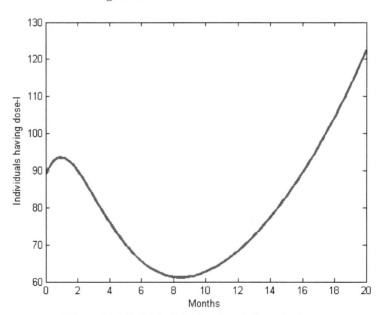

Figure 7.3   Individuals having dose I of vaccination.

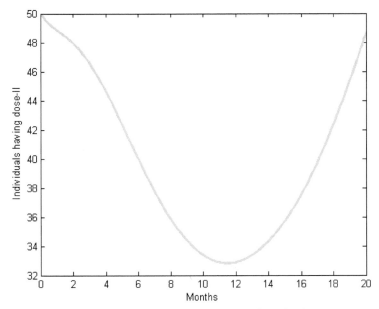

Figure 7.4   Individuals having dose II of vaccination.

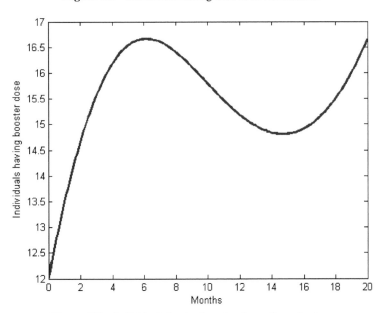

Figure 7.5   Individuals having booster dose of vaccination.

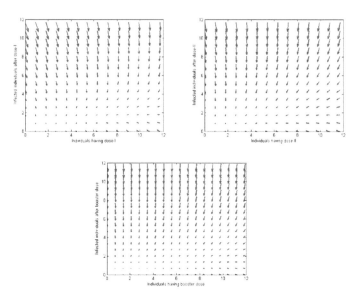

Figure 7.6   Behaviour of infected individual after each dose of vaccination.

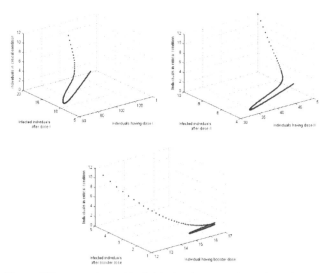

Figure 7.7   Behaviour of individuals after each dose of vaccination.

so does awareness of booster dose and they are negligent in this regard. However, its rate increases after 1 year cause of awareness drives against booster dose.

Figures 7.6 advocates that the number of infected individuals is declining after taking any dose of vaccination. From above figures one can see that the severity of infection after dose I is higher than dose II and booster dose. In addition, booster dose quickly nullifies the infectious effect compared to dose II and dose I.

The phase portraits (Figure7.7) illustrate the effect of vaccination on critically ill individuals. Indication from Figure 7.7 is that those individuals who have taken only dose I compared to dose II and booster dose are more serious in long run.

For dose I: 15.38% of individuals become infected, of which 45% individuals become critical.

For dose II: Out of 14% infected individuals, 42.86% individuals go into critical stage.

For booster dose: 11.76% individuals get infected among only 2% individuals are in critical condition.

Figure 7.8 reflects that the rate at which individuals become infected after taking booster dose is lower and at the same time more individuals recovering after taking booster dose.

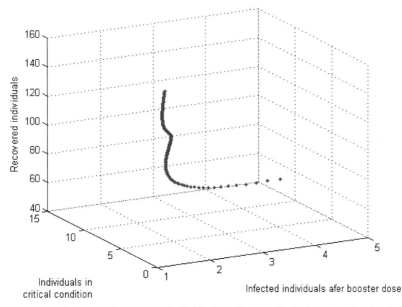

Figure 7.8 Behaviour of recovered individuals and critically ill individuals after booster dose.

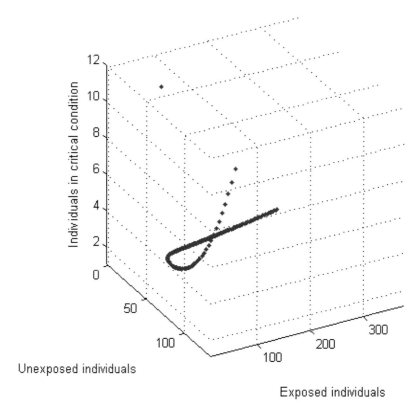

Figure 7.9   Cyclic behaviour of individuals in critical condition.

From Figure 7.9, one can say that unexposed individuals can transfer to exposed individuals. In fact, after recovering, one may get exposed to COVID-19 second time and catch the infection. This loop can continue. Moreover, if not vaccinate one can catch a severe infection and go into critical condition.

7.5 Conclusion

In this mathematical model, the behaviour of infected individuals having the dose I, dose II and booster dose is compared. Additionally, the critical situation of the same individuals is deliberated. We have shown the basic reproduction number of this system. It shows the rate of infection with three types of individuals; (i) the individuals taken dose I of vaccination, (ii) the individuals who have taken dose II of vaccination and (iii) the individuals taken booster dose of vaccination. The basic reproduction number when only susceptible individual is present is 0.3428, which suggests that the system is under control. The basic reproduction number when unexposed individuals are not present in the system, i.e. all individuals are infected with COVID-19 at least once in his lifetime, is 2.9160. This proposes that 291.6% individuals are getting infected with the time which is not in controllable stage. Lastly, the basic reproduction number when every type of individuals is present in the system is 0.0338 that advocates that only 3.38% individuals are getting infection. Comparing these basic reproduction number, it can be concluded that unexposed individuals play vital role for making pandemic controllable. Further, the stability analysis presents that the virus-free stage is asymptotically stable but pandemic stage is not asymptotically stable which can be stable if γ_2 is increased by 1%, that is at least 1% more unexposed individuals should start taking vaccination. These results are supported through numerical simulation that when individuals are start taking booster dose, the infection rate is decreasing from 45% to 2%. Also, the rate of getting into critical condition is decreased. In order to avoid getting exposed to COVID-19 everyone should start his/her vaccination at the earliest and also opt for booster dose. By taking necessary precautions, every individual can make this pandemic to end.

References

[1] A. Korobeinikov, "Global properties of infectious disease models with nonlinear incidence", Bulletin of Mathematical Biology, pp. 1871-1886, 69(6), 2007.

[2] A. Korobeinikov, P. K. Maini, "A Lyapunov function and global properties for SIR and SEIR epidemiological models with nonlinear incidence", Mathematical Biosciences & Engineering, pp. 57, 1(1), 2004.

[3] A. M. Hause, J. Baggs, P. Marquez, T. R. Myers, J. R. Su, P.G. Blanc, D. K. Shay, "Safety monitoring of COVID-19 vaccine booster doses among adults—United States", Morbidity and Mortality Weekly Report, pp. 249, 71(7), September 22, 2021–February 6, 2022.

[4] B. H. Foy, B. Wahl, K. Mehta, A. Shet, G. I. Menon, C. Britto, "Comparing COVID-19 vaccine allocation strategies in India: A mathematical modelling study", International Journal of Infectious Diseases, pp. 431-438, 103, 2021.

[5] C. C. McCluskey, "Lyapunov functions for tuberculosis models with fast and slow progression", Mathematical Biosciences & Engineering, pp. 603, 3(4), 2006.

[6] C. R. MacIntyre, V. Costantino, M. Trent, "Modelling of COVID-19 vaccination strategies and herd immunity, in scenarios of limited and full vaccine supply in NSW, Australia", Vaccine, pp. 2506-2513, 40(17), 2022.

[7] E. D. Moreira Jr, N. Kitchin, X. Xu, S. S. Dychter, S. Lockhart, A. Gurtman, K. U. Jansen, "Safety and Efficacy of a Third Dose of BNT162b2 Covid-19 Vaccine", New England Journal of Medicine, 2022.

[8] J. Croda, O. T. Ranzani, "Booster doses for inactivated COVID-19 vaccines: if, when, and for whom", The Lancet Infectious Diseases, pp. 430-432, 22(4), 2022.

[9] J. P. LaSalle, "The stability of dynamical systems", Siam, 25, 1976.

[10] N. Andrews, J. Stowe, F. Kirsebom, S. Toffa, T. Rickeard, E. Gallagher, J. Lopez Bernal, "Covid-19 vaccine effectiveness against the Omicron (B. 1.1. 529) variant", New England Journal of Medicine, pp. 1532-1546, 386(4), 2022.

[11] N. H. Shah, A. H. Suthar, M. H. Satia, Y. Shah, N. Shukla, J. Shukla, D. Shukla, "Modelling the Impact of Nationwide BCG Vaccine Recommendations on COVID-19 Transmission, Severity and Mortality", In Mathematical Analysis for Transmission of COVID-19, Springer, Singapore, pp. 21-37, 2021.

[12] N. H. Shah, M. H. Satia, F. A. Thakkar, "Effect of Environmental Pollutants on Rain due to Stakeholders", Mathematics in Engineering Sciences: Novel Theories, Technologies, and Applications, CRC Press, pp. 301-327, 2019.

[13] O. Diekmann, J. A. P. Heesterbeek, M. G. Roberts, "The construction of next-generation matrices for compartmental epidemic models", Journal of the Royal Society Interface, pp. 873-885, 7(47), 2009.

[14] P. Bansal, A. Raj, D. M. Shukla, N. Sunder, "COVID-19 vaccine preferences in India", Vaccine, pp. 2242-2246, 40(15), 2022.

[15] S. Mandal, N. Arinaminpathy, B. Bhargava, S. Panda, "India's pragmatic vaccination strategy against COVID-19: a mathematical modelling-based analyses. BMJ open", 11(7), e048874, 2021.

[16] S. Chenchula, P. Karunakaran, S. Sharma, M. Chavan, "Current evidence on efficacy of COVID-19 booster dose vaccination against the Omicron variant: A systematic review", Journal of Medical Virology, 2022.

[17] T. T. Le, Z. Andreadakis, A. Kumar, R. G. Román, S. Tollefsen, M. Saville, S. Mayhew, "The COVID-19 vaccine development landscape. Nat Rev Drug Discov, pp. 305-306, 19(5), 2020.

[18] V. Tolia, R. Renin Singh, S. Deshpande, A. Dave, R. M. Rathod, "Understanding Factors to COVID-19 Vaccine Adoption in Gujarat, India", International Journal of Environmental Research and Public Health, pp. 2707, 19(5), 2022.

8

Impact of Post-COVID-19 Pandemic on Quality Education among School and College Students

Santosh Kumar[1], Surya Kant Pal[2], Isha Sangal[3], Khursheed Alam[4], Mandeep Mittal[5], Mahesh Kumar Jayaswal[6*]

[1,2,4]Department of Mathematics, Sharda School of Basic Sciences and Research, Sharda University, Greater Noida-201306, Uttar Pradesh, India
[1,2,4]The A.H. Siddiqi Centre for Advanced Research in Applied Mathematics and Physics, Sharda University, Greater Noida 201306, Uttar Pradesh, India
[5]Department of Mathematics, Amity Institute of Applied Sciences, Amity University Noida, Noida, Uttar Pradesh-201313, India
[3,6]Department of Mathematics Banasthali Vidyapith, Banasthali Rajasthan-304022, India
E-mail: *corresponding author: Maheshjayaswal17@gmail.com

Abstract

Due to the COVID-19 pandemic, uncertainty has increased in foremost aspects of international and national societies, including educational organizations, like schools and colleges. This uncertainty had paused the functioning of students as it used to, for example, uncertainty has been about how the institute's closure in the last spring affected students' achievements, and likewise, the rapid conversion to the virtual form of learning in the academic years will continue affecting the achievements. Amidst all the uncertainties, there has been an increasing consensus that the lockdown and closures of the schools and colleges due to the COVID-19 have harmed students' academics, mental health and physical health. This chapter deals with how students from schools and colleges feel about the outbreak of

COVID-19 impacting them physically, mentally and academically. In this chapter, we have used a regression model and data collected through a primary source to measure the accuracy of the Corona-positive patients with the worst-case scenario for the country of India. The results show that the COVID-19 is expected to continue to grow with an unpredictable factor and a fear of an uncertain environment as well as several health issues among individuals.

Keywords: Lockdown Effects, Physical Health, Mental Health, Sustainable Development Goals, COVID-19, Anticipative Measures.

8.1 Introduction

Emerging uncertainty due to COVID-19 had put a pause on the functioning of the country. The COVID-19 pandemic has posed several challenges to every stakeholder of education, like students, educators and parents. The children are the most affected by it and they have coped up with their physical and mental conditions. Based on secondary data sources, Ref. [1] explained the effect of sustainable development goals for no poverty and exposed the way the Indian government was willing to achieve these goals till the financial year 2030. Ref. [2] studied the foremost effect on the education of trainees in the radiology environment. Based on the steep drop in inpatient volumes and sequestering of faculty and trainees to maintain the decorum of social distancing, he concluded that experiential learning was affected. Ref. [3] studied the impact of COVID-19 on the psychological state of a group of 584 Chinese youths. The responses to the question about the cognitive status of COVID-19, the overall Health Questionnaire (GHQ-12) and the PTSD Checklist-Civilian Version (PCL-C) were completed. Ref. [4] studied the effect of COVID-19 on overall education. Data collection was done with the primary source together with the structured questionnaires that were administered to 200 respondents comprising students, teachers, parents and policymakers from several countries.

They concluded that COVID-19 harms education and learning disruptions. This also has an impact on access to research facilities and education. Ref. [5] studied the effect of COVID-19 on psychological well-being and psychological state. This research was conducted on Filipino Graduate students in the context of an epidemic. Based on an experiment on mental health problems, Ref. [6] studied the impact of COVID-19 on several factors, like understanding, HCWs' attention and decision-making ability as well as the lasting consequences for the well-being. Ref. [7] studied the impact of COVID-19 on socio-economic conditions through surveys. Ref. [8] studied the effect of COVID-19 on students' experiences and expectations. Ref. [5] studied the impact of COVID-19 on stress anxiety and fear in graduate students and concluded that COVID-19 affected all these factors. Ref. [6] studied the psychological impact of COVID-19 on adults and their children in Italy. Ref. [9] studied the impact of COVID-19 on vulnerability, perceived risk and fear. They concluded that COVID-19 highly affected these factors. Recently, Refs. 10–15] have given significant contributions to the Indian economy and SDGs.

In this chapter, the effect of this changing scenario on physical and mental health as well as academics of different levels of students – school level, undergraduate level and graduate level – has been studied. The undertaking research for an overall evaluation of the impact questionnaire method has been used to get first-hand data and it is divided into three major sections covering all the responses. This chapter also deals with the preventive measures and the way of which the government is dealing with the problems arising from the COVID-19 pandemic.

8.2 Research Gap

- Comparative study of students who are currently pursuing schooling, graduation and postgraduation on how they feel about the pandemic affecting their health, psychology and academics.
- Analysing the differences in the responses, by considering altogether, female and male students of the schools, UG and PG.
- How did the change from the traditional environment to the virtual learning environment affect student psyche, physique, and academics?
- Change in the perspective of thinking about the impact of COVID-19 on school students, undergraduates and postgraduates.

8.3 Research Objectives

- To understand the effect of lockdown on the students' physical health and prove whether it is effective for the students or puts a challenge on their physique.
- To understand about the students' feeling on their academics getting affected by the closures of schools and colleges, and the adoption of virtual platforms of learning.

8.4 Research Hypothesis

(a) Null hypothesis (Ho) – There is no impact of the outbreak of COVID-19 and imposition of lockdown on students' physical health.
Alternative hypothesis (H1) – There is a major impact of the outbreak of COVID-19 and imposition of lockdown on students' physical health.

(b) Null hypothesis (Ho) – There is no impact of the outbreak of COVID-19 and imposition of lockdown on students' mental health.
Alternative hypothesis (H1) – There is a major impact of the outbreak of COVID-19 and imposition of lockdown on students' mental health.

(c) Null hypothesis (Ho) – There is no impact of the outbreak of COVID-19 and imposition of lockdown on students' academics.
Alternative hypothesis (H1) – There is a major impact of the outbreak of COVID-19 and imposition of lockdown on students' academics.

(d) Null hypothesis (Ho) – There is no relationship between adverse effects on health and mental state due to online learning and the effects on the behaviour, such as frequent mood swings.
Alternative hypothesis (H1) – There is a relationship between adverse effects on health and mental state due to online learning and the effects on the behaviour, such as frequent mood swings.

8.5 Methodology

Descriptive statistics and regression analysis approach has been used to analyse and interpret the data. With the help of these tools, we will get to know whether to reject the null hypothesis or accept it. This interpretation is divided into four parts to evaluate four hypothesis statements. Visualization has been performed with the help of a tableau for each question separately to understand the level of deviation and skewness in every question.

Figure 8.1 provides that the majority of the population agreed that the closure of gyms and other outdoor activities due to lockdown affected the physique of individuals. In total 26% of the population strongly agree and 39.33% of the population agree that the physique of individuals has been affected.

Figure 8.2 shows that online learning has affected adversely at most student's physical health, and the total majority of students have reported that they are not able to concentrate properly in their work, and also not being able to make their discussions freely with their teachers and other staffs.

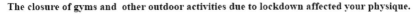

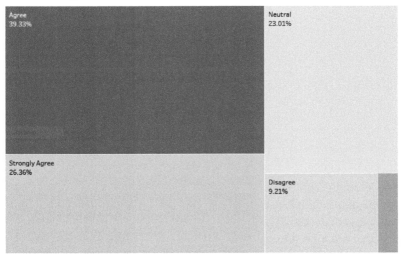

Figure 8.1 The closure of gyms and other outdoor activities due to lockdown affected your physique.

The student's physical health has suffered majorly due to this online mode of learning. They are unable to concentrate on their studies, neither they are interested, which has put an adverse effect on the growth of the students.

Figure 8.3 describes the monotonous daily routine that is causing depression and anxiety in the lives of the students; not only students but adults are also facing the same kind of situation. This has led to many lives with uncertainty; thus, the survey results are also depicting the irregular daily routine which has caused depression problems in the lives of many individuals.

Figure 8.4 describes different genders to whom this lockdown proved to be beneficial, stating that there are hardly few individuals who benefited from the lockdown, and in terms of gender, women were more benefited than men in terms of maintaining their physical health. But the figure shows more of the disagreeing situation for both men and women.

Figure 8.5 discusses about students who are studying in schools, UG or PG courses, and the level of doubt clearance or level of delivery of information for which most of the students disagree and feel uncomfortable with such scenarios of learning and doubts that they face while studying. The information which is being delivered is not up to the level of which teachers used to demonstrate while conducting physical classes, and this leads to an end-time overburden situation.

Figure 8.6 shows the level of confidence of students in schools, under-graduates or postgraduates, and the results depict an agreed situation for schools, undergraduates and for postgraduates who have to seek jobs in the market and require more confidence of which they lack; thus, they strongly agree with the lack of confidence they face due to virtual learning platforms.

Figure 8.7 concludes the following:

- Since the *P*-value of physical and mental health is found to be less than 0.05, we reject the null hypothesis which states that there exists no relation between adverse effects on physical and mental health due to online learning with behaviour changes, such as mood swings.
- From the above ANOVA table, it is found that the significance of the *F*-value is less than 0.05. This indicates rejecting the null hypothesis. This means that not all regression coefficients are zero.

Online mode of learning put an adverse effects on your physical health.

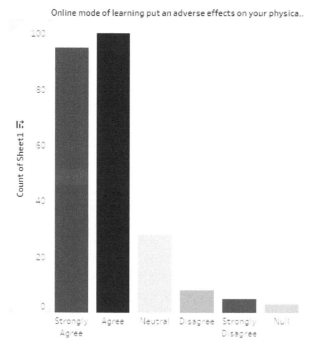

Online mode of learning put an adverse effects on your physica..

Figure 8.2   Online mode of learning put an adverse effect on your physical health.

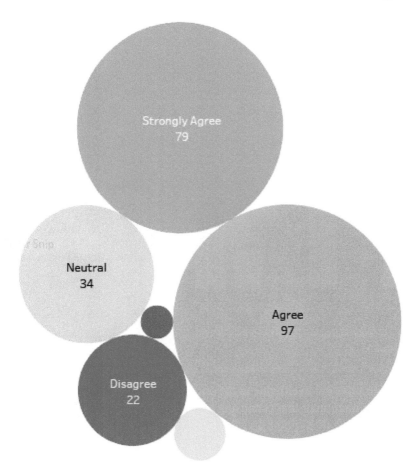

Figure 8.3 Monotonous daily routine, causing depression and anxiety in the lives of the students.

- R^2 is found to be 0.22 which means that 22% of the variation in behaviour is explained by the adverse effects of online learning on physical and mental health.
- Figures 8.1 and 8.2 show that there is a direct relationship between variables as it is sloping upward.

Hence, the regression equation is found to be $Y = 0.92 + 0.16*X1 + 0.39*X2$, where Y denotes the effects on the behaviour, such as causing frequent mood swings due to online learning platform. $X1$ denotes the adverse

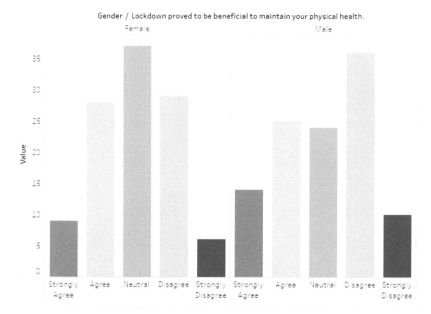

Figure 8.4   Gender/lockdown has proven to be beneficial to maintaining your physical health (male vs. female).

effects on your physical health due to the online mode of learning. $X2$ denotes the screen time increment that is causing mental stress.

Therefore, the null hypothesis (d) is rejected, that is there is a relation between these dependent and independent variables.

From the all above conclusions, it could be demonstrated that there is a major impact of COVID-19 on students.

Figure 8.8 describes whether there is any relation between adverse effects on health and mental state due to online learning and effects on the behaviour, such as frequent mood swings. The results are shown below.

Table 8.1 indicates that the following:

1. The mean value is found to be maximum in Question 6, that is 2.11, followed by Question 2, that is 1.90.
2. The median is found to be the same for all the questions, whereas the mode value of Question 1 is found to be 1 which states that respondents strongly agree with the point that Screen time increment is causing mental stress and for the rest of the questions.

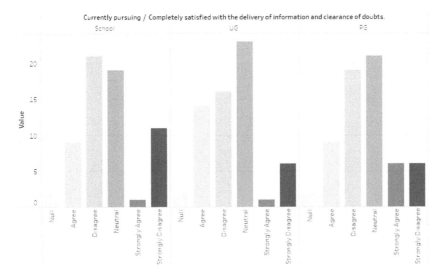

Figure 8.5 Currently pursuing/completely satisfied with the delivery of information and clearance of doubts (school students vs. UG students vs. PG students).

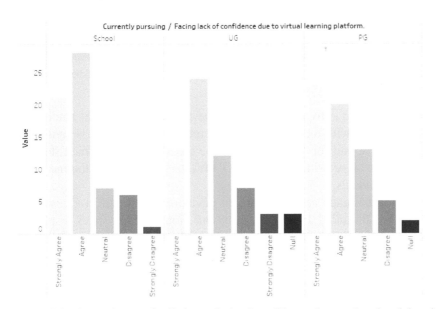

Figure 8.6 Currently pursuing/facing a lack of confidence due to the virtual learning platform (school students vs. UG students vs. PG students).

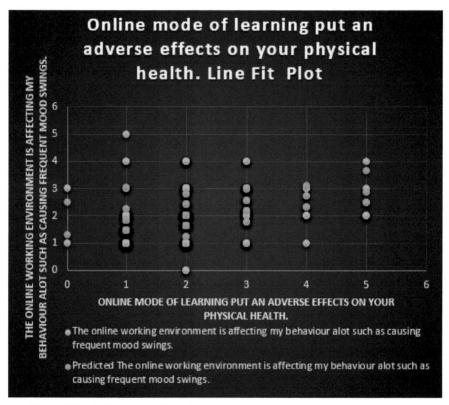

Figure 8.7   Online mode of learning put an adverse effect on your physical health.

3. The mode value is found to be 1 for Question 1 which means students strongly agree with the statement that screen time increment is causing mental stress, and for other questions the mode value is 2 which states respondents agree with the fact that COVID-19 is affecting the mental health of students.

4. Maximum variance and the standard deviation are found in Question 6, that is 1.05 and 1.11, followed by Question 3 which states respondents have different prospects for the question that are the reason their responses have deviated from the mean value. This means that there are people who feel no lack of confidence due to the virtual platform of learning and do not feel depressed and anxious due to a monotonous daily routine.

5. Questions 1 and 4 are found to be highly positively skewed as it is 1.21 and 1.16, respectively, so this demonstrates that data are skewed

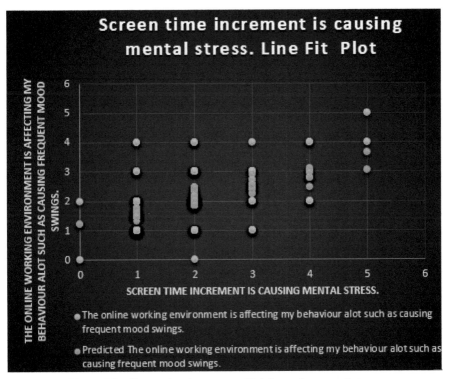

Figure 8.8 Screen time increment which is causing mental stress.

right, meaning the right tail is long relative to the left tail, whereas other question skewness is moderate.

From Table 8.1. it could be seen that all the responses agree with the statement that there is a major effect of pandemic and lockdown on the mental health of students. Hence, the null hypothesis (b) is rejected.

Table 8.2 indicates that the following:

- The mean value is found to be maximum in Question 1, that is 3.18, followed by Questions 2 and 3.
- Median and mode value for Questions 3 and 4 are found to be the same, that is 2, and for Questions 1, 2, and 5 they are calculated as 3, which states that respondents have a neutral response for E-learning to be beneficial for enhancing your academics, satisfied with your grades relating to your efforts or hard work and satisfied with the delivery of information and clearance of doubts. On the other hand, they agree with

Table 8.1   Interpretation of data.

	Question 1 Screen time incre- ment is causing mental stress	Question 2 The unfor- tunate things hap- pening around the world increase the feeling of fear	Question 3 The monotonous daily routine is causing depres- sion and anxiety	Question 4 Pandemic is causing fright of accom- plishment of your goals and plans	Question 5 The online working environ- ment is affecting my behaviour a lot, such as causing frequent mood swings	Question 6 Facing lack of confi- dence due to virtual learning platform
Mean	1.71	1.90	2.05	1.85	1.87	2.11
Median	2	2	2	2	2	2
Mode	1	2	2	2	2	2
Standard devia- tion	0.88	0.87	1.03	0.92	0.89	1.05
Sample vari- ance	0.77	0.76	1.07	0.85	0.79	1.11
Skewness	1.21	0.83	0.86	1.16	0.77	0.67

the statement that lockdown proved to be an effective time to learn new skills (like online courses) and get some time to engage yourself in extracurricular activities apart from your academics.

- Standard deviation and sample variance are found to be quite similar and high for all the questions that state respondents having different prospects for the questions that are the reason their responses have deviated from the mean value.
- All the questions are moderately skewed as the values are lying between -0.5 to 0 and 0 to 0.5.

Table 8.2 Interpretation of data.

	Question 1	Question 2	Question 3	Question 4	Question 5
	E-learning is beneficial for the enhancement of your academics	Satisfied with your grades relating to your efforts or hard work	Lockdown proved to be an effective time to learn new skills (like online courses)	You can get some time to engage yourself in extracurricular activities apart from your academics	Completely satisfied with the delivery of information and clearance of doubts
Mean	3.18	2.98	2.46	2.62	3.16
Median	3	3	2	2	3
Mode	3	3	2	2	3
Standard deviation	1.16	1.12	1.08	1.19	1.14
Sample variance	1.34	1.26	1.16	1.41	1.29
Skewness	-0.32	-0.23	0.59	0.23	-0.45

From Table 8.2, the responses are in favour of the statements; hence, the null hypothesis (c) is rejected, that is there is an effect of the lockdown and the pandemic on the academics of students.

8.6 Limitations of the Study

Research does not apply to primary students below the age of 12 years. Data are collected from the students of metropolitan cities and not from the students of rural areas. This research is applied for the period of the lockdown due to the persistence of COVID-19; therefore, it is not applicable in the normal scenario. Data availability seemed less; hence, the small sample size was considered due to lockdown.

8.7 Future Prospective of Research

The research carried out has contributed well to the state of the effect of COVID-19 on students, so it could be used by other researchers as their secondary data. Research is applicable for the upcoming scenario of lockdown or pandemics. It could be used as a reference to understand the impact of virtual learning platforms on the students' physical health, mental health and academics.

8.8 Conclusion

The above results depict that COVID-19 has created a lot of problems in the country. Also, the quality education has declined for the students, which has led to a downfall in various areas. They are not able to concentrate more on their studies which affects their mental health and reduce their level of confidence. Besides leading to the decline of quality education, COVID-19 has also affected the growth of sustainable development goals. Also, as discussed above, SDG – quality education – has been impacted more by COVID-19. The reason is that students are not getting healthier, and neither have sufficient learning platforms to survive in the upcoming technologies. COVID-19 has also impacted the mental health of the students as they have involved themselves in undesirable activities which have led to fear in their minds. Thus, it is impacting SDG: Good Health and Well-being.

To overcome such situation, for a model of providing remote education to the students, India needs to work a lot in terms of providing updated technology as to further allow students studying remotely, feel a sense of belongingness and similarly perceive the same sense pride or privilege studying in schools or universities, etc., and the teachers as well must be able to provide effective knowledge to their students.

Acknowledgements

The authors are grateful to the handling editor and anonymous reviewers for their constructive comments to enhancing the quality of this chapter.

References

[1] S. Chugh, M. Mathur, P. Rawat, S. K. Pal, "A systematic study on sustainable development goal: In special reference no poverty", Studies in Indian Place Names, 40 (56), pp. 1612-1622, 2020.

[2] J. D. Chertoff, J. G. Zarzour, D. E. Morgan, P. J. Lewis, C. L. Canon, J. A. Harvey, "The early influence and effects of the coronavirus disease (COVID-19) pandemic on resident education and adaptations", Journal of the American College of Radiology, 17(10), pp. 1322-1328, 2019.

[3] L. Liang, H. Ren, R. Cao, Y. Hu, Z. Qin, C. Li, S. Mei, "The effect of COVID-19 on youth mental health", Psychiatric Quarterly, 91(3), pp. 841-852, 2020.

[4] E. M. Onyema, N. C. Eucheria, F. A. Obafemi, S. Sen, F. G. Atonye, A. Sharma, A. O. Alsayed, "Impact of Coronavirus pandemic on education", Journal of Education and Practice, 11(13), pp. 108-121, 2020.

[5] R. M. Oducado, G. Parreño-Lachica, J. Rabacal, "Personal resilience and its influence on COVID-19 stress, anxiety and fear among graduate students", IJERI: International Journal of Educational Research and Innovation, 15, pp. 431-443, 2021.

[6] C. Davico, A. Ghiggia, D. Marcotulli, F. Ricci, F. Amianto, B. Vitiello, "Psychological impact of the COVID-19 pandemic on adults and their children in Italy", Frontiers in Psychiatry, pp. 1-8, 2021.

[7] H. M. Bai, A. Zaid, S. Catrin, K. Ahmed, A. J. Ahmed, "The socio-economic implications of the coronavirus pandemic (COVID-19)", A Review International Journal of Surgery, 8(4), pp. 8-17, 2020.

[8] E. M. Aucejo, J. French, M. P. U. Araya, B. Zafar, "The impact of COVID-19 on student experiences and expectations: Evidence from a survey", Journal of Public Economics, pp. 1-15. 2020.

[9] M. Yıldırım, E. Geçer, Ö. Akgül, "The impacts of vulnerability, perceived risk, and fear on preventive behaviours against COVID-19", Psychology, Health & Medicine, 26(1), pp. 35-43, 2021.

[10] K. Verma, S. K. Pal, V. Kumar, "The impact of micro finance on rural sector and its contribution towards Indian economy", International Journal of Advanced Science and Technology, 29 (3), pp. 1522-1545, 2020.

[11] S. Chugh, G. Vibhu, S. Kanojia, V. Kumar, S. K. Pal, "To study the impact of advancement in E-payment technology used in India", PalArch's Journal of Archaeology of Egypt/ Egyptology, 17 (6), pp. 1617-1635, 2020.

[12] S. K. Pal, N. R. Aderla, MD. S. Rahaman, S. S. Ali, H. Divya, G. N. S. Bandi, "Factors Affecting Impulse Buying on Online Shopping amongst Youth: A Structural Equation Modeling Approach", Turkish Journal of Computer and Mathematics Education, 12(9), pp. 2149-2157, 2021.

[13] S. K. Pal, A. K. Pal, "The impact of increase in Covid-19 cases with exceptional situation to SDG: Good health and Well-being", Journal of Statistics & Management Systems, 24 (1), pp. 209-228, 2021.

[14] U. Sharma, S. K. Pal, "Investigation of slowdown in the Indian economy and suggestive measures to scull out of it", Studies in Indian Place Names, 40 (56), pp. 1623-1630, 2020.

[15] S. Mukherjee, M. M. Baral, V. Chittipaka, S. K. Pal, "Analyzing the factors that will impact the supply chain of the COVID-19 vaccine: A structural equation modeling approach" Journal of Statistics and Management Systems, 25(6), pp. 1-17, 2022.

9

A Mathematical Model for the Dynamics of the Violence Epidemic Spreading due to Misinformation

Arindam Kumar Paul and M. Haider Ali Biswas

Mathematics Discipline, Science Engineering and Technology School,
Khulna University, Khulna-9208, Bangladesh
E-mail: arindam017@gmail.com, mhabiswas@yahoo.com

Abstract

Misinformation spread is considered as one of the most reliable causes of violence and creating disturbance among people. With the rapid development of technology, it has become easier to spread any kind of information and influence others with that information for serving any kind of agenda. In this paper, we represented and analysed the internal mechanism that causes violence, disturbance, etc., due to misinformation spread. Since it is harder to solve this problem without understanding how the system works, mathematical modelling is being used to deal with this problem here. A system of five non-linear ordinary differential equations has been developed to describe the whole system mathematically. All the necessary analysis, like equilibrium, stability, the positivity of the solutions, basic reproduction ratio, etc., has been performed to validate the model. After justifying the equations analytically, we solved the model numerically with different conditions and used the parameters to visualize and understand the system's pattern. Our research differs from previous studies in that we have proposed an entirely new mathematical model, considering each possible cause of the spread of misleading information, and have conducted the whole study in line with the facts. Not only focusing on analytical proofs of equations but also developing a model considering most of the real facts and expressing both the society and online as a medium of spread are the novelties and uniqueness of our model.

Keywords: Mathematical Modelling, Classifying Motive of Information, Misinformation Spread Dynamics, Pattern Identification of Misinformation Spread.

MSC Code: 93A30, 00A71, 65P10, 65P30, 65P40, 92B05.

9.1 Introduction

From the distant past to the present, mathematical science, technology, medical science, sociology, language or human resource development, there is a contribution of distribution and mutual aid–cooperation behind every story.

From ancient times, people would not have been able to do anything significant if they had not shared the knowledge of science or the diseases they had discovered. Nowadays, the most effective way to solve any problem is to take it to the right person or share it with as many people as possible. So, there is no quicker way to solve any problem than to share it now and in the future. But like all great discoveries or good things, there are good and bad aspects [1]. No matter how tremendous the discovery is, there will always be the opportunity to do something wrong by using it.

The easiest way to accomplish any kind of present purpose is to share it in a beautifully distorted way and present it in such a way for the development of all. In addition, the current World Wide Web and Internet connectivity have made it easier to do such things. Through the Internet time, every person can connect with anyone else in any part of the world at any and spend the least amount of time and effort. Over the last 10 years, most of the violence has been caused by deliberately misrepresenting any information or object. Suppose people see something in front of their eyes that needs to be solved very quickly and is very useful for many others. Then they try to solve it by sharing it with everyone regarding empathy or humanity [2].

Since this kind of strategy of spreading information and creating violence is directly connected with human psychology and behaviour, it is hard to predict or defend violence frequently occurring for only accepting and sharing misinformation and fake news over the Internet. Those who are intended to create violence and use misinformation spread as a strategy are very creative. They use their ingenuity to design a strategy and lead the whole operation of creating violence. The popularity of social media amplifies this process at a higher rate. The age of reading print newspapers will be in the museum very quickly if people keep accepting online media instead of print media [2]. Since it is impossible to prevent sharing information and knowledge among people now or in the future, it is the most challenging task to deal with misinformation spreading. Another major constraint while preventing misinformation spread through the Internet and online media is time. It is possible to make anything possible within the shortest time before the detection process of misinformation. Those who are intended to spread misinformation and alloys are intelligent and target a particular population for spreading and fulfilling their objective by using the psychology of common people [3]. Researchers and scientists have been focusing on this issue recently and trying to solve it using different technologies and strategies. To develop the detection and prevention processes of spreading misinformation, the world undeniably needs mathematicians' and statisticians' help. The very complex nature and a new strategy for every occurrence made the whole process very complex. To find a solution first, it is very necessary to understand the problem as well as the inner mechanisms of the whole system of fake news or information diffusion. That is why engineers and other experts need the validation and justification of mathematicians regarding specific strategies. There are plenty of technologies and hardware in the deployment phase for deploying any kind of solution and activating instantly, by using artificial intelligence and data mining algorithms for spreading misinformation and fake information diffusion. Contrarily, the intended individuals are using artificial intelligence as a medium for spreading their misinformation [4].

Mathematical modelling, simulation and analysis add a very ancient base to deal with any system. Modelling is the best technique for understanding and cracking the inner dynamics and interactions [5]. Many processes have already been initiated to deal with this issue with mathematical and statistical modelling. Some models have been developed considering different types of phenomena, statistics, characteristics, demography, psychology and other properties. Very specific to generalized scenarios have been considered for understanding and solving this problem. A huge analysis has been done on particular countries or areas because of the similarities among the population of those areas.

9.2 Previous Studies

Rumours refer to unconfirmed information and are spread in certain ways rumour-spreading model can be traced back to 1960. Daley and Kendall's "stochastic rumours" paper borrowed the infectious disease model and proposed the rumour propagation model for the first time. Later, researchers called the rumour-spreading model in the author's name for the D–K model.

The D–K model analyses the problem of rumours based on a random process and divides people in the process of spreading rumours into three categories: people who have never heard of rumours, people who spread rumours and people who have heard rumours but did not spread them. Later, in [6], authors modified the D–K model and proposed the M–K model, in which rumours spread directly to other people through the communicator [7]. It is found that the D–K model and the M–K model are only applicable to rumour propagation models in small-scale social networks. But social networks in daily life are large in scale and have the characteristics of small world and scale-free.

Zanetti studied the spread of rumours on the small-world network considering the topology. Moreno studied the M–K model on scale-free networks and recorded the interaction between the network topology and the rules of the M–K model. By combining the M–K model with the SIR model in [8], authors further studied the rumour propagation state of complex networks in general, took the forgetting mechanism into account and proposed a rumour propagation model with an *immune mechanism* and *forgetting mechanism.* They introduced the forgetting and memory mechanism in the SI model and analysed the influence of information dissemination on the scale-free network using numerical simulation.

The results showed that the forgetting memory mechanism might lead to the termination of dissemination. According to the characteristics that people will forget and then remember for a short time, they become communicators when they are remembered. The authors [7] call this kind of people who temporarily forget to spread rumours when they encounter a communicator and start spreading rumours a hibernator, call this mechanism a memory mechanism and propose a hibernator, immune mechanism and forgetting mechanism.

In [9], the authors abstracted the objective process of rumour propagation and established a mathematical model to understand the influence of various related factors, such as transmission and mortality rates. In addition to individual factors, social factors have been acknowledged as indispensable ingredients of the rumour diffusion process. As revealed by many researchers, the social reinforcement effect is one of the key factors in the rumour-spreading process on which social networks proposed a rumour propagation model considering the negative and positive double social reinforcement mechanisms on the individual propagation behaviour [6]. Zhang et al. [10]

presented a novel model based on the *generation function* and *cavity method* developed from the statistical physics of the disordered system.

A new variant of the SEIZ model was proposed by Isea and Lonngren [11], assuming a new population type called sceptics who can stop the spreading process by creating awareness. After that study, many types of research were done based on these assumptions and by adding other assumptions as well. Some models using the Markov chain, game theory and other problem stick models were developed for understanding and predicting the trends of rumour spreading [12].

9.3 An Overview on Misinformation

There are many sources from which any kind of misleading information may produce. Generally, people prioritize the source of information while they interact with that information. We have found three types of sources from where any kind of this information may produce:

- From news media.
- From a reputed person.
- From a group of people.

The initial population and contact rate with other populations of each category are different. News media and reputed people rarely share any kind of information, but they are connected with too many people. That is why the contact rate becomes very high when any kind of information comes from any news media or reported person. Generally, people believe in the news media and the person who has a very good reputation or a person they like or follow. When a group of people tries to intentionally spread any kind of misleading information, the initial population of intended people is larger, but they are not connected with mass people. That is why generally, people do not believe in them but then again believe in news media and any kind of reputed source [7].

In most countries, the lion's share of the total population lives in rural areas, and verifying and validating any information is not a very easy task for them [15]. For spreading misinformation, the rural population is the perfect target worldwide [21].

We have considered two types of misinformation spread. The initial population of the source is assumed accordingly – dissemination of two types of misinformation divided forget the importance of information and the purpose of the disseminator. No significant or long-term plan is required to disseminate such information, and most of the dissemination intends to be very insignificant. Because it is not very important, many people or organizations do not think much before sharing such information again. As a result, the information spreads very fast and disappears very quickly.

9.3.1 Intention is not creating violence

- Disclosure of information for marketing of any product.
- Spreading information increases the reputation of an individual or a brand.
- Spreading information for fun.
- To excite a person or persons without any reason. To attract a particular person or organization.
- Releasing information in a self-centred way to establish oneself.
- Dissemination of information at random without any purpose.

In the case of the second type of proliferation, the information is essential, or some of the information can make a significant difference depending on the situation. If such information obtains from other known or verified sources, people avoid it after viewing it. However, such information comes from a well-known and verified source. If the information presents logically, most people share it somehow, realizing the importance of the information so that others can find out.

9.3.2 Intention is to create violence

- Information related to a country or region's political, religious or other issues.
- Information about the spread of any disease.
- Information about war, rebellion or any fight.
- People see this kind of information as a blessing for any ethnic group.

9.4 Mathematical Model of Spreading Misinformation

A variant of the SEIZ model considered that some people forget information, so some information is always lost [12]. Another SEIZ model added a new population type Z, defined as sceptics or protestors who generally do not believe in anything without verifying and protesting if defined in information misleading. Later in this paper, the authors considered that when the exposed population contacted the sceptic population, *some of them* became part of the sceptic during the spread is running [13]. After analysing all the papers explaining the dynamics of spreading misleading information, we differentiated violence-intended misleading information spread and general misleading information spread by each other by studying their properties and behaviour. We also considered a population type responsible for the spread of violence-intended misleading information and denoted them as the source population of spread [18].

Here, we have designed a mathematical model containing five compartments. The model describes the dynamics of spreading misinformation among general Internet users. This model describes not only how information differs from information but also how information can be differentiated from the spread of misinformation.

We have considered Internet users as the population who can spread misinformation but do not have that misinformation yet. We have defined this population by the symbol **S**. We thought of another compartment named the exposed population, which is not instantly spreading any information or misinformation. We have added a population type intended to spread misinformation to fulfil their bad agenda of creating violence and other suspicious activities. We denoted them by the symbol **C**. Then we considered another population type that is concerned about misinformation and not being infected without justifying any kind of information. We are calling them the sceptic population. Moreover, the last compartment contains the population who had already spread this information without thinking about it.

When the general Internet users communicate with the intended population, some of the population **S** are infected and start spreading misinformation. Some of them take a little bit of time before spreading or sharing that information. When the general users communicate with the infected individual, some enter the exposed population, and some instantly share that information without any justification [2].

Here are some specialities and unique objectives of our model:

- The spread of misinformation aimed to create violence is considered in this model. Another misinformation spread is described by the same model but using different initial populations and contact rates.
- The spreader population is divided into two categories, one is the intended spreader and another is the purposeless spreader. The second type population is considered as being used as a *medium* by the first type.

- A susceptible population is considered fixed because the intended population targets a group of populations or a specific region to spread misinformation.
- People connected to the Internet are considered the only medium for spreading misinformation.
- It is considered that both the sceptic and intended populations are decreasing by a certain rate because of personal and social problems.
- Some people rarely use the Internet or social media. But when they verbally get in touch with others' misinformation, they start joining the susceptible population.

Figure 9.1 visualizes how this type of population interacts with others and spreads or tries to spread misinformation to others.

- **The Internet user (S):** Basically, all Internet users fall into this category. Such people include all kinds of educated and uneducated people.
- **Exposed population (E):** This type of community comprises people who do not share any information immediately after learning it. They take some time to verify whether or not to share it. After that, they can share it, or many can decide to verify.
- **Intended spreader (C):** In this category, there are mainly those whose purpose is to spread violence by spreading misleading information.
- **General spreader (I):** This category includes those who share any information without proper verification so as not to create violence. Many from such included populations may have deliberately decided to spread misleading information again.
- **Skeptic population (Z):** This category includes those who make decisions after verifying and reasoning appropriately after seeing or knowing any information. Many within this population category can encourage others to raise awareness about misleading information.

By following [12, 13] and modifying them according to the description and model diagram, the dynamical behaviour of the misinformation spreading system can be expressed by eqn (9.1)–(9.5):

$$\frac{d}{dt}S(t) = r - \gamma C(t) S(t) - \beta I(t) S(t) - bS(t) Z(t), \qquad (9.1)$$

$$\frac{d}{dt}E(t) = -E(t) + (1-p)\beta I(t) S(t) + (1-l) bS(t) Z(t), \qquad (9.2)$$

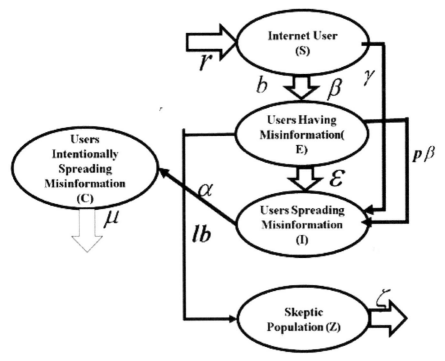

Figure 9.1   Diagram of misinformation spread model.

$$\frac{d}{dt}C\left(t\right)=\alpha C\left(t\right)I\left(t\right)-\mu C\left(t\right),\tag{9.3}$$

$$\frac{d}{dt}I\left(t\right)=E\left(t\right)-\alpha C\left(t\right)I\left(t\right)+\gamma C\left(t\right)S\left(t\right)+\beta pI\left(t\right)S\left(t\right),\tag{9.4}$$

$$\frac{d}{dt}Z\left(t\right)=blS\left(t\right)Z\left(t\right)-\xi Z\left(t\right),\tag{9.5}$$

where the initial conditions are

$$S\left(0\right)=S_0\geq0,E\left(0\right)=E_0\geq0,C\left(0\right)=C_0\geq0,I(0)$$
$$=I_0\geq0,Z(0)>Z_0.$$

The above model has to be analysed to describe the dynamics of the criminal population in Bangladesh. The objective of this analysis is to find and verify the criminal population of our country for any specified time duration for that a closed set has been considered as
$\Omega=\left\{(S(t),E(t),C(t),I(t),Z(t))\in R_+^5\right\}$ with the initial conditions.

In this model, initially, the parameter b and β are the most important. It's because when general people contact people who can decide or verify whether any content is intentional or not, at first, they do not share that information instantly. Secondly, some started analysing the news for information and revealing its truth. To start this activity, they join the *sceptic population* and make others aware of this kind of misleading information by lb rate.

In the case of general users, contacting such people who are spreading any misleading information unintentionally, some of them try to understand the scenario and take some time to clarify the situation. Another share of general people starts sharing instantly any kind of information without verifying at a $(1 - p)\beta$ rate.

If the source of information is directly contacting general users, they (general users) are convinced by their presentation, logic or anything and start sharing the information instantly by γ rate.

9.4.1 Equilibrium analysis

For calculating the spread-free equilibrium, we've considered $\frac{dS}{dt} \neq 0$ and $\frac{dE}{dt} = 0, \frac{dC}{dt} = 0, \frac{dI}{dt} = 0, \frac{dZ}{dt} = 0$; then by solving for S, E, C, I, and Z we get,

$$E^0 \left(S^*, E^*, C^*, I^*, Z^* \right) = \left(\tfrac{r}{\gamma}, \quad 0, \quad 1, \quad 0, \quad 0 \right). \tag{9.6}$$

By considering each state equal to zero $\frac{dS}{dt} = \frac{dE}{dt} = \frac{dC}{dt} = \frac{dI}{dt} = \frac{dZ}{dt} = 0$ and solving for state variables S, E, C, I and Z we get the system's equilibrium state $(S^*, E^*, C^*, I^*, Z^*)$ which is

$$\left(\frac{\xi}{bl}, -\frac{\mu\xi \left(\beta\gamma\xi + b\beta\mu p - \beta\gamma p\xi - b\beta l\mu \right)}{\alpha bl \left(b\mu - \gamma\xi \right)}, \frac{\beta\mu\xi}{\alpha \left(b\mu - \gamma\xi \right)}, \frac{\mu}{\alpha}, -\frac{\beta\mu^2}{\alpha b\mu - \alpha\gamma\xi} \right).$$
$$\tag{9.7}$$

9.4.2 Stability analysis

At first, we investigate the stability at the equilibrium point E by proving Theorem 9.1.

9.4.2.1 Stability at $E^0 \left(S^*, E^*, C^*, I^*, Z^* \right)$
Theorem 9.1. The spread-free equilibrium point $E^0 \left(S^*, E^*, C^*, I^*, Z^* \right)$ of the model (9.1)–(9.5) is asymptotically stable if the eigenvalues of the Jacobean matrix are negative [19].

Proof. To establish the stability analysis of the model at the equilibrium point E^* $(S^*, E^*, C^*, I,^* Z^*)$. The model is as follows:

$$J = \begin{bmatrix} -C\gamma - I\beta - Zb & 0 & -S\gamma & -S\beta & -Sb \\ I\beta r + Zb\pi & - & 0 & S\beta r & Sb\pi \\ 0 & 0 & -\mu & \alpha & 0 \\ C\gamma + I\beta p & S\gamma & S\beta p - \alpha & 0 \\ Zbl & 0 & 0 & 0 & Sbl - \xi \end{bmatrix}. \quad (9.8)$$

Substituting the value of S, E, C, I and Z from eqn (9.6) in eqn (9.8), we get

$$J = \begin{Vmatrix} 0 & 0 & 0 & 0 & 0 \\ 0 & - & 0 & 0 & 0 \\ 0 & 0 & -\mu & 0 & 0 \\ 0 & & 0 & 0 & 0 \\ 0 & 0 & 0 & 0 & -\xi \end{Vmatrix}. \quad (9.9)$$

Subtracting an identity matrix from j and taking the determinant of eqn (9.9), we get

$$\|J - \lambda I\| = \begin{Vmatrix} -\lambda & 0 & 0 & 0 & 0 \\ 0 & -- \lambda & 0 & 0 & 0 \\ 0 & 0 & -\lambda - \mu & 0 & 0 \\ 0 & & 0 & -\lambda & 0 \\ 0 & 0 & 0 & 0 & -\lambda - \xi \end{Vmatrix} = 0, \quad (9.10)$$

$$- \lambda^2 \left(+ \lambda \right) \left(\lambda + \mu \right) \left(\lambda + \xi \right) = 0. \quad (9.11)$$

Solving eqn (9.11) for λ we get

$$\lambda_1 = 0, \lambda_2 = 0, \lambda_3 = -, \lambda_4 = -\mu, \lambda_5 = -\xi.$$

Since all λ values are negative or $\lambda \leq 0$ here, our model is asymptotically stable for the spread-free equilibrium point.

9.4.2.2 Stability at E^* $(S^*, E^*, C^*, I^*, Z^*)$
Substituting the value of equilibriums from E^* $(S^*, E^*, C^*, I^*, Z^*)$ in eqn (9.12) and subtracting the identity matrix from it, we get

$$\|J - \lambda I\| \quad (9.12)$$

$$= \begin{pmatrix} \frac{b\beta\mu^2}{\alpha b\mu - \alpha\gamma\xi} - \frac{\beta\mu}{\alpha} - \lambda - \frac{\beta\gamma\mu\xi}{\alpha(b\mu - \gamma\xi)} & 0 & -\frac{\gamma\xi}{bl} & -\frac{\beta\xi}{bl} & -\frac{\xi}{l} \\ \frac{b\beta\mu^2(l-1)}{\alpha b\mu - \alpha\gamma\xi} - \frac{\beta\mu(p-1)}{\alpha} & -- \lambda & 0 & -\frac{\beta\xi(p-1)}{bl} & -\frac{\xi(l-1)}{l} \\ 0 & 0 & -\lambda & \frac{\beta\mu\xi}{b\mu - \gamma\xi} & 0 \\ \frac{\beta\mu p}{\alpha} + \frac{\beta\gamma\mu\xi}{\alpha(b\mu - \gamma\xi)} & \frac{\gamma\xi}{bl} - \mu & \frac{\beta p\xi}{bl} - \frac{\beta\mu\xi}{b\mu - \gamma\xi} - \lambda & 0 \\ - \frac{b\beta l\mu^2}{\alpha b\mu - \alpha\gamma\xi} & 0 & 0 & 0 & -\lambda \end{pmatrix} = 0.$$

Taking the determinant of eqn (9.12) and then solving it for λ, we found the conditions for which all the eigenvalues will have a negative real part:

$$b^2\mu^2 + \gamma^2\xi^2 \neq 2b\gamma\mu\xi, \alpha \neq 0, b \neq 0, l \neq 0, b\mu \neq \gamma\xi.$$

If the above three conditions hold, we will get the real part as negative of the eigenvalues of the Jacobean matrix. Since the eigenvalues are too long to write here, we calculated the stability conditions in terms of the inequality of the combination of parameters [20]. Since all values of λ are negative or $\lambda \leq 0$ under these conditions, our model is asymptotically stable for the endemic equilibrium point.

9.4.3 Basic reproduction bumber

The basic reproduction number is a threshold parameter for the stability of spread-free equilibrium. If the number of spreader individuals increases at a high rate, misinformation spread prevails for an extended period. Otherwise, the spread will disappear within the shortest possible time. Now to find the basic reproduction number of the model (9.1)–(9.5), we have to apply the next-generation matrix method [14]. Let us consider that the prevalence rate of spread is F, and the rate of transferring into and out of any class is V. Since the misinformation travels from and to the exposed (E) and spreader (I), we are considering only eqn (9.2) and (9.4) for calculating the basic reproduction number for this system:

$$\frac{d}{dt}E(t) = -E(t) + (1-p)\beta I(t)S(t) + (1-l)bS(t)Z(t), \quad (9.13)$$

$$\frac{d}{dt}I(t) = E(t) - \alpha C(t)I(t) + \gamma C(t)S(t) + \beta pI(t)S(t). \quad (9.14)$$

Using these two equations, we get that transition and transmission matrix

$$F = \begin{bmatrix} 0 & S\beta \\ 0 & 0 \end{bmatrix}, \quad (9.15)$$

$$V = \begin{bmatrix} - & -S\beta(p-1) \\ & S\beta p - C\alpha \end{bmatrix}. \quad (9.16)$$

Taking the inverse of the matrix V and multiplying with F, we get

$$FV^{-1} = \begin{bmatrix} -\frac{S^*\beta}{C^*\alpha - S^*\beta} & -\frac{S^*\beta}{C^*\alpha - S^*\beta} \\ 0 & 0 \end{bmatrix}. \quad (9.17)$$

By determining all the eigenvalues of the Jacobean matrix obtained from (9.10) and substituting the values of S^*, C^* in (9.18), we get

$R_0 = \frac{\beta r}{\beta r - \alpha \gamma}$. (9.18)Numerically substituting all the values and plotting for different ranges, we get the impact of different parameters on maximizing or minimizing spread [21].

Figure 9.2 indicates that the basic reproduction ratio decreases with the increment of β, while the ratio is equal to or above 1.

Figure 9.3 indicates that the basic reproduction ratio decreases with the increment of r. But the ratio is above 1 since the spread will not stop.

From Figure 9.4, we can see that it is impossible to minimize spread in terms of the basic reproduction number. But with the increment of the parameters r and β, the basic reproduction number becomes 1. Γ and α are primarily responsible for increasing the spread. Figures 9.1–9.4 show a significant outcome of the system. Since $R_0 < 1$, the spread will minimize very fast; otherwise, the spread will increase rapidly. In this case, $R_0 \geq 1$ for every possible individual parameter.

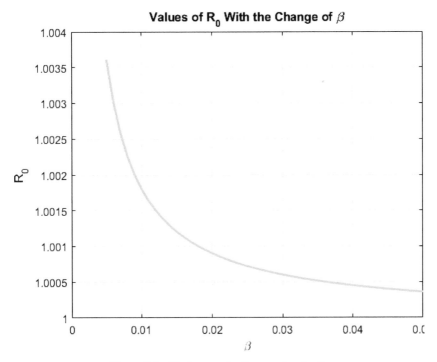

Figure 9.2 Basic reproduction number for β.

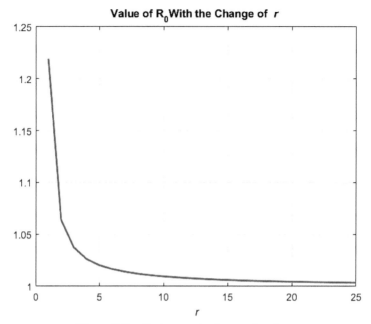

Figure 9.3   Basic reproduction number for r.

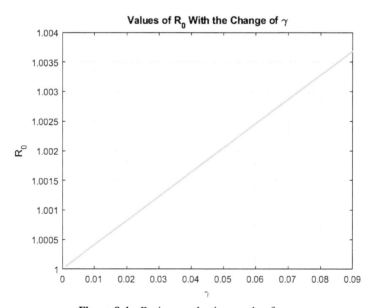

Figure 9.4   Basic reproduction number for γ.

9.5 Numerical Simulations

Numerically solving the model is crucial to connect an analytical model and justify the model's outcome in reality. No practical solutions to any problems can be provided without numerical solutions. These solutions and diagrams give the analytical findings strong validity.

We have simulated our mathematical model numerically for finding the pattern of misinformation spread. Since we started with a hypothesis and from statistical data and reports, we have validated our hypothesis to set the initial value and parameter values accordingly. Initial values for the state variables and parameters used in the model are taken from Table 9.1.

We have used one of the previous research papers to determine the value of parameters as it is not possible to verify whether it is a rumour or a fact as soon as any information starts spreading [2, 4, 13, 12]. A survey has been conducted about spreading rumours to know the general point of view. Moreover, we used *Python* programming language to extract information about various published news and their authenticity or purpose. The information has been taken from various websites and news media. The value of the initial standard parameter has been determined using these three methods.

Initial values of the population for all the state variables are the same except the intended misinformation spreaders because it is the key point to differentiate information from misinformation. For further information spread, we have considered the initial population of the source as 100–200 people with a lower contact rate with other people [14]. In this case, the contact rate of violence-intended misinformation spreader with other people is considered as 0.001–0.005 for visualizing the outcome and difference with the general intentional spread. For the susceptible, exposed, spreader and sceptic populations, the initial population is considered as $S_0 = 600 - 100, E_0 = 100 - 200, I_0 = 20 - 50, Z_0 = 20 - 50$. Since all the Internet users can initially participate in spreading, we considered a fixed population in a region who are already targeted as the medium of spread by the intended population. Exposed are initially bigger because people generally do not care about anything before seeing it from different sources. The spreader and sceptic populations are considered the same. But a logical population is rare. They do not believe everything they see online and do not share actually. Those who do not share are also considered sceptics.

In the case of intentional information spread, we have considered the number of source population from 5 to 15 with a higher rate of contact with other people. All the other parameters and initial values remain the

Table 9.1   Parameter values.

Symbols	Description	Values	Unit
β	S–I contact rate	0.02	Day^{-1}
b	S–Z contact rate	0.03	Day^{-1}
γ	S–C contact rate	0.01–0.005	Day^{-1}
ε	Incubation rate	0.3	Day^{-1}
ζ	Decay rate from the sceptic population	0.01	Day^{-1}
μ	Decay rate from criminal population	0.3	Day^{-1}
α	Transmission rate from infected to criminals	0.002	Day^{-1}
p	Transmission rate from susceptible to infected	0.2	Day^{-1}
l	Transmission rate from S to Z	0.5	Day^{-1}

same during the simulation. In case of intentional misinformation spread, the contact rate is considered as 0.01–0.05 in the simulation to understand the outcome.

To solve the model numerically, we used MATLAB 2020a as the simulation software. We also used the ODE45 solver to solve the ordinary differential equation system with the specified parameter values and initial conditions. The change of state trajectories for different parameter values has been shown in numerical simulation.

If the source of information is any news media or a very reputable person, their initial population is relatively small. But they have connected with many people who believe in them or their words. That is why the initial population is considered only 10, and the other initial population is the same here. For this case, we consider $\gamma = 0.03$ and $C_0 = 10$.

From Figure 9.5, we can see that if the misleading news comes from any source which is known as authentic, then the contact rate is very high. The number of spreaders who are spreading that news unintentionally grows within a concise time, and some of them join with the source population or act like the source population after a few times. When the intention of spreading any misinformation is not creating violence, we consider the initial source population as 200 and the contact rate with other people as 0.0003.

Let us consider that misleading news is information, coming from a group of people and not from any media or reputable person. The population of the group is much more than news media for high-profile people. But this group of people is not known as authentic, and they are not connected with too many people so that they can spread the news quickly. From every figure, we can see that those who are sharing or spreading news without knowing its base are multiplying the first time, but the rate of spread is coming down concisely.

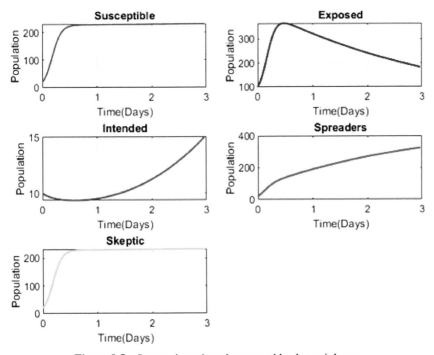

Figure 9.5 State trajectories when spread leads to violence.

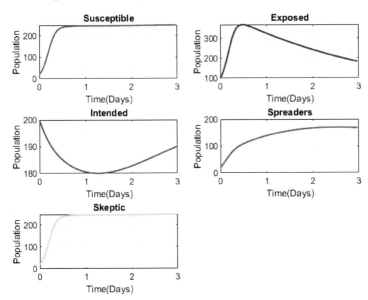

Figure 9.6 State trajectories when spread does not lead to violence.

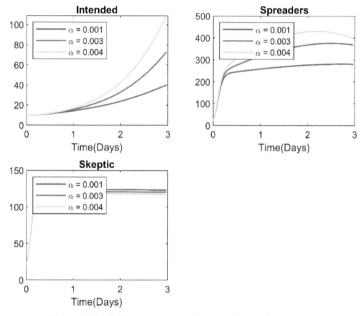

Figure 9.7   Different state trajectories for variation of α.

So any news media or any authentic profile on social media is the best option for them. If the news source is a media or high-profile person, the spread rate becomes higher and higher, and it is very tough to control quickly. That is why these types of misleading information often lead to violence.

Providing the best strategies to solve this problem, we have simulated our model for searching the states we should control so that the whole system will be under control. We have investigated the performance of parameters on different states and how every state variable changes with respect to other state variables and time.

9.5.1 State Trajectories of Intentional Spread

The different value of alpha, starting from 0.001 to 0.005, plays a significant role in increasing the intended spreader. When the general spreaders are communicating or contacting intended personnel for their motive after being influenced by the intended population, some join as a source population who had always intended to spread misleading information and create violence.

From Figure 9.8, it is clear that l is mainly liable for increasing the sceptic population as well as preventing transmission from being exposed to the

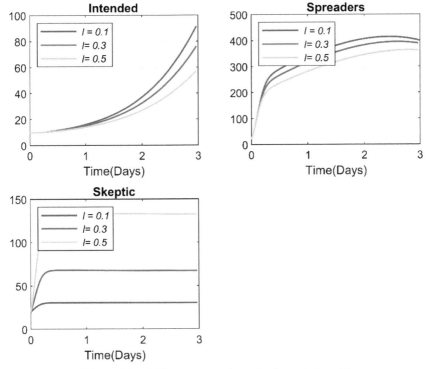

Figure 9.8 Different state trajectories for variation of *l*.

spreader population. If the transmission rate from the exposed population to the sceptic population becomes high, the spreader population becomes lower without any interference. By this time, many spreaders join as intended spreaders.

From Figure 9.9, it is clear that γ is the main reason for instantly spreading any misleading information like a diffusion process. As many Internet *user population* contact the *intended population* directly, they believe in them and start spreading without verifying or knowing about the information properly.

From Figure 9.10, we see that parameter *b* does just the opposite of what β does. If the value of the *b* is big, the spreader population will be minimized. And the intended population will be minimized too by a significant rate.

The variables β, *b* and γ are the contact rate with the susceptible population and exposed, sceptic and intended populations. From the above figures, we can see that, at first, the selectable population diverted and joined the exposed population at a significant rate. Then how many populations will

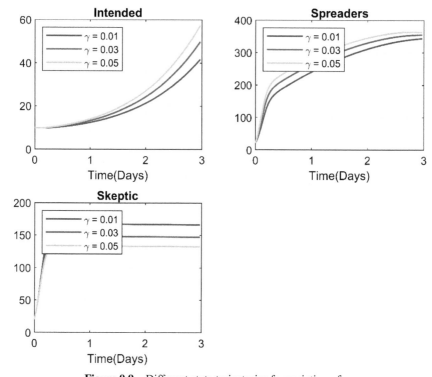

Figure 9.9   Different state trajectories for variation of γ.

start sharing the information at some time or will not share and protest against the misleading information depends on the transmission rate from exposure to infected and exposed to sceptic population. Some people will join the spreaders and start spreading just after taking some time.

9.5.2 State Trajectories for Spread with Other Intentions

If we simulate our model with a higher initial number of intentional spreaders and a lower contact rate with susceptible population S, we get the trajectories for different parameter values, and the simulation result is given below. For these cases $\gamma = 0.0002$ and $C_0 = 200$.

If the motive of the spread is not creating violence, the estimated population decreases after a certain period, and the increment stops naturally. The intended or source population of the spread also decreases.

When we look at the spreader population, we can see that for different parameter values, the spreader population becomes higher within a short time.

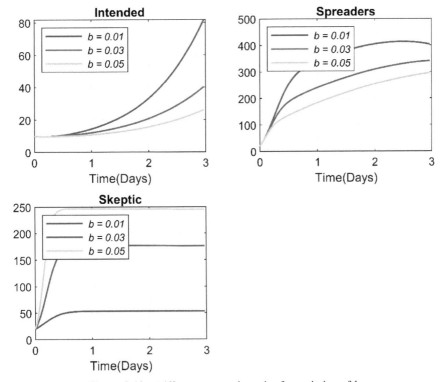

Figure 9.10 Different state trajectories for variation of b.

In addition, after two days, the spreader population decreases and almost becomes zero after three days. If the case was violence-intended information spread, then for higher contact rate with the intended source population, the population growth with the high rate and the general spread does not decrease so easily.

9.5.3 Bifurcation analysis

Bifurcation theory is a branch of mathematical study, that is mathematics that deals with the changes in the qualitative or topological structure of a family of curves, for example, the integral curves of a vector fields' family, and solutions of a family of differential equations. Bifurcation diagrams help us visualize and understand how the system's qualitative or topological change in behaviour is caused by a small change in the parameter values (the bifurcation parameters).

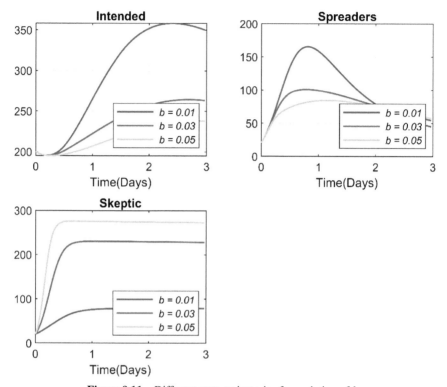

Figure 9.11   Different state trajectories for variation of b.

Here we have performed a numerical bifurcation analysis to understand the system's *global behaviour*. Considering all the initial values as the same as above, we investigated the properties and behaviour of the system for changing parameters. Most of the parameters liable for transmitting information are being considered here.

For $\beta = 0 \rightarrow 1$, $l = 0 \rightarrow 1$, $b = 0 \rightarrow 1$ *and* $r = 5 \rightarrow 25$, we can see that the states change a little, and the highest value of each state is under a certain limit. Furthermore, all the states converge to a fixed point with the increment of time; hence, the system remains stable for any changes in those parameters.

With the increment of time, only the sceptic population continues increasing because, gradually, individuals get the actual information and become aware of misinformation. The other two states, *intended* and *spreader*, stay the same with the increment of sceptic population after a certain time.

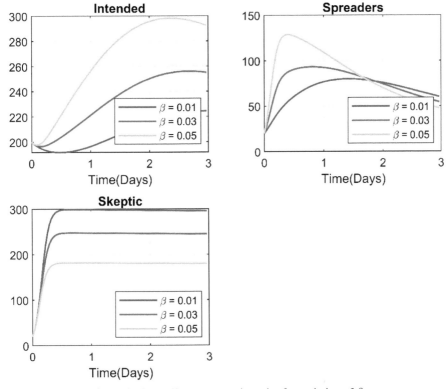

Figure 9.12 Different state trajectories for variation of β.

For different values, $\alpha = 0 \rightarrow 1$, we can see that the spreader population is increasing after crossing the value of 0.003. For the values < 0.003, the states are changing but not crossing the limit.

Since μ is the removal rate from the intended population, we investigated the threshold of μ. From the figure, it is clear that for $\mu \leq 0.13$, the intended population is being increased, but if $\mu \geq 0.13$, the intended population starts decreasing by a significant rate. The spreader population also decreases and reaches its lowest.

9.6 Results and Discussion

From the figures and analysis, we can see that the *contact rate is higher* when any news content comes from a news media or any highly deported person. That is why we have simulated the model for the different contact

Nonlinear behaviour of system for β, I, b, r

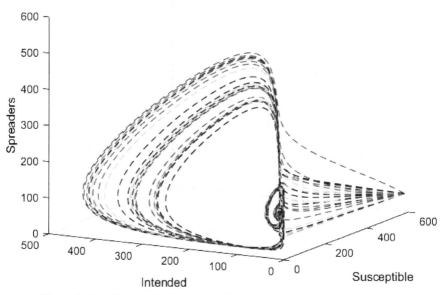

Figure 9.13   Changes of state trajectories with time for different parameters.

rates with news media and with other people. From Figures 9.7 to 9.10, it is clear that when the source is any kind of news media or authentic people, they are influencing the news content comes a very short time, and the kind of misinformation is growing up without any constraint for a long time.

Again we can see that when the source of any information is not any kind of news media, they are being contacted by a smaller number of people. So that the news or information receives fewer people and decreases after a specific time.

It can be said that if someone spreads any kind of news or misinformation intentionally to create violence, the number of people affected by the misinformation increases at a higher rate. Since some of the spreaders join with the intentional spreaders, the number of intentional misinformation spreaders becomes higher than the last time.

It indicates that after intentionally misleading information spreads, the source population who want to spread misinformation deliberately increases. This extra source population will amplify the spread of the next spread of misinformation at a significant rate.

Bifurcation Diagram for the System

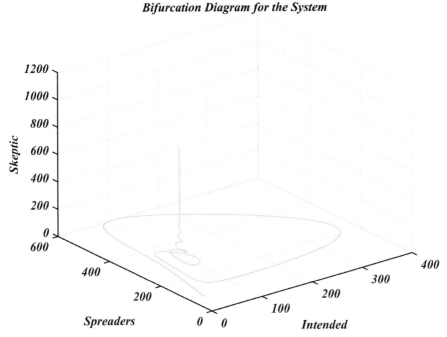

Figure 9.14 Changes of state trajectories with time for different parameters.

Finally, we have summarized our findings altogether below:

- The general spreader population automatically decreases after a certain time so that they can be minimized with the least effort. The number of intended criminals increases because some of the general spreaders start joining with them and help increase the population.
- The intended criminals are liable for every misleading information spread that leads to violence. That is why we have to focus on minimizing the intended criminals.
- When we investigated the spreader population for different parameters, we found that the time duration was the same as intended, but the spreader population can be minimized with the increase of the sceptic population. The parameters α, b, β, and 1 significantly influence the number of people in both spreader and intended populations.
- From the above analysis and simulation, we decided to control the whole scenario before day number 2. We also decided to control the intended and spreader populations as soon as possible.

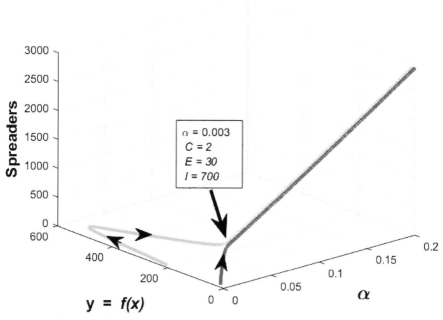

Figure 9.15   Changes of state trajectories with time for $\alpha = 0 \rightarrow 0.2$.

- Figures 9.13–9.16 depicts the system change for the change in parameters. From those figures, we can see, except for α and μ, for all other combinations of parameter values, the solution of the system is bounded. Besides, we found a limit cycle there. But only state Z increases continuously with time for any cases. All other states show a converging property for different values of parameters.
- Any machine learning model can be developed for predicting the spread type considering our study as a base [16]. Our study gives the data scientists a clue about variable/indicator selection, the importance of indicators and the behaviour of different indicators on the outcome.

9.7 Conclusion

In this study, we have considered a very realistic case scenario by considering communication through online media and person-to-person communication

Threshold value of μ for minimizing the Intended

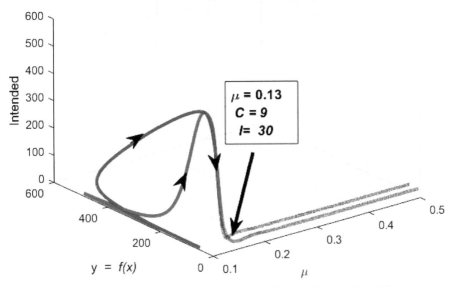

Figure 9.16 Changes of state trajectories with time for $\mu = 0 \rightarrow 0.5$.

as a medium of misinformation spread. The whole analysis and simulation give a transparent picture of the internal mechanism of spreading misinformation and the control techniques.

The control strategies can be determined easily from the findings of our study. Mathematical models generally give an approximation about the system, which can be valid in practice or not. And all the models use the least assumptions of the system. Here we developed the model by considering a very specific case of misinformation spread. And our result leads towards the underlying patterns of misinformation spread with different types of motives. The pattern of violence-intended approaches and general approaches are visually different. Selecting the initial values of the intended population for different agendas gives us a very fruitful outcome in practice.

In this study, we have ignored all the unnecessary and irreverent assumptions we noticed in some previous studies only for analytical purposes. This gives our model the extra power to reveal reality. All the findings are similar to the statistical data and the logic of spreading misinformation.

References

[1] K. M. a. J. Caroline Fisher, "How Asia gets its news: News consumption trends 20162020 – News in Asia," 2021. [Online]. Available: https://newsinasia.jninstitute.org/chapter/how-asia-gets-its-news-news-consumption-trends-2016-2020/.[AccessedApril2022].

[2] B. Guerin and Y. Miyazaki, "Analyzing Rumors, Gossip, and Urban Legends Through Their Conversational Properties," *The Psychological Record,* vol. 56, pp. 23-33, 2006.

[3] P. Suciu, "Spotting Misinformation On Social Media Is Increasingly Challenging," August 2021. [Online]. Available: https://www.forbes.com/sites/petersuciu/2021/08/02/spotting-misinformation-on-social-media-is-increasingly-challenging/?sh=3bd08ab2771c.

[4] X. Zhao and J. Wang, "Dynamical Model about Rumor Spreading with Medium," *Discrete Dynamics in Nature and Society,* vol. 2013, p. 2013, 2013.

[5] M. H. A. Biswas, P. R. Dey, M. S. Islam, and S. Mandal, "Mathematical Model Applied to Green Building Concept for Sustainable Cities Under Climate Change," *Journal of Contemporary Urban Affairs,* vol. 6, no. 1, pp. 36-50, 2021.

[6] M. Nekovee, Y. Moreno, G. Bianconi, and M. Marsili, "Theory of rumour spreading in complex social networks," *Physica A: Statistical Mechanics and its Applications,* vol. 374, no. 1, pp. 457-470, 2007.

[7] MailChimp, "How To Identify Fake News on Social Media," April 2022. [Online]. Available: https://mailchimp.com/resources/fake-news-on-social-media/.

[8] L. Wei, G. Jiao and C. Xu, "Message Spreading and Forget–Remember Mechanism on a Scale-Free Network," *Chinese Physics Letters,* vol. 25, no. 6, pp. 2303-2306, 2008.

[9] Y. Liu, C. Zeng and Y. Luo, "Dynamics of a New Rumor Propagation Model with the Spread of Truth," *Applied Mathematics,* vol. 9, no. 5, pp. 536-549, 2018.

[10] W. Zhang, H. Deng and X. Li, "Dynamics of the Rumor-Spreading Model with Control Mechanism in Complex Network," *Journal of Mathematics,* 2022.

[11] R. Isea and K. E. Lonngren, "A New Variant of the SEIZ Model to Describe the Spreading of a Rumor," *International Journal of Data Science and Analysis,* vol. 3, no. 4, pp. 28-33, 2017.

[12] A. Mathur and C. P. G., "Dynamic SEIZ in Online Social Networks: Epidemiological Modeling of Untrue Information," *International Journal of Advanced Computer Science and Applications,* vol. 11, no. 7, pp. 577-585, 2020.

[13] A. K. Paul and M. H. A. Biswas, "Modeling the Dynamics of Spreading Rumors and Fake News through Online and Social Media," in *2nd International Conference on Industrial and Mechanical Engineering and Operations*, Dhaka, December 12, 2019.

[14] O. Diekmann, J. A. P. Heesterbeek and M. G. Roberts, "The construction of next-generation matrices for compartmental epidemic models," *Journal of The Royal Society Interface,* p. 873–885, 2019.

[15] I. T. Union, "Individuals using the Internet (% of population) | Data," 2022. [Online]. Available: https://data.worldbank.org/indicator/IT.NET. USER.ZS.

[16] A. K. Paul and M. H. A. Biswas, "Violence Detection: Introducing a Machine Learning Based Novel Method," 3rd International Conference on Industrial and Mechanical Engineering and Operations Management (IMEOM), Dhaka, Bangladesh, Dhaka, February 08, 2021.

[17] A. K. Paul, M. S. Khatun, and M. H. A. Biswas, "Modeling and Optimal Control Applied to Reduce the Effects of Greenhouse Gases Emitted from the Coal-based Power Plant in Bangladesh (Revised Manuscript Submitted for Publication)," *Hellyon,* April 2022.

[18] H. Mahmoud, "A model for the spreading of fake news," *Journal of Applied Probability,* vol. 57, no. 1, pp. 332-342, 2020.

[19] M. H. A. Biswas, "On the Evolution of AIDS/HIV Treatment: An Optimal Control Approach," *Current HIV Research,* vol. 12, no. 1, pp. 1-12, 2014.

[20] M. H. A. Biswas, M. A. Islam, S. Akter, S. Mandal, M. S. Khatun, S. A. Samad, A. K. Paul, and M. R. Khatun, "Modelling the Effect of Self-Immunity and the Impacts of Asymptomatic and Symptomatic Individuals on COVID-19 Outbreak," *Computer Modeling in Engineering & Sciences,* vol. 125, no. 3, pp. 1133-1060, 2021.

[21] "Ranking by Population (Rural)," 2022. [Online]. Available: https://datacommons.org/ranking/Count_Person_Rural/Country/?h=country%2FBGD.[AccessedApril2022].

10

Diseased Predator–Prey Model Incorporating Herd Behaviour in Prey: A Study Under an Alternative Food Source Scenario

Sumit Kaur Bhatia[1], Riju Chaudhary[1], and Devyanshi Bansal[2]

[1]Amity Institute of Applied Sciences, Amity University Uttar Pradesh, Noida, India
[2]Amity School of Engineering and Technology, Amity University Uttar Pradesh, Noida, India
E-mail: sumit2212@gmail.com; riju.chaudhary@gmail.com; rchaudhary@amity.edu; devyanshi.bansal@s.amity.edu

Abstract

A diseased prey predator model with herd behaviour exhibited by prey population has been analysed in the current study. The modified functional response of Holling type-II, where the prey-density term is given by its square root, has been incorporated into the model. Details of the boundedness of solutions, conditions for the existence of points of equilibria, and their stability analyses have all been studied. Global stability and Hopf bifurcation have also been presented. The biomass conversion parameter has been shown to play a crucial impact in the system's dynamics. Analytical findings has been substantially supported by numerical simulation of the proposed model, which have been done using MATLAB. Biological implications of the results have also been discussed.

Keywords: Prey Predator Model, Stability Analysis of Equilibrium points, Square Root Functional Responses, Transmissible Diseases in Predator, Hopf Bifurcation.

10.1 Introduction

The interaction between diverse species and the natural environment defines ecosystems. The pioneering work of Lotka [1] and Volterra [2] paved way for other researchers to explore the system of prey predator interaction. The interaction where an organism consumes another living organism termed as 'predation' has a major effect on population dynamics. 'Functional response', one of several elements in prey predator models, is critical in describing a prey predator interaction. Functional response was defined by the authors in [8, 13, 14, 15, 16, 17, 18, 20] and [21] as an assessment of the predation rate's relative responsiveness to changes in prey density at diverse prey populations. The capacity of prey to flee, the shape of the prey habitat, and the time it takes the predator to conquer and consume prey before moving on to the next hunt are all factors that influence functional response. The functional responses of Holling type II and Holling type III ([13, 14]), Tanner ratio dependency ([3]) and Beddington Dengelis ([4]) have all been investigated in the literature. Cosner at el [5] introduced different spatial-patterns a prey predator model can develop due to various functional responses. The Holling type II functional response shows that a saturation function is a better fit for describing food intake. It is predicated on the concept that a single individual can only eat until its stomach is empty. Following is the mathematical form of Holling type II functional response:

$$F\left(\wp\right) = \frac{\varpi x}{1 + T_h \varpi \wp},$$

where ϖ is the search efficiency of the predator for prey, $\wp\left(t\right)$ is the prey density at time t and T_h denotes the predators's mean handling time for each prey.

Within the context of an ecological system, different species employ different strategies, such as clustering, refuging and so on, in order to acquire food and defend themselves. Herd behaviour refers to the act of individuals within a group behaving in a coordinated manner. Herd behaviour can be caused by a number of different factors which has been studied by different authors in the past. When it comes to protecting themselves from predators, prey species can benefit from grouping together in herds. When it comes to defeating prey, it is unquestionably more effective for the predator species to work together in a group. Several prey predator models with herd behaviour have been studied [9–12]. The authors in [6, 7] studied a predator prey model in which prey species exhibit herd behaviour and the predator interacts with the prey along the outside corridor of the herd of prey. The use of square root term suggests that the interaction is happening along the boundary of the population. Recent searches deal with more complex situations with the aim of better understanding reality. In order to capture the memory effect, fractional order can be considered as has been suggested by [22–27]. In [19, 28] time delay has been incorporated to better comprehend the complex dynamics of the system.

In this study, a prey-predator model, where predator population suffers from disease and the prey exhibits herd behaviour has been studied. The mathematical model is introduced in the second section, followed by the boundedness and positivity of the solutions. Criteria for existence of equilibrium points have been obtained in the following section. Stability study is carried out for all the points of equilibria, and stability near the origin is investigated from a novel perspective using non-linear analysis, resulting in more realistic results. Global stability has also been discussed, followed by Hopf bifurcation. In the next section, we verify our analytical results numerically using MATLAB. Finally, in the last section, the work's conclusion has been presented.

10.2 Mathematical Model Formulation

In this chapter, let X be the prey population density which exhibits herd behaviour. Also, we have considered that the predator is infected with a disease. Let Y_1, Y_2 be the population density of susceptible and infected predators, respectively. In the absence of predator, the prey population has been assumed to grow logistically. The intrinsic growth rate and carrying capacity of the prey population have been denoted by r and K, respectively. The susceptible predator Y_1 has another food source rather than X so we introduce logistic growth term in susceptible predator population and let r_1 be the intrinsic growth rate of the susceptible predator. Also, let K_1 be the carrying capacity of the predator population which is shared by both susceptible predators and infected predators. The infection in the predator is spread out according to the law of mass of action and its rate is denoted by λ. d denotes the death rate of infected predator. Biomass conversion rate of prey to predator population has been denoted by e. As suggested by Braza [7], square root of prey population should be considered as it is best suited for the situation in which the prey population forms a herd to find better food resources and to defend itself from the predators. Keeping Holling type II functional response in consideration, functional response has been taken as follows:

$$F\left(x\right) = \frac{\alpha x}{1 + T_h \alpha x}.$$

Thus, we consider the following model under the above-mentioned observations:

$$\frac{dX}{dT} = rX\left(1 - \frac{X}{K}\right) - \frac{\alpha\sqrt{X}Y_1}{1 + T_h\alpha\sqrt{X}}, \tag{10.1}$$

$$\frac{dY_1}{dT} = r_1Y_1\left(1 - \frac{Y_1 + Y_2}{K_1}\right) + \frac{e\alpha\sqrt{X}Y_1}{1 + T_h\alpha\sqrt{X}} - \lambda Y_1 Y_2, \tag{10.2}$$

$$\frac{dY_2}{dT} = \lambda Y_1 Y_2 - dY_2, \tag{10.3}$$

where predator's search efficiency for prey is denoted by α. After rescaling, the system given by eqn (10.1)–(10.3), is reduced to the system given by eqn (10.4) and (10.5):

$$\frac{dx}{dt} = x\left(1 - x\right) - \frac{ay_1\sqrt{x}}{1 + b\sqrt{x}}, \tag{10.4}$$

$$\frac{dy_1}{dt} = r_ny_1\left[1 - y_1 - y_2\right] + \frac{c\sqrt{x}y_1}{1 + b\sqrt{x}} - my_1y_2, \tag{10.5}$$

$$\frac{dy_2}{dt} = my_1y_2 - ny_2, \tag{10.6}$$

with re-scaling variables

$$x = \frac{X}{K}, y_1 = \frac{Y_1}{K_1}, y_2 = \frac{Y_2}{K_1}, t = rT, a = \frac{\alpha K_1}{\sqrt{K}r}, b = T_h\alpha\sqrt{K},$$

$$c = \frac{e\alpha\sqrt{K}}{r}, m = \frac{\lambda K_1}{r}, n = \frac{d}{r}, r_n = \frac{r_1}{r}.$$

10.3 Qualitative Analysis of the Model

Theorem 10.1. All solutions of the system (10.4) and (10.5) with positive initial conditions remain positive for all $t \geq 0$.

Proof. From eqn (10.4), it can be seen that $\frac{dx}{dt} \geq x(1 - x) \implies x(t) > 0 \, \forall \, t \geq 0$

From eqn (10.5), we can see that,

$$\frac{dy_1}{y_1} = r_n \left[1 - y_1 - y_2\right] + \frac{c\sqrt{x}}{1 + b\sqrt{x}} - my_2$$

Therefore,

$$y(t) = y(0) \exp\left(\int r_n \left[1 - y_1(w) - y_2(w)\right] + \frac{c\sqrt{x}}{1 + b\sqrt{x}} - my_2(w)\right) dw$$

$> 0, \forall\, t \geq 0.$

Positivity of $y_2(t)$ can be proved on similar lines. Hence, the theorem. □

Theorem 10.2. Solutions of the system (10.4) and (10.5) with positive initial conditions are uniformly bounded.

Proof. Let $W = cx + ay_1$. Then,

$$\frac{dW}{dt} + W = cx\,(1 - x) + ar_n y_1 \,(1 - y_1 - y_2) + cx + ay_1$$
$$\leq 2c + ar_n{}^2 + 2ar_n + a = \mu(let).$$

From [28], we get, $0 < W < \mu + \dfrac{(x(0), y_1(0))}{e^t}$ and therefore for $t \to \infty$, we have $0 < W < \mu$

Now consider,

$$\frac{dy_1}{dt} + \frac{dy_2}{dt} = r_n y_1 \left[1 - y_1 - y_2\right] + \frac{c\sqrt{x}\,y_1}{1 + b\sqrt{x}} - ny_2$$

Similarly, we can prove that $0 < y_1(t) + y_2(t) < \dfrac{(r_n + c + n)\mu}{n}$.

Solutions of the system (10.4) and (10.5) with positive initial conditions are confined in the region given below:

$$\left\{(x, y_1, y_2) \in R_+^3 : 0 < W < \mu, 0 < y_1(t) + y_2(t) < \frac{(r_n + c + n)\mu}{n}\right\}$$

Hence, the theorem. □

10.3.1 Existence of points of equilibria

Existence of point of equilibria for the system given by eqn (10.4)–(10.6) has been obtained, in this section:

$$E_0\,(0,0,0)\,, E_1\,(1,0,0)\,, E_2\left(x, \frac{c\sqrt{x}}{r_n\,(1 + b\sqrt{x})} + 1, 0\right),$$

$$E_3\left(x, \frac{n}{m}, \frac{r_n\left(1 - \frac{n}{m}\right) + \frac{can}{m\left(1 + b\sqrt{x}\right)^2 (1 - x)}}{m + r_n}\right)$$

It is easy to see that $E_0\,(0,0,0)$, $E_1\,(1,0,0)$.

Evaluating E_2:

From eqn (10.5) we get, $y_1 = \dfrac{c\sqrt{x}}{r_n\left(1+b\sqrt{x}\right)} + 1$, where $\sqrt{x} = p$ and p is root of the equation

$$\gamma_1 p^5 + \gamma_2 p^4 + \gamma_3 p^3 + \gamma_4 p^2 + \gamma_5 p + \gamma_6 = 0 \text{ and } \gamma_1 = b^2 r_n, \gamma_2 = 2br_n, \gamma_3 = r_n - b^2 r_n, \gamma_4 = -2br_n, \gamma_5 = ac - r_n + abr_n, \gamma_6 = ar_n.$$

Evaluating E_3:

From eqn (10.4), we can find $\dfrac{x}{\sqrt{x}} = \dfrac{ay_1}{\left(1+b\sqrt{x}\right)\left(1-x\right)}$.

Now from eqn (10.5) :$y_1 = \dfrac{r_n\left(1-\frac{n}{m}\right) + \frac{can}{m\left(1+b\sqrt{x}\right)^2(1-x)}}{m+r_n}$.

Hence, we obtain $E_3\left(x, \dfrac{n}{m}, \dfrac{r_n\left(1-\frac{n}{m}\right) + \frac{can}{m\left(1+b\sqrt{x}\right)^2(1-x)}}{m+r_n}\right)$

$E_3\left(x, \dfrac{n}{m}, \dfrac{r_n\left(1-\frac{n}{m}\right) + \frac{can}{m\left(1+b\sqrt{x}\right)^2(1-x)}}{m+r_n}\right)$ will exist if $\dfrac{r_n \lambda K_1}{r_n - ca} > d$

where x can
be obtained from the equation $p^8 q_1 + p^7 q_2 + p^6 q_3 + p^5 q_4 + p^4 q_5 + p^3 q_6 + p^2 q_7 + q_8 = 0$ where $\sqrt{x} = p$ where $q_1 = b^2 m^2$, $q_2 = 2bm^2$, $q_3 = m^2 - 2b^2 m^2$, $q_4 = -4m^2 b$, $q_5 = b^2 m^2$, $q_6 = 2bm^2$, $q_7 = m^2$, $q_8 = -a^2 n^2$.

10.3.2 Stability analysis

Model system's stability around equilibrium points is studied in this section. We will first calculate the Jacobian matrix of the system which is as follows:

$$J = \begin{pmatrix} (1-2x) - \dfrac{ay_1}{2\sqrt{x}\left(1+b\sqrt{x}\right)^2} & \dfrac{-a\sqrt{x}}{1+b\sqrt{x}} & 0 \\ \dfrac{cy_1}{2\sqrt{x}\left(1+b\sqrt{x}\right)^2} & r_n\left(1-y_2-2y_1\right) + \dfrac{c\sqrt{x}}{1+b\sqrt{x}} - my_2 & -y_1(r_n+m) \\ 0 & my_2 & my_1 - n \end{pmatrix}$$

10.3.2.1 The prey-only equilibrium
This point can also be referred to as the predator-free equilibrium.

$$JE_1 = \begin{pmatrix} -1 & \frac{-a}{1+b} & 0 \\ 0 & r_n + \frac{c}{1+b} & 0 \\ 0 & 0 & -n \end{pmatrix}$$

The characteristic equation obtained is: $(\lambda + 1) \left[\lambda - \left(r_n + \frac{c}{1+b} \right) \right] (\lambda + n)$.

The corresponding eigen-values are $-1, -n, r_n + \frac{c}{1+b}$. Since $r_n + \frac{c}{1+b} > 0$, the system at E_1 is always unstable.

Remark: Due to the presence of an alternate food source for the predator, the prey-only equilibrium point is always unstable.

10.3.2.2 Dynamical behavior near the trivial equilibrium point

The variational matrix at the origin consists of indeterminate term due to the functional response. So, the system is not linearisable at the origin. Rescaling the variable x, say $x = z^2$, will remove the square root singularity, but this can also mask the true nature of the system.

In this case, a local non-linear analysis, as suggested by Braza [7], is the ideal way to highlight the effect of the square root term. This would uncover the singular dynamics near origin. For stability analysis, it is best to neglect the effect of infected predator as the susceptible predator is extremely small near the origin.

We assume $x(t) << 1, x^2$ or higher order term vanishes, $1 + b\sqrt{x} \approx 1, c\sqrt{x}y_1 \approx y_1$.

Then, the system becomes:

$$\frac{dx}{dt} = x - ay_1\sqrt{x}$$
$$\frac{dy_1}{dt} = -\mu y_1$$
$$\text{where} \mu = (m - r_n - 1)$$

Origin is a saddle point, if $\dfrac{dx}{dt} \approx x$ and $x = O(y)$. Also $\sqrt{x}y_1$ may be larger or the same size as x.

Assuming $y_1 \ll 1$ and $x = O(y_1^s(0))$ so that $y_1(t) = y_1(0)e^{(-\mu t)}$ and $y_1(0) \ll 1$. Three distinguishing cases have been formed: $s = 1, s = 2$ and $s > 2$.

Case $s = 1$: Saddle behaviour near the origin is noted.
Case $s > 2$: Since $x \ll y_1\sqrt{x}$, the system reduces to

$$\frac{dx}{dt} = -ay_1\sqrt{x}$$
$$\frac{dy_1}{dt} = -\mu y_1$$

Now we have

$$\frac{dy_1}{dx} = \frac{\mu}{a\sqrt{x}} \tag{10.7}$$

Integrating eqn (10.7) from $y_1(0)$ to y_1 and $x_0 y_1^{s(0)}$ to x.

$$y_1 - y_1(0) = \frac{2\mu}{a}\left(\sqrt{x} - \sqrt{x_0 y_1^s(0)}\right) \tag{10.8}$$

The curve represented by eqn (10.8) is part of a parabola that starts at $(x, y_1(0)) = \left(x_0 y_1^{s(0)}, y_1(0)\right)$ and terminates on the y_1 axis at $y_1 = y_1(0)\frac{-2\mu}{a}(-x_0 y_1^s(0)) > 0$, after which the predator y_1 declines to zero because of $\frac{dy_1}{dt} = -\mu y_1$.

This means that if the prey population is smaller than the predator's, the prey will become extinct first, followed by the predator.

Case $s = 2$: As in the $s > 2$ case above, here also, at some positive value of y_1 and $x = 0$, the trajectory terminates. Therefore, it is a saddle point.

The stability condition of other equilibrium points of this model system has been stated in the following theorems:

10.3.2.3 Disease-free equilibrium point

Theorem 10.3. The equilibrium point $E_2\left(x, \frac{c\sqrt{x}}{r_n(1+b\sqrt{x})} + 1, 0\right)$ is stable if $r_1 < r$, $\lambda K_1 < d$ and $\lambda K_1 ea\sqrt{K} > r_1 b(d - \lambda K_1)$

Proof. Here,

$$J_{E_2} =$$
$$\begin{pmatrix} (1-2x) - \frac{ay_1}{2\sqrt{x}\left(1+b\sqrt{x}\right)^2} & \frac{-a\sqrt{x}}{1+b\sqrt{x}} & 0 \\ \frac{cy_1}{2\sqrt{x}\left(1+b\sqrt{x}\right)^2} & r_n(1-2y_1) + \frac{c\sqrt{x}}{1+b\sqrt{x}} & -y_1(r_n+m) \\ 0 & 0 & my_1 - n \end{pmatrix}$$

One of the eigenvalues of the characteristic equation is $\lambda_1 = n - my_1$ and other eigenvalues are given by roots of the following equation:

$$\lambda^2 - \lambda\,(p+q) + \left(pq + \frac{acy_1}{2\,(1+b\sqrt{x})^3}\right) = 0$$

where $p = 1 - 2x - \frac{ay_1}{2\sqrt{x}(1+b\sqrt{x})^2}$, $q = r_n\,(1 - 2y_1) + \frac{c\sqrt{x}}{1+b\sqrt{x}}$.

Therefore, if $y_1 < \dfrac{n}{m}$ and $p + q < 0$. It can be easily seen that $y_1 < \dfrac{n}{m}$ if $\lambda K_1 < d$ and $\lambda K_1 ea\sqrt{K} > r_1 b(d - \lambda K_1)$ and $p + q < 0$ is possible if $r_1 < r$, then the equilibrium is stable otherwise unstable.

Remark: According to the above theorem, a disease-free equilibrium point is stable if the predator's intrinsic growth rate due to an alternate food source is less than the growth rate of the prey, and if the infection rate times the predator's carrying capacity due to an alternate food source is less than the death rate of the infected predator. □

10.3.2.4 Endemic equilibrium point

Theorem 10.4. The equilibrium point $E_3\left(x, \dfrac{n}{m}, y_2\right)$ will be asymptotically stable if the following conditions are satisfied:

(1) $a_1 + b_1 < 0$

(2) $a_1 b_1 + (r_n + m)\,ny_2 + \dfrac{acn}{2m(1+b\sqrt{x})^3} > 0$

(3) $-(a_1 + b_1)\left[a_1 b_1 + (r_n + m)\,ny_2 + \dfrac{acn}{2m(1+b\sqrt{x})^3}\right] + a_1\,(r_n + m)\,ny_2 > 0$

Proof. The Jacobian matrix at the point E_3 is as follows:

$$\begin{pmatrix} (1-2x) - \frac{an}{2m\sqrt{x}(1+b\sqrt{x})^2} & \frac{-a\sqrt{x}}{1+b\sqrt{x}} & 0 \\ \frac{cn}{2m\sqrt{x}(1+b\sqrt{x})^2} & r_n\left(1 - 2\frac{n}{m} - y_2\right) + \frac{c\sqrt{x}}{1+b\sqrt{x}} - my_2 & -\frac{n}{m}(r_n + m) \\ 0 & my_2 & 0 \end{pmatrix}$$

The characteristic equation obtained from the above Jacobian is:

$$\lambda^3 - (a_1 + b_1)\,\lambda^2 + \left[a_1 b_1 + (r_n + m)\,ny_2 + \frac{acn}{2m(1+b\sqrt{x})^3}\right]\lambda - a_1\,(r_n + m)\,ny_2 = 0$$

where $a_1 = (1-2x) - \dfrac{an}{2m\sqrt{x}(1+b\sqrt{x})^2}$, $b_1 = r_n\left(1 - 2\frac{n}{m} - y_2\right) + \dfrac{c\sqrt{x}}{1+b\sqrt{x}} - my_2$

and $y_2 = \dfrac{r_n\left(1 - \frac{n}{m}\right) + \dfrac{can}{m\left(1+b\sqrt{x}\right)^2(1-x)}}{m + r_n}$

$E_3\left(x, \frac{n}{m}, y_2\right)$ is stable when:

(1) $a_1 + b_1 < 0$,

(2) $a_1 b_1 + (r_n + m)\, n y_2 + \dfrac{acn}{2m(1 + b\sqrt{x})^3} > 0$,

(3) $-(a_1 + b_1)\left[a_1 b_1 + (r_n + m)\, n y_2 + \dfrac{acn}{2m(1 + b\sqrt{x})^3}\right] + a_1\,(r_n + m)\, n y_2 > 0$.

$\square$

10.3.3 Transversality condition for Hopf bifurcation

For the value c, we will prove that the given system experiences a Hopf bifurcation around the endemic equilibrium point.

Theorem 10.5. The system enters into Hopf-bifurcation around the endemic equilibrium point E_3 for the parameter $c = \tilde{\jmath}$ if, $\varsigma_1(\tilde{\jmath}) > 0$, $\varsigma_1(\tilde{\jmath})\varsigma_2(\tilde{\jmath}) - \varsigma_3(\tilde{\jmath}) = 0$, $[\varsigma_1(\tilde{\jmath})\varsigma_2(\tilde{\jmath})]' < \varsigma_3'(\tilde{\jmath})$ hold.

Proof. Let c be the biomass conversion rate. We assume that the interior point E_5 is asymptotically stable. Our interest is on the parameter c that whether E_5 loses its stability with the change in the parameter c, that is, we assume c as the bifurcation parameter, then there exists a critical value of $\tilde{\jmath}$ such that, $\varsigma_1(\tilde{\jmath}) > 0$, $\varsigma_1(\tilde{\jmath})\varsigma_2(\tilde{\jmath}) - \varsigma_3(\tilde{\jmath}) = 0,)[\varsigma_1(\tilde{\jmath})\varsigma_2(\tilde{\jmath})]' < \varsigma_3'(\tilde{\jmath})$. In order for Hopf bifurcation to occur, we must have the following:

$$(\lambda_2(\tilde{\jmath}) + \varsigma_2(\tilde{\jmath}))\,(\lambda(\tilde{\jmath}) + \varsigma_1(\tilde{\jmath})) = 0. \tag{10.9}$$

The corresponding eigen-values are $\lambda_1(\tilde{\jmath}) = i\sqrt{\varsigma_2(c^*)}$, $\lambda_2(\tilde{\jmath}) = -i\sqrt{\varsigma_2(c^*)}$, $\lambda_3(\tilde{\jmath}) < 0$.

We will check if Hopf bifurcation is likely to occur at $c = \tilde{\jmath}$, for this to happen we will verify transversality criteria, that is, $\left[\dfrac{d\mathrm{Re}\,(\lambda\,(c))}{dc}\right]_{c=\tilde{\jmath}} \neq 0$

The roots for all c are: $\lambda_1 = \tau_1\,(c) + i\tau_2\,(c)$, $\lambda_2 = \tau_1\,(c) - i\tau_2\,(c)$, $\lambda_3 = -\sigma_1\,(c)$.

Now, the transversality condition is verified by substituting $\lambda_j = \tau_1\,(c) \pm i\tau_2\,(c)$ in equation (10.9) and then by calculating the derivative, we get

$$F\,(c)\,\tau_1{}'\,(C) - G\,(c)\,\tau_2{}'\,(c) + H\,(c) = 0, \tag{10.10}$$

$$G\left(c\right)\tau_1{}'\left(C\right)+F\left(c\right)\tau_2{}'\left(c\right)+I\left(c\right)=0, \tag{10.11}$$

where

$F\left(c\right)=3\tau_1^2\left(c\right)+2\sigma_1\left(c\right)\tau_1\left(c\right)+\sigma_2\left(c\right)-3\tau_2{}^2\left(c\right),$

$G\left(c\right)=6\tau_1\left(c\right)\tau_2\left(c\right)+2\sigma_1\left(c\right)\tau_2\left(c\right),$

$H\left(c\right)=\tau_1^2\left(c\right)\sigma_1'\left(c\right)+\sigma_2'\left(c\right)\tau_1\left(c\right)+\sigma_1'\left(c\right)-\sigma_1'\left(c\right)v^2\left(c\right),$

$I\left(c\right)=2\tau_1\left(c\right)\tau_2\left(c\right)\sigma_1'\left(c\right)+\sigma_2'\left(c\right)\tau_2\left(c\right).$

We know that $\tau_1\left(\tilde{\jmath}\right)=0$, $\tau_2\left(\tilde{\jmath}\right)=\sqrt{\sigma_2\left(\tilde{\jmath}\right)}$ which results in

$F\left(\tilde{\jmath}\right)=-2\sigma_2\left(\tilde{\jmath}\right),$

$G\left(\tilde{\jmath}\right)=2\sigma_2\sqrt{\sigma_2\left(\tilde{\jmath}\right)},$

$H\left(\tilde{\jmath}\right)=\sigma_3'-\sigma_1'\left(\tilde{\jmath}\right)\sigma_2\left(\tilde{\jmath}\right),$

$I\left(\tilde{\jmath}\right)=\sigma_2'\left(\tilde{\jmath}\right)\sqrt{\sigma_2\left(\tilde{\jmath}\right)}.$

We now determine the value for $\tau_1'\left(\tilde{\jmath}\right)$ using equations (10.10) and (10.11) we get,

$$\left[\frac{d\mathrm{Re}\left(\lambda_j\left(c\right)\right)}{dc}\right]_{c=\tilde{\jmath}}=\frac{\sigma_3'(\tilde{\jmath})-\sigma_1'(\tilde{\jmath})\sigma_2(\tilde{\jmath})-\sigma_1(\tilde{\jmath})\sigma'(\tilde{\jmath})}{\sigma_1^2(\tilde{\jmath})+\sigma_2(\tilde{\jmath})}>0. \tag{10.12}$$

If $\left[\sigma_1\left(\tilde{\jmath}\right)\sigma_1\left(\tilde{\jmath}\right)\right]'-\sigma_3'\left(\tilde{\jmath}\right)<0$ then it clearly implies that transervality condition holds. Therefore, at $c=\tilde{\jmath}$, Hopf bifurcation occurs. □

10.3.4 Global stability analysis

In this section, will study the global stability of our system's endemic equilibrium point.

Theorem 10.6. If, $y_1\sqrt{x^*}>y_1^*\sqrt{x}$; $y_1x^*>y_1^*x$ and $x\sqrt{x^*}<x^*\sqrt{x}$ hold, then around the endemic equilibrium point E_3, the system is globally asymptotically stable.

Proof. Assume the following:

$$V(x,y_1,y_2)=p_0\left(x-x^*-x^*ln\frac{x}{x^*}\right)+p_1\left(y_1-y_1^*-y_1^*ln\frac{y_1}{y_1^*}\right)$$

$$+p_2\left(y_2-y_2^*-y_2^*ln\frac{y_2}{y_2^*}\right), \tag{10.13}$$

where p_0, p_1, p_2 are the constants which are greater than 0, whose values will be determined in the following steps.

At the equilibrium (x,y_1^*,y_2^*), the function V is zero and on D, the function V is positive.

Consider

$$\frac{dV}{dt} = p_0 \left(\frac{x - x^*}{x} \right) \frac{dx}{dt} + p_1 \left(\frac{y_1 - y_1^*}{y_1} \right) \frac{dy_1}{dt} + p_2 \left(\frac{y_2 - y_2^*}{y_2} \right) \frac{dy_2}{dt}.$$
(10.14)

This implies

$$\frac{dV}{dt} = p_0(x - x^*) \left((1 - x) - \frac{\alpha y_1}{\sqrt{x} + bx} \right) + p_1(y_1 - y_1^*)$$

$$\left(r_n(1 - y_1 - y_2) + \frac{cx}{\sqrt{x} + bx} - my_2 \right) + + p_2(y_2 - y_2^*)(my_1 - n).$$
(10.15)

Solving it further we get

$$\frac{dV}{dt} = -p_0(x-x^*)^2 - p_0\alpha(x-x^*) \left(\frac{y_1\sqrt{x^*} - y_1^*\sqrt{x} + b(y_1 x^* - y_1^* x)}{\sqrt{xx^*} + bx\sqrt{x^*} + bx^*\sqrt{x} + b^2 xx^*} \right)$$
$$-p_1 r_n(y_1 - y_1^*)^2 - p_1 r_n(y_1 - y_1^*)(y_2 - y_2^*) + p_1 c(y_1 - y_1^*)$$
$$\left(\frac{x\sqrt{x^*} - x^*\sqrt{x}}{\sqrt{xx^*} + bx\sqrt{x^*} + bx^*\sqrt{x} + b^2 xx^*} \right) - p_1 m(y_1 - y_1^*)(y_2 - y_2^*) + p_2 m(y_1 - y_1^*)(y_2 - y_2^*).$$

Now, choosing $p_0 = 1$, $p_1 = 1$, and $p_2 = 1$, then the above equation reduces to:

$$\frac{dV}{dt} = -(x-x^*)^2 - \alpha(x-x^*) \left(\frac{y_1\sqrt{x^*} - y_1^*\sqrt{x} + b(y_1 x^* - y_1^* x)}{\sqrt{xx^*} + bx\sqrt{x^*} + bx^*\sqrt{x} + b^2 xx^*} \right) - r_n(y_1 - y_1^*)^2$$

$$-r_n(y_1 - y_1^*)(y_2 - y_2^*) + c(y_1 - y_1^*) \left(\frac{x\sqrt{x^*} - x^*\sqrt{x}}{\sqrt{xx^*} + bx\sqrt{x^*} + bx^*\sqrt{x} + b^2 xx^*} \right).$$

Now if, $y_1\sqrt{x^*} > y_1^*\sqrt{x}$; $y_1 x^* > y_1^* x$ and $x\sqrt{x^*} < x^*\sqrt{x}$, then we can see that, $\frac{dV}{dt}$ will be a negative definite function, since all the coefficients in the above equation will become negative.

Then by Lyapunov's direct method, E_3^* is globally asymptotically stable.

10.4 Numerical Simulation

Numerical simulations help validate the analytical findings. As a result, we have used MATLAB to run a variety of numerical simulations. For the set of parameters $a = 0.02$, $b = 0.85$, $r_n = 0.2$, $c = 0.5$, trajectories approach disease-free equilibrium point $(0.97802, 2.0, 0)$, which is depicted in Figure 10.1. Also, it can be checked that disease-free equilibrium point's existence conditions are satisfied.

Figure 10.2 shows that the endemic equilibrium point $(0.99517265, 0.45462029, 0.02370219)$ is locally asymptotically stable for the set of parameters $a = 0.02$, $b = 0.89$, $c = 0.8$, $r_n = 0.1890$. Conditions for stability of endemic equilibrium point for these can be easily verified. From Figure 10.3, it can be seen that if there is no infection in the system, then for the same value of parameters, system is stable and trajectories approach equilibrium point $E_2(0.97419609, 2.34180845)$. It can be noted that in the absence of infection, equilibrium value of predator is more than in the case of system with infection.

It can be seen from Figures 10.4 and 10.5 that c is the bifurcation parameter. As shown in Figures 10.4 and 10.5, the biomass conversion rate plays a significant role in the dynamics of the diseased prey predator system. In Figure 10.5 , we see that in infected system, as we decrease the value of c (biomass conversion rate), to $c = 0.0005$ oscillations decrease and we observe a stable behaviour around endemic equilibrium point, whereas by increasing c to 0.07, oscillations increase and system enters Hopf bifurcation as can be seen from Figure 10.4. This perhaps implies that if more food is available then system may not remain stable.

From Figure 10.6, it is observed that if alternate food source is not provided then the system is unstable around endemic equilibrium point. Thus, it is seen that alternate food source and biomass conversion factors can alter the dynamics of the model.

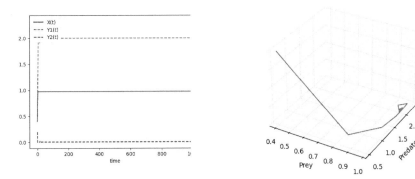

Figure 10.1 $a = 0.02, b = 0.85, c = 0.5, r_n = 0.2$.

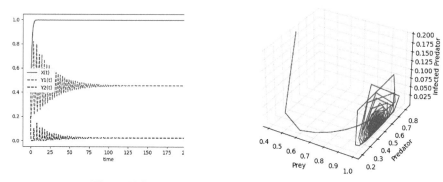

Figure 10.2 $a = 0.02, b = 0.89, c = 0.8, r_n = 0.1890.$

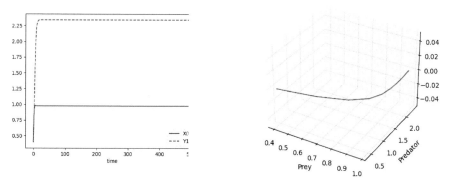

Figure 10.3 $a = 0.02, b = 0.89, c = 0.8, r_n = 0.1890.$

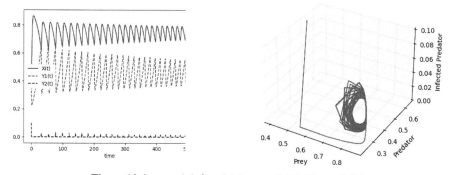

Figure 10.4 $a = 0.8, b = 0.79, r_n = 0.00049, c = 0.07.$

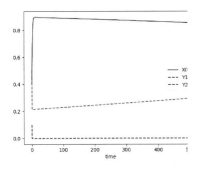

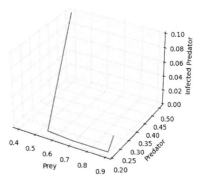

Figure 10.5 $a = 0.8, b = 0.79, r_n = 0.00049, c = 0.0005.$

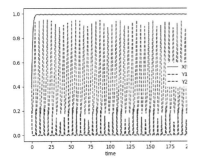

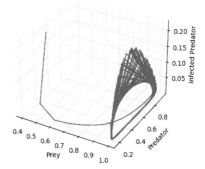

Figure 10.6 $a = 0.02, b = 0.89, c = 0.8, r_n = 0.$

10.5 Conclusion

We discussed a model with a diseased predator and a Holling type II functional response in which the prey-density term is given by its square root due to prey herd behaviour in this study. Logistic growth for prey population is considered. It has been proved that the system's solutions are always positive and bounded. Existence of four equilibrium points has also been established. The model's stability was investigated around equilibrium points, and it was discovered that if alternate food source is available, then the prey-only equilibrium point exhibits unstablity. We also proved that disease-free equilibrium point is stable if intrinsic growth rate of predator due to alternate food source is less than growth rate of prey and also if infection rate times carrying capacity of predator due to alternate food source is less than the death rate of infected predator. We also obtained the conditions under which endemic equilibrium point is stable. Due to the square root term, the variational matrix obtained at the trivial equilibrium was observed to be indeterminate. Consequently, approximation method, as suggested by Braza [7], was used to analyse the stability at origin. It was assumed that the prey population density was extremely small near the origin, thus the higher-order terms were excluded and a more stable path was used to analyse the point. We observed that if the prey population is small then it first goes to extinction followed by the predator population. Further, biomass conversion rate was taken as the bifurcation parameter, as it dictated the dynamics of the model. It was observed that by increasing the biomass conversion rate, system stabilizes around endemic equilibrium point. We also, observed that if alternate food source is not provided then even for small biomass conversion rate system is stable around endemic equilibrium point. Thus, by controlling the alternate food source to predator disease can be controlled.

References

[1] A. Lotka. Elements of physical biology. Williams and Wilkins Company, Baltimore, 1925.

[2] V. Volterra, Fluctuations in the abundance of species considered mathematically, Nature. 118(1926), 558-560.

[3] Z. Lianga and H. Pan, Qualitative analysis of a ratio-dependent Holling-Tanner model, Journal of Mathematical Analysis and Applications, Vol 334(2), (2007), 954-964.

[4] Tzy-Wei Hwang, Global Analysis of the predator-prey system with Beddington - DeAngelis functional response, Jounal of Mathematical Analysis and Applications, Vol 281(1)(2003),395-(401).

[5] Chris Cosner, Donald L. DeAngelis, and Jerald S. Ault and Donald B. Olson, Effects of Spatial Grouping on the Functional Response of Predators. Theoretical Population Biology, 56(1999), 65-75.

[6] Valerio Ajraldi, Marta Pittavino, Ezio Venturio, Modelling herd behaviour in population systems, Nonlinear Analysis: Real World Applications, 12 (2011), 2319-2338.

[7] Peter A. Braza, Predator-prey dynamics with square root functional responses, Nonlinear Analysis: Real World Applications, 13(2012), 1837-1843.

[8] C. Holling. The components of predation as revealed by a study of small-mammal predation of the European pine sawfly. The Canadian Entomologist, 1959, 91(5):293-320.

[9] S. Belvisi, E. Venturino. Anecoepidemic model with diseased predators and prey group defense. Simulation Modelling Practice and Theory, 2013, 34: 144-155.

[10] E. Venturino, S. Petrovskii. Spatiotemporal behavior of a prey-predator system with a group defense for prey. Ecological Complexity, 2013, 14: 37-47.

[11] C. Xu, S. Yuan. Stability and Hopf bifurcation in a delayed predator-prey system with herd behavior. Abstract and Applied Analysis, 2014, 8: 568943, 2014.

[12] S. Yuan, C. Xu, et al. Spatial dynamics in a predator-prey model with herd behavior. Chaos: An Interdisciplinary Journal of Nonlinear Science, 2013,23: 033102.

[13] C. Holling. Some characteristics of simple types of predation and parasitism. The Canadian Entomologist, 1959, 91(7):385-398.

[14] C. Holling. The functional response of invertebrate predators to prey density. Memoirs of the Entomological Society of Canada, 1966, 98(48): 5-86.

[15] M. Haque and E. Venturino, An eco-epidemiological model with disease in predator: the ratio-dependent case, Mathematical Methods in the Applied, Sciences, 30(14)(2007), 1791-1809.

[16] J. Murray. Mathematical Biology. Springer-Verlag, New York, 2002.

[17] D. Xaio, S. Ruan. Global analysis in a predator-prey system with non-monotonic functional response. SIAM Journal on Applied Mathematics, 2001, 64(4):1445-1472.

[18] P. Abrams, L. Ginzburg. The nature of predation: prey dependent,ratio dependent or neither? Trends in Ecology and Evolution, 2000, 15(8):337-341.

[19] Y. Kuang. Delay differential equations:with applications in population dynamics. Academic Press, Boston, 1933.

[20] H. Freedman. Deterministic mathematical models in population ecology. Marcel Dekker, New York, 1980.

[21] W. Herbert, A. Hethcotea, W. Wangb, L. Hanc, and Z. Mad, A predator-prey model with infected prey, Theoretical Population Biology, 66(2004), 259-268.

[22] Tukur Abdulkadir Sulaiman, Mehmet Yavuz, Hasan Bulut, Haci Mehmet Baskonus, Investigation of the fractional coupled viscous Burgers' equation involving Mittag-Leffler kernel, Physica A: Statistical Mechanics and its Applications, 527, 2019, 121-126.

[23] Sulaiman, Tukur Abdulkadir, Haci Mehmet Baskonus, and Hasan Bulut. "Optical solitons and other solutions to the conformable space-time fractional complex Ginzburg-Landau equation under Kerr law nonlinearity." Pramana, 91, 4 2018, 1-8.

[24] Yavuz, Mehmet, et al., The Schrödinger-KdV equation of fractional order with Mittag-Leffler nonsingular kernel, Alexandria Engineering Journal 60.2 (2021): 2715-2724.

[25] Sulaiman, Tukur Abdulkadir, Hasan Bulut, and Haci Mehmet Baskonus, Optical solitons to the fractional perturbed NLSE in nano-fibers, Discrete & Continuous Dynamical Systems-S 13, 3, 2020, 925.

[26] Sulaiman, Tukur Abdulkadir, et al., Optical solitons to the fractional perturbed Radhakrishnan-Kundu-Lakshmanan model, Optical and Quantum Electronics 50, 10, 2018, 1-10.

[27] Yusuf, Abdullahi, et al., Modeling the effect of horizontal and vertical transmissions of HIV infection with Caputo fractional derivative, Chaos, Solitons & Fractals 145 (2021): 110794.

[28] Jaradat, Imad, et al., Analytic simulation of the synergy of spatial-temporal memory indices with proportional time delay, Chaos, Solitons & Fractals 156 (2022): 111818.

11

A COVID-19-related Atangana–Baleanu Fractional Model for Unemployed Youths

Albert Shikongo

Department of Computing, Mathematical and Statistical Science
Faculty of Agriculture, Engineering and Natural Sciences,
University of Namibia, Namibia
E-mail: ashikongo@unam.na

Abstract

In this study, a system of ordinary differential equations (ODEs) modelling the dynamics of unemployment of youths in our society is derived. Due to the burden of COVID-19 across all boards of economic activities, the dynamics are extended to a system of Atangana–Baleanu fractional (ABF) dynamics. The ABF dynamics are analysed, and certain conditions are established with respect to existence of unique solution, eradication of unemployment and persistence of unemployment. These conditions are reserved for further discussion within the implementation phase of the AB fractional dynamics. Since the analytic solution for the dynamics is not available, a numerical method for discretizing the Atangana–Baleanu fractional derivative in the Caputo sense, is derived, analysed for convergence, implemented, and the results are presented for further discussion.

Keywords: COVID-19, Atangana–Baleanu Fractional, Unemployment, Uniqueness of Solution, Stability Analysis.

MSC Code: 104A40, 00A09, 34A05, 34C11, 34D05, 65L05, 65M06.

11.1 Introduction

Unemployment occurs when someone is willing and able to work but does not have a paying job. Thus, this constitutes a rate of unemployment people as the percentage of people in a labour force who are unemployed. This includes people who are not in a paid job, but who are actively looking for work. Consequently, in the event of natural disasters such as outbreak of disease such as COVID-19 [18], the United Nation labour experts, mentioned that the economic crisis caused by the COVID-19 pandemic is expected to contribute to global unemployment of more than 200 million people in 2022, particularly with women and youth workers worst-hit. Thus, with the impact of COVID-19 ([6, 24]) very well known to the author, this implies that most countries' per capita real gross domestic product (GDP) in long-term and short-term is indeed negatively affected.

Thus, in this chapter, we consider unemployment as one of the social evils apart from diseases [23], crime and corruption. Hence, in the current fraternity it suffices first to highlight the most recent development in terms of deriving mathematical dynamics to deal with the surge of unemployment rates, particularly among youths.

In [7], an in-depth background of unemployment dynamics are also presented, a simple, realistic and useful mathematical dynamics for unemployment is derived. Hence, an optimal control problem is formulated and solved, which provides some non-trivial and interesting conclusions.

A non-linear mathematical dynamic for the challenge of unemployment in terms of unemployed, employed and available vacancies is derived in [1]. They [1] also mentioned that their model resembles the situation in some countries where the support of the government reaches a certain limited level where the rate of creating new jobs becomes constant and can no longer be proportional to the number of unemployed due to limited financial and economic resources.

An effect of the action of the government and private sector to control unemployment without any delay is analysed in [21], coupled with the effect of attempt of self-employment made by unemployed persons.

An unemployment model of differential equations with time delay, taking into consideration the role of government for the support of vacancies creation is derived and analysed in [22].

In view of the above studies, one can see that nothing much has been done in the direction of modelling the dynamics of the unemployed youths, even though youth unemployment is on the rise contrary to the availability of equal opportunities among men and women. Thus, in the next section, we are proposing a mathematical dynamic with the aim of illustrating (with regard to the outbreak of natural disasters such COVID-19 pandemic) the current status of affairs in most countries.

11.2 The Conceptual Model

In this section, the dynamics governing the proposed model is partitioned into six compartments. The compartments are learners ($\mathcal{L}$), employees ($\mathcal{E}$), small and medium-sized enterprises ($\mathcal{S}$), tertiary ($\mathcal{T}_r$), corporate ($\mathcal{C}$) and unemployed ($\mathcal{U}$). Hence, the six compartments are as follows:

- Learners ($\mathcal{L}$): in secondary education.
- Employees ($\mathcal{E}$): employed for wages or salary, at non-executive level.
- Small and medium-sized enterprises ($\mathcal{S}$): owning small and mid-size enterprises (SMEs).
- Tertiary ($\mathcal{T}_r$): in education for above school age, including college, university and vocational courses.
- Corporate ($\mathcal{C}$): owning large companies.
- Unemployed ($\mathcal{U}$): without a paid job but available to work.

The probabilities of members' interactions for the above compartments are as follows.

- Learners ($\mathcal{L}$): The compartment is increasing with the rates Υ and ς due to the influx of pupils from primary education and some of those previously dropped out of secondary education, respectively. However, the compartment is decreasing by the rates of those getting employed at the rate of ω, establishing their SMEs at the rate of κ, joining tertiary education at the rate of β, dropping out of school and/or those completed their secondary education but unable to get admission in tertiary institutions, employment opportunity or establish an SME, at the rate of $\eta_\mathcal{L}$ and natural death at the rate of α.
- Employed ($\mathcal{E}$): This compartment is increasing with the rate of ω due to learners getting employed, ν students and/or graduates from tertiary education, ρ by the unemployed ones getting employed, and decreasing by the rates of those establishing their SMEs at the rate of ϖ, joining tertiary education at the rates of ϱ, drop outs and/or losing employment due to some circumstances at the rate of $\eta_\mathcal{E}$ and natural death at the rate of α.

- SMEs ($\mathcal{S}$): This compartment is increasing by recruiting from learners and employees compartments with the rates κ and ϖ, respectively, and is decreasing at the rates of those successfully turn their SMEs into corporate at the rate of σ, unsuccessful SMEs at the rate of η_S and by natural death at the rate of α.
- Tertiary ($\mathcal{T}_r$): The compartment is increasing with the contact rates from the learners, employees and unemployed youths with the contact rates of β, ϱ and γ, respectively. It is decreasing with the contact rates of those getting employed at the rates of ν, dropouts and/or graduates completing their studies due to lack of employment opportunities, and natural death at the rates of η_{T_r} and α, in that order.
- Corporate ($\mathcal{C}$): The compartment is increasing at the rate of σ due to those grow their SMEs into corporate. However, the compartment is decreasing at the rates of η_C and α due to the impact of factors such as mismanagement or COVID-19 and natural death, respectively.
- Unemployed ($\mathcal{U}$): This compartment is increasing due to the dropouts from all the above-mentioned compartments at the rate of $\eta_L, \eta_\mathcal{E}, \eta_S, \eta_{T_r}, \eta_C$ and decreasing at the rates of $\varsigma, \rho, \gamma, \alpha$ due to those going back to secondary education system, getting employed, joining tertiary education and natural death, in that order.

Compartmental, the above interactions among members are depicted in Figure 11.1, and mathematically, the basic dynamics for the youths with varying population size in a homogeneously mixing population are as follows

$$
\left.
\begin{aligned}
\frac{d\mathcal{L}}{dt} &= \Upsilon\mathcal{N} + \varsigma\frac{\mathcal{U}\mathcal{L}}{\mathcal{N}} - \omega\mathcal{E} - \kappa\mathcal{S} - \beta\mathcal{T}_r - \eta_\mathcal{L}\mathcal{L} - \alpha\mathcal{L}, \\
\frac{d\mathcal{E}}{dt} &= \omega\mathcal{E} + \nu\frac{\mathcal{E}\mathcal{T}_r}{\mathcal{N}} + \rho\frac{\mathcal{U}\mathcal{E}}{\mathcal{N}} - \varpi\mathcal{S} - \varrho\mathcal{T}_r - \eta_\mathcal{E}\mathcal{E} - \alpha\mathcal{E}, \\
\frac{d\mathcal{S}}{dt} &= \kappa\mathcal{S} + \varpi\mathcal{S} - \sigma\mathcal{C} - \eta_S\mathcal{S} - \alpha\mathcal{S}, \\
\frac{d\mathcal{T}_r}{dt} &= \beta\mathcal{T}_r + \varrho\mathcal{T}_r + \gamma\frac{\mathcal{U}\mathcal{T}_r}{\mathcal{N}} - \nu\frac{\mathcal{E}\mathcal{T}_r}{\mathcal{N}} - \eta_{T_r}\mathcal{T}_r - \alpha\mathcal{T}_r, \\
\frac{d\mathcal{C}}{dt} &= \sigma\mathcal{C} - \eta_C\mathcal{C} - \alpha\mathcal{C}, \\
\frac{d\mathcal{U}}{dt} &= \eta_\mathcal{L}\mathcal{L} + \eta_\mathcal{E}\mathcal{E} + \eta_S\mathcal{S} + \eta_{T_r}\mathcal{T}_r \\
&\quad + \eta_C\mathcal{C} - \varsigma\frac{\mathcal{U}\mathcal{L}}{\mathcal{N}} - \rho\frac{\mathcal{U}\mathcal{E}}{\mathcal{N}} - \gamma\frac{\mathcal{U}\mathcal{T}_r}{\mathcal{N}} - \alpha\mathcal{U},
\end{aligned}
\right\}
$$

$$(11.2.1)$$

$\forall\, t \in [a, b]$ such that $0 = a < b \in \mathbb{N}_0$ and $\mathcal{N}(t) := \mathcal{L}(t) + \mathcal{E}(t) + \mathcal{S}(t) + \mathcal{T}_r(t) + \mathcal{C}(t) + \mathcal{U}(t)$, denotes the total population for the youths at time t. The system in eqn (11.2.1) is a system of first-order, non-linear ordinary

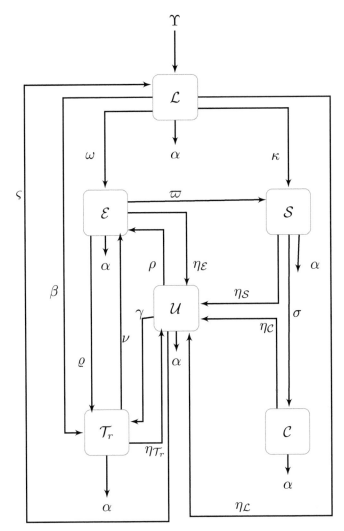

Figure 11.1   Graphical representation of the flow of youths among the compartments in eqn (11.2.1).

differential equations (ODEs), therefore, initial conditions

$$\left.\begin{array}{l} \mathcal{L}(a) = \mathcal{L}_a, \ \mathcal{E}(a) = \mathcal{E}_a, \ \mathcal{S}(a) = \mathcal{S}_a, \\ \mathcal{T}_r(a) = \mathcal{T}_{ra}, \ \mathcal{C}(a) = \mathcal{C}_a, \ \mathcal{U}(a) = \mathcal{U}_a, \\ \text{where, } \mathcal{L}_a \geq 0, \ \mathcal{E}_a \geq 0, \ \mathcal{S}_a \geq 0, \ \mathcal{T}_{ra} \geq 0, \ \mathcal{C}_a \geq 0, \ \mathcal{U}_a \geq 0, \end{array}\right\} (11.2.2)$$

are required. We refer to the system of initial valued problems (IVPs) in eqns (11.2.1) and (11.2.2) as LESTCU and/or unemployed youths. It follows from the total population that

$$\frac{d\mathcal{N}}{dt} = (\Upsilon - \alpha)\mathcal{N},$$

from which one finds that

$$\mathcal{N}(t) = \exp(\Upsilon - \alpha)t. \tag{11.2.3}$$

Considering the proportions of individuals in six compartments in eqn (11.2.1) as

$$l = \frac{\mathcal{L}}{\mathcal{N}}, \quad e = \frac{\mathcal{E}}{\mathcal{N}}, \quad s = \frac{\mathcal{S}}{\mathcal{N}}, \quad t_r = \frac{\mathcal{T}_r}{\mathcal{N}}, \quad c = \frac{\mathcal{C}}{\mathcal{N}}, \quad u = \frac{\mathcal{U}}{\mathcal{N}},$$

and retaining the original notations as in equation (11.2.1) to avoid unnecessary complications. It is trivial to show that l, e, s, t_r, c, u, satisfies the following system of first order, non-linear ODEs,

$$\left. \begin{aligned}
\frac{d\mathcal{L}}{dt} &= \Upsilon(1 - \mathcal{L}) + \varsigma \mathcal{U}\mathcal{L} - \omega\mathcal{E} - \kappa\mathcal{S} - \beta\mathcal{T}_r - \eta_{\mathcal{L}}\mathcal{L}, \\
\frac{d\mathcal{E}}{dt} &= \omega\mathcal{E} + \nu\mathcal{E}\mathcal{T}_r + \rho\mathcal{U}\mathcal{E} - \varpi\mathcal{S} - \varrho\mathcal{T}_r - \eta_{\mathcal{E}}\mathcal{E} - \Upsilon\mathcal{E}, \\
\frac{d\mathcal{S}}{dt} &= \kappa\mathcal{S} + \varpi\mathcal{S} - \sigma\mathcal{C} - \eta_{\mathcal{S}}\mathcal{S} - \Upsilon\mathcal{S}, \\
\frac{d\mathcal{T}_r}{dt} &= \beta\mathcal{T}_r + \varrho\mathcal{T}_r + \gamma\mathcal{U}\mathcal{T}_r - \nu\mathcal{E}\mathcal{T}_r - \eta_{\mathcal{T}_r}\mathcal{T}_r - \Upsilon\mathcal{T}_r, \\
\frac{d\mathcal{C}}{dt} &= \sigma\mathcal{C} - \eta_{\mathcal{C}}\mathcal{C} - \Upsilon\mathcal{C}, \\
\frac{d\mathcal{U}}{dt} &= \eta_{\mathcal{L}}\mathcal{L} + \eta_{\mathcal{E}}\mathcal{E} + \eta_{\mathcal{S}}\mathcal{S} \\
&\quad + \eta_{\mathcal{T}_r}\mathcal{T}_r + \eta_{\mathcal{C}}\mathcal{C} - \varsigma\mathcal{U}\mathcal{L} - \rho\mathcal{U}\mathcal{E} - \gamma\mathcal{U}\mathcal{T}_r - \Upsilon\mathcal{U},
\end{aligned} \right\} \tag{11.2.4}$$

subject to the restriction $l + e + s + t_r + c + u = 1$, and initial conditions in eqn (11.2.2).

Since, fractional calculus is advantageous over the ordinary integer order in the explanation of real-world problems, as well as in the modelling of real phenomena, then various types of fractional-order or non-local derivatives were proposed to deal with the reduction of an ordinary derivative. For instance, based on a power–law, a fractional derivative was introduced in [20], followed by the fractional derivative utilizing the exponential kernel. To overcome the locality of the kernel, a version of a fractional derivative equipped with memory, with the aid of a generalized Mittag–Leffler function (MLF) as a non-singular kernel was introduced in [2].

However, most of the studies describing the unemployment and employment dynamics are of Cauchy type of a system of classical differential equations [8, 14, 15, 16, 17]. Due to the great interest in studying the behavior of solution for some systems using fractional differential equations involving the Atangana–Baleanu operator by several author [12], then for this purpose of investigating several real-world phenomena such as in [9] and in [26, 27, 28] have been experimented through the fractional operators. However, there is no literature available on fractional dynamics for unemployment and employment dynamics. Therefore, due to its features to overcome the locality of the kernel, the dynamics in eqn (11.2.4) are extended to a system of non-local Atangana–Baleanu–Caputo ($\mathbb{ABC}$) derivative operator with $0 < \delta \leq 1$ as,

$$
\left.
\begin{aligned}
{}^{\mathbb{ABC}}\mathbb{D}^{\delta}_{a+}[\mathcal{L}(t)] &= \Upsilon(1-\mathcal{L}) + \varsigma\mathcal{U}\mathcal{L} - \omega\mathcal{E} \\
&\quad -\kappa\mathcal{S} - \beta\mathcal{T}_r - \eta_{\mathcal{L}}\mathcal{L}, \\
{}^{\mathbb{ABC}}\mathbb{D}^{\delta}_{a+}[\mathcal{E}(t)] &= \omega\mathcal{E} + \nu\mathcal{E}\mathcal{T}_r + \rho\mathcal{U}\mathcal{E} - \varpi\mathcal{S} - \varrho\mathcal{T}_r \\
&\quad -\eta_{\mathcal{E}}\mathcal{E} - \Upsilon\mathcal{E}, \\
{}^{\mathbb{ABC}}\mathbb{D}^{\delta}_{a+}[\mathcal{S}(t)] &= \kappa\mathcal{S} + \varpi\mathcal{S} - \sigma\mathcal{C} - \eta_{\mathcal{S}}\mathcal{S} - \Upsilon\mathcal{S}, \\
{}^{\mathbb{ABC}}\mathbb{D}^{\delta}_{a+}[\mathcal{T}_r(t)] &= \beta\mathcal{T}_r + \varrho\mathcal{T}_r + \gamma\mathcal{U}\mathcal{T}_r \\
&\quad -\nu\mathcal{E}\mathcal{T}_r - \eta_{\mathcal{T}_r}\mathcal{T}_r - \Upsilon\mathcal{T}_r, \\
{}^{\mathbb{ABC}}\mathbb{D}^{\delta}_{a+}[\mathcal{C}(t)] &= \sigma\mathcal{C} - \eta_{\mathcal{C}}\mathcal{C} - \Upsilon\mathcal{C}, \\
{}^{\mathbb{ABC}}\mathbb{D}^{\delta}_{a+}[\mathcal{U}(t)] &= \eta_{\mathcal{L}}\mathcal{L} + \eta_{\mathcal{E}}\mathcal{E} + \eta_{\mathcal{S}}\mathcal{S} + \eta_{\mathcal{T}_r}\mathcal{T}_r \\
&\quad +\eta_{\mathcal{C}}\mathcal{C} - \varsigma\mathcal{U}\mathcal{L} - \rho\mathcal{U}\mathcal{E} - \gamma\mathcal{U}\mathcal{T}_r - \Upsilon\mathcal{U},
\end{aligned}
\right\} \quad (11.2.5)
$$

subject to the same initial conditions in eqn (11.2.2).

The rest of the chapter is structured as follows: Section 11.3 deals with the mathematical analysis, whereas in Section 11.4, the discretization of the $\mathbb{ABC}$ derivative operator and analysis for its convergence and stability are dealt with. The implementation of the discretized $\mathbb{ABC}$ derivative operator and discussion for the numerical results are presented in Section 11.5, and Section 11.6 concludes the chapter.

11.3 Mathematical Analysis

In this section, existence and uniqueness of bounded solution, positivity of solution, equilibrium points and stability conditions for the dynamics in eqn (11.2.5) are established.

11.3.1 Existence of solution

Let

$$
\left.\begin{aligned}
\Pi_1 &:= \Upsilon(1 - \mathcal{L}) + \varsigma\mathcal{UL} - \omega\mathcal{E} - \kappa\mathcal{S} - \beta\mathcal{T}_r - \eta_{\mathcal{L}}\mathcal{L}, \\
\Pi_2 &:= \omega\mathcal{E} + \nu\mathcal{E}\mathcal{T}_r + \rho\mathcal{UE} - \varpi\mathcal{S} - \varrho\mathcal{T}_r - \eta_{\mathcal{E}}\mathcal{E} - \Upsilon\mathcal{E}, \\
\Pi_3 &:= \kappa\mathcal{S} + \varpi\mathcal{S} - \sigma\mathcal{C} - \eta_{\mathcal{S}}\mathcal{S} - \Upsilon\mathcal{S}, \\
\Pi_4 &:= \beta\mathcal{T}_r + \varrho\mathcal{T}_r + \gamma\mathcal{UT}_r - \nu\mathcal{E}\mathcal{T}_r - \eta_{\mathcal{T}_r}\mathcal{T}_r - \Upsilon\mathcal{T}_r, \\
\Pi_5 &:= \sigma\mathcal{C} - \eta_{\mathcal{C}}\mathcal{C} - \Upsilon\mathcal{C}, \\
\Pi_6 &:= \eta_{\mathcal{L}}\mathcal{L} + \eta_{\mathcal{E}}\mathcal{E} + \eta_{\mathcal{S}}\mathcal{S} + \eta_{\mathcal{T}_r}\mathcal{T}_r + \eta_{\mathcal{C}}\mathcal{C} - \varsigma\mathcal{UL} \\
&\quad - \rho\mathcal{UE} - \gamma\mathcal{UT}_r - \Upsilon\mathcal{U}.
\end{aligned}\right\} \tag{11.3.6}
$$

Then, in view of eqn (11.2.5) and (11.3.6), one has

$$
\left.\begin{aligned}
{}^{\mathbb{ABC}}\mathbb{D}_{a+}^{\delta}[\mathcal{L}(t)] &= \Pi_1, \\
{}^{\mathbb{ABC}}\mathbb{D}_{a+}^{\delta}[\mathcal{E}(t)] &= \Pi_2, \\
{}^{\mathbb{ABC}}\mathbb{D}_{a+}^{\delta}[\mathcal{S}(t)] &= \Pi_3, \\
{}^{\mathbb{ABC}}\mathbb{D}_{a+}^{\delta}[\mathcal{T}_r(t)] &= \Pi_4, \\
{}^{\mathbb{ABC}}\mathbb{D}_{a+}^{\delta}[\mathcal{C}(t)] &= \Pi_5, \\
{}^{\mathbb{ABC}}\mathbb{D}_{a+}^{\delta}[\mathcal{U}(t)] &= \Pi_6.
\end{aligned}\right\} \tag{11.3.7}
$$

Taking the integral form on both sides of eqn (11.2.5), one finds

$$
\left.\begin{aligned}
\Pi_1(t) - \Pi_1(a) &= \frac{1-\delta}{\mathbb{ABC}[\delta]}\Pi_1(t,\mathcal{L}) + \frac{\delta}{\mathbb{ABC}[\delta]}\int_a^t (t-\tau)^{\delta-1}\Pi_1(\tau,\mathcal{L})d\tau, \\
\Pi_2(t) - \Pi_2(a) &= \frac{1-\delta}{\mathbb{ABC}[\delta]}\Pi_2(t,\mathcal{E}) + \frac{\delta}{\mathbb{ABC}[\delta]}\int_a^t (t-\tau)^{\delta-1}\Pi_2(\tau,\mathcal{E})d\tau, \\
\Pi_3(t) - \Pi_3(a) &= \frac{1-\delta}{\mathbb{ABC}[\delta]}\Pi_3(t,\mathcal{S}) + \frac{\delta}{\mathbb{ABC}[\delta]}\int_a^t (t-\tau)^{\delta-1}\Pi_3(\tau,\mathcal{S})d\tau, \\
\Pi_4(t) - \Pi_4(a) &= \frac{1-\delta}{\mathbb{ABC}[\delta]}\Pi_4(t,\mathcal{T}_r) + \frac{\delta}{\mathbb{ABC}[\delta]}\int_a^t (t-\tau)^{\delta-1}\Pi_4(\tau,\mathcal{T}_r)d\tau, \\
\Pi_5(t) - \Pi_5(a) &= \frac{1-\delta}{\mathbb{ABC}[\delta]}\Pi_5(t,\mathcal{C}) + \frac{\delta}{\mathbb{ABC}[\delta]}\int_a^t (t-\tau)^{\delta-1}\Pi_5(\tau,\mathcal{C})d\tau, \\
\Pi_6(t) - \Pi_6(a) &= \frac{1-\delta}{\mathbb{ABC}[\delta]}\Pi_6(t,\mathcal{U}) + \frac{\delta}{\mathbb{ABC}[\delta]}\int_a^t (t-\tau)^{\delta-1}\Pi_6(\tau,\mathcal{U})d\tau.
\end{aligned}\right\} \tag{11.3.8}
$$

Theorem 11.3.1. The kernels Π_i, (for $i = 1, 2, 3, 4, 5, 6$) satisfy the Lipschitz conditions and contraction, if there exists a constant h_i, such that $0 \le h_i < 1$.

Proof. Let $\mathcal{L}$ and $\mathcal{L}^*$ denotes two functions. Then, for Π_1, one finds that

$$\|\Pi_1(t,\mathcal{L}) - \Pi_1(t,\mathcal{L}^*)\| = \| - \Upsilon\,(\mathcal{L} - \mathcal{L}^*) + \varsigma\mathcal{U}\,(\mathcal{L} - \mathcal{L}^*) - \eta_{\mathcal{L}}\,(\mathcal{L} - \mathcal{L}^*)\|,$$

$$\leq (\Upsilon + \varsigma\|\mathcal{U}\| + \eta_{\mathcal{L}})\,\|\mathcal{L} - \mathcal{L}^*\|,$$

$$\leq (\Upsilon + \varsigma C_{\mathcal{U}} + \eta_{\mathcal{L}})\,\|\mathcal{L} - \mathcal{L}^*\| = h_1\|\mathcal{L} - \mathcal{L}^*\|,$$

where $h_1 := \Upsilon + \varsigma C_{\mathcal{U}} + \eta_{\mathcal{L}}$ and $\|\mathcal{U}\| \leq C_{\mathcal{U}}$, denotes bounded function. Similarly, one finds that

$$\left.\begin{array}{l} \|\Pi_2(t,\mathcal{E}) - \Pi_2(t,\mathcal{E}^*)\| \leq h_2\|\mathcal{E} - \mathcal{E}^*\|, \\[4pt] \|\Pi_3(t,\mathcal{S}) - \Pi_3(t,\mathcal{S}^*)\| \leq h_3\|\mathcal{S} - \mathcal{S}^*\|, \\[4pt] \|\Pi_4(t,\mathcal{T}_r) - \Pi_4(t,\mathcal{T}_r^*)\| \leq h_4\|\mathcal{T}_r - \mathcal{T}_r^*\|, \\[4pt] \|\Pi_5(t,\mathcal{C}) - \Pi_5(t,\mathcal{C}^*)\| \leq h_5\|\mathcal{C} - \mathcal{C}^*\|, \\[4pt] \|\Pi_6(t,\mathcal{U}) - \Pi_6(t,\mathcal{U}^*)\| \leq h_6\|\mathcal{U} - \mathcal{U}^*\|, \end{array}\right\}$$

where

$$\left.\begin{array}{l} h_2 := \omega + \nu C_{\mathcal{T}_r} + \rho C_{\mathcal{U}} + \eta_{\mathcal{E}} + \Upsilon, \\[4pt] h_3 := \kappa + \varpi + \eta_{\mathcal{S}} + \Upsilon, \\[4pt] h_4 := \beta + \varrho + \gamma C_{\mathcal{U}} + \nu C_{\mathcal{E}} + \eta_{\mathcal{T}_r} + \Upsilon, \\[4pt] h_5 := \sigma + \eta_{\mathcal{C}} + \Upsilon, \\[4pt] h_6 := \varsigma C_{\mathcal{L}} + \rho C_{\mathcal{E}} + \gamma C_{\mathcal{T}_r} + \Upsilon, \end{array}\right\}$$

in which $\|\mathcal{L}\| \leq C_{\mathcal{L}}, \|\mathcal{T}_r\| \leq C_{\mathcal{T}_r}, \|\mathcal{E}\| \leq C_{\mathcal{E}}$ denote bounded functions. $\qquad\square$

Theorem 11.3.2. Let Theorem 11.3.1 holds. If, for $i = 1, 2, 3, 4, 5, 6$,

$$\left(1 - \frac{1 - \delta}{\mathrm{ABC}[\delta]} - \frac{T^\sigma}{\mathrm{ABC}[\delta]\Gamma(\delta)}\right) h_i < 1,$$

then the solution to the dynamics in eqn (11.2.5) exists.

Proof. In view of the system in eqn (11.3.8), one finds the successive difference

$$
\begin{aligned}
\chi_{1n}(t) &= \mathcal{L}_n(t) - \mathcal{L}_{n-1}(t) = \tfrac{1-\delta}{\mathbb{ABC}[\delta]}\left(\Pi_1(t,\mathcal{L}_{n-1}) - \Pi_1(t,\mathcal{L}_{n-2})\right)\\
&\quad + \tfrac{\delta}{\mathbb{ABC}[\delta]}\tfrac{1}{\Gamma(\delta)}\int_a^t (t-\tau)^{\delta-1}\left(\Pi_1(\tau,\mathcal{L}_{n-1}) - \Pi_1(\tau,\mathcal{L}_{n-2})\right)d\tau,\\
\chi_{2n}(t) &= (\mathcal{E})_n(t) - (\mathcal{E})_{n-1}(t) = \tfrac{1-\delta}{\mathbb{ABC}[\delta]}\left(\Pi_2(t,(\mathcal{E})_{n-1}) - \Pi_2(t,(\mathcal{E})_{n-2})\right)\\
&\quad + \tfrac{\delta}{\mathbb{ABC}[\delta]}\tfrac{1}{\Gamma(\delta)}\int_a^t (t-\tau)^{\delta-1}\left(\Pi_2(\tau,(\mathcal{E})_{n-1}) - \Pi_2(\tau,(\mathcal{E})_{n-2})\right)d\tau,\\
\chi_{3n}(t) &= \mathcal{S}_n(t) - \mathcal{S}_{n-1}(t) = \tfrac{1-\delta}{\mathbb{ABC}[\delta]}\left(\Pi_3(t,\mathcal{S}_{n-1}) - \Pi_3(t,\mathcal{S}_{n-2})\right)\\
&\quad + \tfrac{\delta}{\mathbb{ABC}[\delta]}\tfrac{1}{\Gamma(\delta)}\int_a^t (t-\tau)^{\delta-1}\left(\Pi_3(\tau,\mathcal{S}_{n-1}) - \Pi_3(\tau,\mathcal{S}_{n-2})\right)d\tau,\\
\chi_{4n}(t) &= \mathcal{T}_{r(n)}(t) - \mathcal{T}_{r(n-1)}(t) = \tfrac{1-\delta}{\mathbb{ABC}[\delta]}\left(\Pi_4(t,\mathcal{T}_{r(n-1)}) - \Pi_4(t,\mathcal{T}_{r(n-2)})\right)\\
&\quad + \tfrac{\delta}{\mathbb{ABC}[\delta]}\tfrac{1}{\Gamma(\delta)}\int_a^t (t-\tau)^{\delta-1}\left(\Pi_4(\tau,\mathcal{T}_{n-1}) - \Pi_4(\tau,\mathcal{T}_{n-2})\right)d\tau,\\
\chi_{5n}(t) &= (\mathcal{C})_n(t) - (\mathcal{C})_{n-1}(t) = \tfrac{1-\delta}{\mathbb{ABC}[\delta]}\left(\Pi_5(t,(\mathcal{C})_{n-1}) - \Pi_5(t,(\mathcal{C})_{n-2})\right)\\
&\quad + \tfrac{\delta}{\mathbb{ABC}[\delta]}\tfrac{1}{\Gamma(\delta)}\int_a^t (t-\tau)^{\delta-1}\left(\Pi_5(\tau,(\mathcal{C})_{n-1}) - \Pi_5(\tau,(\mathcal{C})_{n-2})\right)d\tau,\\
\chi_{6n}(t) &= \mathcal{U}_n(t) - \mathcal{U}_{n-1}(t) = \tfrac{1-\delta}{\mathbb{ABC}[\delta]}\left(\Pi_6(t,\mathcal{U}_{n-1}) - \Pi_6(t,\mathcal{U}_{n-2})\right)\\
&\quad + \tfrac{\delta}{\mathbb{ABC}[\delta]}\tfrac{1}{\Gamma(\delta)}\int_a^t (t-\tau)^{\delta-1}\left(\Pi_6(\tau,\mathcal{U}_{n-1}) - \Pi_6(\tau,\mathcal{U}_{n-2})\right)d\tau,
\end{aligned}
\tag{11.3.9}
$$

with the initial conditions as reported in eqn (11.2.2).

Eqn (11.3.9) implies,

$$
\begin{aligned}
\mathcal{L}_n(t) &= \sum_{i=1}^n \chi_{1i}(t),\ (\mathcal{E})_n(t) = \sum_{i=1}^n \chi_{2i}(t),\ \mathcal{S}_n(t) = \sum_{i=1}^n \chi_{3i}(t),\\
\mathcal{T}_{r(n)}(t) &= \sum_{i=1}^n \chi_{4i}(t),\ (\mathcal{C})_n(t) = \sum_{i=1}^n \chi_{5i}(t),\ \mathcal{U}_n(t) = \sum_{i=1}^n \chi_{6i}(t).
\end{aligned}
$$

Taking the norm for the system in eqn (11.3.9) and using Theorem 11.3.1, one finds

$$
\begin{aligned}
\|\chi_{1n}(t)\| &\le h_1 \tfrac{1-\delta}{\mathbb{ABC}[\delta]}\|\chi_{1(n-1)}(t)\| + \tfrac{\delta}{\mathbb{ABC}[\delta]}\tfrac{h_1}{\Gamma(\delta)}\int_a^t (t-\tau)^{\delta-1}\|\chi_{1(n-1)}(t)\|d\tau,\\
\|\chi_{2n}(t)\| &\le h_2 \tfrac{1-\delta}{\mathbb{ABC}[\delta]}\|\chi_{2(n-1)}(t)\| + \tfrac{\delta}{\mathbb{ABC}[\delta]}\tfrac{h_2}{\Gamma(\delta)}\int_a^t (t-\tau)^{\delta-1}\|\chi_{2(n-1)}(t)\|d\tau,\\
\|\chi_{3n}(t)\| &\le h_3 \tfrac{1-\delta}{\mathbb{ABC}[\delta]}\|\chi_{3(n-1)}(t)\| + \tfrac{\delta}{\mathbb{ABC}[\delta]}\tfrac{h_3}{\Gamma(\delta)}\int_a^t (t-\tau)^{\delta-1}\|\chi_{3(n-1)}(t)\|d\tau,\\
\|\chi_{4n}(t)\| &\le h_4 \tfrac{1-\delta}{\mathbb{ABC}[\delta]}\|\chi_{4(n-1)}(t)\| + \tfrac{\delta}{\mathbb{ABC}[\delta]}\tfrac{h_4}{\Gamma(\delta)}\int_a^t (t-\tau)^{\delta-1}\|\chi_{4(n-1)}(t)\|d\tau,\\
\|\chi_{5n}(t)\| &\le h_5 \tfrac{1-\delta}{\mathbb{ABC}[\delta]}\|\chi_{5(n-1)}(t)\| + \tfrac{\delta}{\mathbb{ABC}[\delta]}\tfrac{h_5}{\Gamma(\delta)}\int_a^t (t-\tau)^{\delta-1}\|\chi_{5(n-1)}(t)\|d\tau,\\
\|\chi_{6n}(t)\| &\le h_6 \tfrac{1-\delta}{\mathbb{ABC}[\delta]}\|\chi_{6(n-1)}(t)\| + \tfrac{\delta}{\mathbb{ABC}[\delta]}\tfrac{h_6}{\Gamma(\delta)}\int_a^t (t-\tau)^{\delta-1}\|\chi_{6(n-1)}(t)\|d\tau.
\end{aligned}
$$

Thus,

$$
\left.\begin{aligned}
\|\chi_{1n}(t)\| &\le \|\mathcal{L}_n(a)\| h_1 \left(\frac{1-\delta}{\mathbb{ABC}[\delta]} + \frac{\delta}{\mathbb{ABC}[\delta]} \frac{T^\sigma}{\Gamma(\delta)} \right)^n, \\
\|\chi_{2n}(t)\| &\le \|(\mathcal{E})_n(a)\| h_2 \left(\frac{1-\delta}{\mathbb{ABC}[\delta]} + \frac{\delta}{\mathbb{ABC}[\delta]} \frac{T^\sigma}{\Gamma(\delta)} \right)^n, \\
\|\chi_{3n}(t)\| &\le \|\mathcal{S}_n(a)\| h_3 \left(\frac{1-\delta}{\mathbb{ABC}[\delta]} + \frac{\delta}{\mathbb{ABC}[\delta]} \frac{T^\sigma}{\Gamma(\delta)} \right)^n, \\
\|\chi_{4n}(t)\| &\le \|\mathcal{T}_{r(n)}(a)\| h_4 \left(\frac{1-\delta}{\mathbb{ABC}[\delta]} + \frac{\delta}{\mathbb{ABC}[\delta]} \frac{T^\sigma}{\Gamma(\delta)} \right)^n, \\
\|\chi_{5n}(t)\| &\le \|\mathcal{C}_n(a)\| h_5 \left(\frac{1-\delta}{\mathbb{ABC}[\delta]} + \frac{\delta}{\mathbb{ABC}[\delta]} \frac{T^\sigma}{\Gamma(\delta)} \right)^n, \\
\|\chi_{6n}(t)\| &\le \|\mathcal{U}_n(a)\| h_6 \left(\frac{1-\delta}{\mathbb{ABC}[\delta]} + \frac{\delta}{\mathbb{ABC}[\delta]} \frac{T^\sigma}{\Gamma(\delta)} \right)^n.
\end{aligned}\right\} \quad (11.3.10)
$$

Hence, the system in eqn (11.2.5) possesses a continuous solution. This implies that the solution to the system in eqn (11.2.5) exists. Consequently, one verifies that eqn (11.3.10) constitutes the solution to the system in eqn (11.2.2)–(11.2.5) by assuming that

$$
\left.\begin{aligned}
\mathcal{L}(t) - \mathcal{L}(a) &= \mathcal{L}_n(t) - Q_{1n}(t), \\
\mathcal{E}(t) - \mathcal{E}(a) &= \mathcal{E}_n(t) - Q_{2n}(t), \\
\mathcal{S}(t) - \mathcal{S}(a) &= \mathcal{S}_n(t) - Q_{3n}(t), \\
\mathcal{T}_r(t) - \mathcal{T}_r(a) &= \mathcal{T}_{r(n)}(t) - Q_{4n}(t), \\
\mathcal{C}(t) - \mathcal{C}(a) &= \mathcal{C}_n(t) - Q_{5n}(t), \\
\mathcal{U}(t) - \mathcal{U}(a) &= \mathcal{U}_n(t) - Q_{6n}(t).
\end{aligned}\right\}
$$

Thus,

$$
\begin{aligned}
\|Q_{1n}(t)\| &\le \left\| \frac{1-\delta}{\mathbb{ABC}[\delta]} \left(\Pi_1(t,\mathcal{L}) - \Pi_1(t,\mathcal{L}_{n-1}) \right) \right. \\
&\quad + \frac{\delta}{\mathbb{ABC}[\delta]} \frac{1}{\Gamma(\delta)} \left. \int_a^t (t-\tau)^{\delta-1} \left(\Pi_1(\tau,\mathcal{L}) - \Pi_1(\tau,\mathcal{L}_{n-1}) \right) d\tau \right\|, \\
&\le h_1 \left(\frac{1-\delta}{\mathbb{ABC}[\delta]} + \frac{T^\sigma}{\mathbb{ABC}[\delta]\Gamma(\delta)} \right) \|\mathcal{L} - \mathcal{L}_{n-1}\|,
\end{aligned}
$$

where successive iterations yield

$$
\|Q_{1n}(t)\| \le \left(\frac{1-\delta}{\mathbb{ABC}[\delta]} + \frac{t_a^\sigma}{\mathbb{ABC}[\delta]\Gamma(\delta)} \right)^{n+1} h_1^{n+1},
$$

which implies that $\|Q_{1n}(t)\| \to 0$ as $n \to \infty$. Similarly, as $n \to \infty$, one easily verifies that

$$
\|Q_{2n}(t)\| \to 0, \|Q_{3n}(t)\| \to 0, \|Q_{4n}(t)\| \to 0, \|Q_{5n}(t)\| \to 0, \|Q_{6n}(t)\| \to 0.
$$

$\square$

11.3.2 Uniqueness of solution

Then the following hypothesis, holds.

Theorem 11.3.3. Let $\mathcal{L}^*, \mathcal{E}^*, \mathcal{S}^*, \mathcal{T}_r^*, \mathcal{C}^*, \mathcal{U}^*$ denote another solution to the proposed dynamics in eqn (11.2.5). Then,

$$\mathcal{L}(t) = \mathcal{L}^*(t), \ \mathcal{E}(t) = \mathcal{E}^*(t), \ \mathcal{S}(t) = \mathcal{S}^*(t), \ \mathcal{T}_r(t) = \mathcal{T}_r^*(t), \ \mathcal{C}(t) = \mathcal{C}^*(t),$$
$$\mathcal{U}(t) = \mathcal{U}^*(t).$$

Proof. In view of eqn (11.3.9), one obtains

$$\|\mathcal{L}(t) - \mathcal{L}^*(t)\| \leq \frac{1-\delta}{\mathbb{ABC}[\delta]} \|\Pi_1(t, \mathcal{L}) - \Pi_1(t, \mathcal{L}^*)\|$$

$$+ \frac{\delta}{\mathbb{ABC}[\delta]} \frac{1}{\Gamma(\delta)} \int_a^t (t-\tau)^{\delta-1} \|\Pi_1(\tau, \mathcal{L}) - \Pi_1(\tau, \mathcal{L}^*)\| d\tau,$$

$$\leq h_1 \left(\frac{1-\delta}{\mathbb{ABC}[\delta]} + \frac{t^\sigma}{\mathbb{ABC}[\delta]\Gamma(\delta)} \right) \|\mathcal{L}(t) - \mathcal{L}^*(t)\|,$$

from which one finds

$$\|\mathcal{L}(t) - \mathcal{L}^*(t)\| \left(1 - \frac{1-\delta}{\mathbb{ABC}[\delta]} - \frac{t^\sigma}{\mathbb{ABC}[\delta]\Gamma(\delta)} \right) h_1 \leq 0.$$

In view of Theorem 11.3.2, one obtains that

$$\left(1 - \frac{1-\delta}{\mathbb{ABC}[\delta]} - \frac{t^\sigma}{\mathbb{ABC}[\delta]\Gamma(\delta)} \right) h_1 > 0,$$

which implies that $\mathcal{L}(t) - \mathcal{L}^*(t) = 0$. Similarly, one can easily verifies that

$$\mathcal{E}(t) - \mathcal{E}^*(t) = 0, \ \mathcal{S}(t) - \mathcal{S}^*(t) = 0, \ \mathcal{T}_r(t) - \mathcal{T}_r^*(t) = 0, \ \mathcal{C}(t) - \mathcal{C}^*(t) = 0,$$
$$\mathcal{U}(t) - \mathcal{U}^*(t) = 0.$$

$\square$

11.3.3 Positive solution

In this section, the positivity invariant region Ω, with regard to the dynamics in eqn (11.2.5) is established.

Theorem 11.3.4. The set

$$\Omega := \left\{ (\mathcal{N}, \mathcal{E}, \mathcal{S}, \mathcal{T}_r, \mathcal{C}, \mathcal{U}) \in \mathbb{R}_+^6, : 0 < \mathcal{N}(t) \leq 1 \right\},$$

is positively invariant with respect to the ABC-fractional dynamics in eqn (11.2.5).

Proof. It suffices to note that from eqn (11.2.5), one finds that

$$^{ABC}\mathbb{D}_{a+}^{\delta}[\mathcal{N}(t)] + \Upsilon\mathcal{N}(t) > \Upsilon.$$

Hence, the results follows easily. ☐

11.3.4 Equilibrium points

At the equilibrium point(s) one finds that

$$\left.\begin{array}{l}^{ABC}\mathbb{D}_{a+}^{\delta}[\mathcal{L}(t)] =^{ABC}\mathbb{D}_{a+}^{\delta}[\mathcal{E}(t)] =^{ABC}\mathbb{D}_{a+}^{\delta}[\mathcal{S}(t)] = 0, \\[2mm] ^{ABC}\mathbb{D}_{a+}^{\delta}[\mathcal{T}_r(t)] =^{ABC}\mathbb{D}_{a+}^{\delta}[\mathcal{C}(t)] =^{ABC}\mathbb{D}_{a+}^{\delta}[\mathcal{U}(t)] = 0. \end{array}\right\}$$

Thus, the following results follow.

Theorem 11.3.5. Let $\mathcal{Q} := (\mathcal{L}^*, \mathcal{E}^*, \mathcal{S}^*, \mathcal{T}_r^*, \mathcal{C}^*, \mathcal{U}^*)$ denotes the equilibrium point for the system in eqn (11.2.5). Then,

$$\left.\begin{array}{l}\mathcal{Q} = \left(\frac{\rho-(\eta\varepsilon+\Upsilon-w)}{\rho}, 0, 0, 0, 0, \frac{\eta\varepsilon+\Upsilon-w}{\rho}\right), \\[2mm] \& \\[2mm] \mathcal{Q} = \left(\frac{\rho-(\eta\varepsilon+\Upsilon-w)}{\rho}, \frac{\rho\beta+\rho\varrho+\gamma\eta\varepsilon+\gamma\Upsilon-\gamma w-\rho\eta\mathcal{T}_r-\rho\Upsilon}{\nu\rho}, 0, 0, \frac{\kappa+\varpi-\eta s-\Upsilon}{\sigma}, \\[2mm] \frac{\eta\varepsilon+\Upsilon-w}{\rho}\right), \end{array}\right\}$$

are the equilibrium points for the phenomena in eqn (11.2.5).

Proof. The proof of Theorem 11.3.5 follows trivially. ☐

Lemma 11.3.1. The equilibrium points in Theorem 11.3.5 are positive if the following conditions hold.

(a) $\mathcal{L}^* > 0$, if $\rho + w > \eta\varepsilon + \Upsilon$.
(b) $\mathcal{E}^* > 0$, if $\rho\beta + \rho\varrho + \gamma\eta\varepsilon + \gamma\Upsilon > \gamma w + \rho\eta\mathcal{T}_r + \rho\Upsilon$.
(c) $\mathcal{C}^* > 0$, if $\kappa + \varpi > \eta s + \Upsilon$.
(d) $\mathcal{U}^* > 0$, if $\eta\varepsilon + \Upsilon > w$.

Proof. The proof of Lemma 11.3.1 follows trivially. ☐

11.3.5 Stability analysis

In order to determine stability condition for the system in eqn (11.2.5), we use the Jacobian matrix $J_{ij} := \frac{\partial\Pi_i}{\partial x_j}$, for $i = 1, 2, 3, 4, 5, 6$ and

$j = 1, 2, 3, 4, 5, 6$. Thus, the non-zero entries of the Jacobian matrix J_{ij} evaluated at the equilibrium points $(\mathcal{L}^*, \mathcal{E}^*, \mathcal{S}^*, \mathcal{T}_r^*, C^*, \mathcal{U}^*)$ are

$$
\begin{aligned}
J_{11} &= -\Upsilon + \sigma\mathcal{U}^* - \eta_{\mathcal{L}}; J_{12} = -\omega; J_{13} = -\kappa; J_{14} \\
&= -\beta; J_{16} = \varsigma\mathcal{L}^*; \\
J_{22} &= \omega + \nu\mathcal{T}_r^* + \rho\mathcal{U}^* - \eta_{\mathcal{E}} - \Upsilon; J_{23} = -\varpi; J_{24} = -\varrho; \\
J_{33} &= \kappa + \varpi - \eta_{\mathcal{S}} - \Upsilon;, J_{35} = -\sigma \\
J_{42} &= -\nu\mathcal{T}_r^*; J_{44} = \beta + \rho + \gamma\mathcal{U}^* - \eta_{\mathcal{T}_r} - \Upsilon; J_{46} = \gamma\mathcal{T}_r^* \\
J_{55} &= \delta - \eta_C - \Upsilon; \\
J_{61} &= -\varsigma\mathcal{U}^*; J_{62} = \eta_{\mathcal{L}}\eta_{\mathcal{E}}\mathcal{L}^* - \rho\mathcal{U}^*; J_{63} = \eta_{\mathcal{S}}; J_{64} = \eta_{\mathcal{T}_r} \\
&\quad -\gamma\mathcal{U}^*; J_{65} = \eta_C; \\
J_{66} &= -\varsigma\mathcal{L}^* - \gamma\mathcal{T}_r - \Upsilon,
\end{aligned}
\tag{11.3.11}
$$

from which one finds the characteristic equation as

$$\lambda^5 + a_4\lambda^4 + a_3\lambda^3 + a_2\lambda^2 + a_1\lambda + a_0 = 0, \tag{11.3.12}$$

where, the explicit expressions for the a_k, $\forall\, k = 0, 1, 2, 3, 4$, terms are omitted due to their long expressions.

Theorem 11.3.6. The characteristic equation in (11.3.12) is asymptotic stable if all the $a_k > 0$, $\forall\, k = 0, 1, 2, 3, 4$ [11].

Proof. In view of Lemma 11.3.1, the proof of Theorem 11.3.6 follows easily. □

11.4 Derivation of the Numerical Method

Let $N \in \mathbb{N} < \infty$, such that $k = \frac{b-a}{N}$ denotes a uniform time step–size. Consequently, let $t_n = nk$ for $n = 0, 1, 2, \ldots, N, \forall\, t_n \in [a, b]$. Then the $\mathbb{ABC}$ derivative operator with $0 < \delta \le 1$ in eqn (11.2.5) is discretized as follows. Due to rigorous derivation for the numerical method, it suffices to consider one of the compartments from the system in (11.2.5) as follows

$$
\left.
\begin{aligned}
\mathbb{ABC}\mathbb{D}_{a^+}^\delta[\mathcal{L}(t)] &= f(t, \mathcal{L}), \\
\mathcal{L}(a) &= \mathcal{L}_a,
\end{aligned}
\right\}
\tag{11.4.13}
$$

where $f(t, \mathcal{L}) = \Upsilon(1 - \mathcal{L}) + \varsigma \mathcal{U} \mathcal{L} - \omega \mathcal{E} - \kappa \mathcal{S} - \beta \mathcal{T}_r - \eta_{\mathcal{L}} \mathcal{L}$. The remaining compartments are deduced analogously.

Integrating equation in (11.4.13) one obtains

$$\mathcal{L}(t) - \mathcal{L}(a) = \frac{1 - \delta}{\mathbb{ABC}(\delta)} f(t, \mathcal{L}(t)) + \frac{\delta}{\mathbb{ABC}(\delta)\Gamma(\alpha)}$$

$$\int_a^t (t - \tau)^{\delta-1} f(t, \mathcal{L}(\tau)) d\tau. \qquad (11.4.14)$$

At any time t_{n+1}, eqn (11.4.14) becomes

$$\mathcal{L}(t_{n+1}) - \mathcal{L}(a) = \frac{1 - \delta}{\mathbb{ABC}(\delta)} f(t_n, \mathcal{L}_n) + \frac{\delta}{\mathbb{ABC}(\delta)\Gamma(\alpha)}$$

$$\int_a^{t_{n+1}} (t_{n+1} - \tau)^{\delta-1} f(t, \mathcal{L}(t)) dt, \qquad (11.4.15)$$

and at time t_n, eqn (11.4.14) is equivalent to

$$\mathcal{L}(t_n) - \mathcal{L}(a) = \frac{1 - \delta}{\mathbb{ABC}(\delta)} f(t_{n-1}, \mathcal{L}_{n-1}) + \frac{\delta}{\mathbb{ABC}(\delta)\Gamma(\alpha)}$$

$$\int_a^{t_n} (t_n - \tau)^{\delta-1} f(t, \mathcal{L}(t)) dt. \qquad (11.4.16)$$

Subtracting eqn (11.4.16) from eqn (11.4.15), one obtains

$$\mathcal{L}(t_{n+1}) - \mathcal{L}(t_n) = \frac{1-\delta}{\mathbb{ABC}(\delta)} \left(f(t_n, \mathcal{L}_n) - f(t_{n-1}, \mathcal{L}_{n-1}) \right)$$

$$+ A_{\delta,1} - A_{\delta,2},$$

where

$$A_{\delta,1} = \frac{\delta}{\mathbb{ABC}(\delta)\Gamma(\alpha)} \int_a^{t_{n+1}} (t_{n+1} - \tau)^{\delta-1} f(t, \mathcal{L}(t)) dt, $$

$$\left. \right\} \qquad (11.4.17)$$

$$A_{\delta,2} = \frac{\delta}{\mathbb{ABC}(\delta)\Gamma(\alpha)} \int_a^{t_n} (t_n - \tau)^{\delta-1} f(t, \mathcal{L}(t)) dt, $$

in which the Lagrange interpolation for $f(t, \mathcal{L}(t))$ [5] yields

$$f(t, \mathcal{L}(t)) \approx \frac{t - t_{n+1}}{t_n - t_{n+1}} f(t_n, \mathcal{L}_n)$$

$$+ \frac{t - t_{n-1}}{t_{n-1} - t_n} f(t_{n-1}, \mathcal{L}_{n-1}). \qquad (11.4.18)$$

Eqn (11.4.18) and [4] imply that eqn (11.4.17) becomes

$$
\left.
\begin{aligned}
A_{\delta,1} &= \frac{f(t_n,\mathcal{L}_n)}{\mathrm{ABC}(\delta)\Gamma(\delta)k}\left(\frac{t_{n+1}^{\delta+1}}{\delta+1} - t_{n-1}t_{n+1}^{\delta}\right) \\[6pt]
&\quad - \frac{f(t_{n-1},\mathcal{L}_{n-1})}{\mathrm{ABC}(\delta)\Gamma(\delta)k}\left(\frac{t_{n+1}^{\delta+1}}{\delta+1} - t_n t_{n+1}^{\delta}\right), \\[6pt]
A_{\delta,2} &= \frac{f(t_n,\mathcal{L}_n)}{\mathrm{ABC}(\delta)\Gamma(\delta)k}\left(\frac{t_n^{\delta+1}}{\delta+1} - t_{n-1}t_n^{\delta}\right) \\[6pt]
&\quad - \frac{f(t_{n-1},\mathcal{L}_{n-1})}{\mathrm{ABC}(\delta)\Gamma(\delta)k}\left(\frac{t_n^{\delta+1}}{\delta+1} - t_n^{\delta+1}\right).
\end{aligned}
\right\}
\tag{11.4.19}
$$

Eqn (11.4.19) implies that eqn (11.4.17) simplifies to

$$
\left.
\begin{aligned}
\mathcal{L}(t_{n+1}) &= \mathcal{L}(t_n) + \frac{1-\delta}{\mathrm{ABC}(\delta)}\left(f(t_n,\mathcal{L}_n) - f(t_{n-1},\mathcal{L}_{n-1})\right) \\[6pt]
&\quad + \frac{f(t_n,\mathcal{L}_n)k^{\delta}}{\mathrm{ABC}(\delta)\Gamma(\delta)}\left(\frac{(n+1)^{\delta+1}}{\delta+1} - (n-1)(n+1)^{\delta}\right. \\[6pt]
&\quad \left. - \frac{n^{\delta+1}}{\delta+1} + (n-1)n^{\delta}\right) - \frac{f(t_{n-1},\mathcal{L}_{n-1})k^{\delta}}{\mathrm{ABC}(\delta)\Gamma(\delta)} \\[6pt]
&\quad \left(\frac{(n+1)^{\delta+1}}{\delta+1} - n(n+1)^{\delta} - \frac{n^{\delta+1}}{\delta+1} + n^{\delta+1}\right).
\end{aligned}
\right\}
$$

Let

$$
\Psi_{1n} := \frac{(n+1)^{\delta+1}}{\delta+1} - (n-1)(n+1)^{\delta} - \frac{n^{\delta+1}}{\delta+1} + (n-1)n^{\delta}
$$

$$
\Psi_{2n} := \frac{(n+1)^{\delta+1}}{\delta+1} - n(n+1)^{\delta} - \frac{n^{\delta+1}}{\delta+1} + n^{\delta+1}.
$$

Then eqn (11.4.20) becomes

$$
\left.
\begin{aligned}
\mathcal{L}(t_{n+1}) &= \mathcal{L}(t_n) + \frac{1-\delta}{\mathrm{ABC}(\delta)}\left(f(t_n,\mathcal{L}_n) - f(t_{n-1},\mathcal{L}_{n-1})\right) \\[6pt]
&\quad + \frac{f(t_n,\mathcal{L}_n)k^{\delta}}{\mathrm{ABC}(\delta)\Gamma(\delta)}\Psi_{1n} - \frac{f(t_{n-1},\mathcal{L}_{n-1})k^{\delta}}{\mathrm{ABC}(\delta)\Gamma(\delta)}\Psi_{2n},
\end{aligned}
\right\}
$$

which implies that the numerical approximation for the dynamics in eqn (11.2.5) is given by

$$
\begin{aligned}
\mathcal{L}(t_{n+1}) &= \mathcal{L}(t_n) + \tfrac{1-\delta}{\mathbb{ABC}(\delta)}\left(f(t_n,\mathcal{L}_n) - f(t_{n-1},\mathcal{L}_{n-1})\right) \\
&\quad + \tfrac{f(t_n,\mathcal{L}_n)k^\delta}{\mathbb{ABC}(\delta)\Gamma(\delta)}\Psi_{1n} - \tfrac{f(t_{n-1},\mathcal{L}_{n-1})k^\delta}{\mathbb{ABC}(\delta)\Gamma(\delta)}\Psi_{2n},
\end{aligned}
$$

$$
\begin{aligned}
\mathcal{E}(t_{n+1}) &= \mathcal{E}(t_n) + \tfrac{1-\delta}{\mathbb{ABC}(\delta)}\left(f(t_n,\mathcal{E}_n) - f(t_{n-1},\mathcal{E}_{n-1})\right) \\
&\quad + \tfrac{f(t_n,\mathcal{E}_n)k^\delta}{\mathbb{ABC}(\delta)\Gamma(\delta)}\Psi_{1n} - \tfrac{f(t_{n-1},\mathcal{E}_{n-1})k^\delta}{\mathbb{ABC}(\delta)\Gamma(\delta)}\Psi_{2n},
\end{aligned}
$$

$$
\begin{aligned}
\mathcal{S}(t_{n+1}) &= \mathcal{S}(t_n) + \tfrac{1-\delta}{\mathbb{ABC}(\delta)}\left(f(t_n,\mathcal{S}_n) - f(t_{n-1},\mathcal{S}_{n-1})\right) \\
&\quad + \tfrac{f(t_n,\mathcal{S}_n)k^\delta}{\mathbb{ABC}(\delta)\Gamma(\delta)}\Psi_{1n} - \tfrac{f(t_{n-1},\mathcal{S}_{n-1})k^\delta}{\mathbb{ABC}(\delta)\Gamma(\delta)}\Psi_{2n},
\end{aligned}
$$

$$
\begin{aligned}
\mathcal{T}_r(t_{n+1}) &= \mathcal{T}_r(t_n) + \tfrac{1-\delta}{\mathbb{ABC}(\delta)}\left(f(t_n,\mathcal{T}_{rn}) - f(t_{n-1},\mathcal{T}_{rn-1})\right) \\
&\quad + \tfrac{f(t_n,\mathcal{T}_{rn})k^\delta}{\mathbb{ABC}(\delta)\Gamma(\delta)}\Psi_{1n} - \tfrac{f(t_{n-1},\mathcal{T}_{rn-1})k^\delta}{\mathbb{ABC}(\delta)\Gamma(\delta)}\Psi_{2n},
\end{aligned}
$$

$$
\begin{aligned}
\mathcal{C}(t_{n+1}) &= \mathcal{C}(t_n) + \tfrac{1-\delta}{\mathbb{ABC}(\delta)}\left(f(t_n,\mathcal{C}_n) - f(t_{n-1},\mathcal{C}_{n-1})\right) \\
&\quad + \tfrac{f(t_n,\mathcal{C}_n)k^\delta}{\mathbb{ABC}(\delta)\Gamma(\delta)}\Psi_{1n} - \tfrac{f(t_{n-1},\mathcal{C}_{n-1})k^\delta}{\mathbb{ABC}(\delta)\Gamma(\delta)}\Psi_{2n},
\end{aligned}
$$

$$
\begin{aligned}
\mathcal{U}(t_{n+1}) &= \mathcal{U}(t_n) + \tfrac{1-\delta}{\mathbb{ABC}(\delta)}\left(f(t_n,\mathcal{U}_n) - f(t_{n-1},\mathcal{U}_{n-1})\right) \\
&\quad + \tfrac{f(t_n,\mathcal{U}_n)k^\delta}{\mathbb{ABC}(\delta)\Gamma(\delta)}\Psi_{1n} - \tfrac{f(t_{n-1},\mathcal{U}_{n-1})k^\delta}{\mathbb{ABC}(\delta)\Gamma(\delta)}\Psi_{2n},
\end{aligned}
$$

$$\tag{11.4.20}$$

for each $n = 0, 1, 2, \ldots, N$, and with initial conditions as in (11.2.2), where,

$$
\begin{aligned}
f(t_n,\mathcal{E}_n) &= \omega\mathcal{E}_n + \nu\mathcal{E}_n\mathcal{T}_{rn} + \rho\mathcal{U}_n\mathcal{E}_n - \varpi\mathcal{S}_n - \varrho\mathcal{T}_{rn} - \eta_\mathcal{E}\mathcal{E}_n - \Upsilon\mathcal{E}_n, \\
f(t_n,\mathcal{S}_n) &= \kappa\mathcal{S}_n + \varpi\mathcal{S}_n - \sigma\mathcal{C}_n - \eta_s\mathcal{S}_n - \Upsilon\mathcal{S}_n, \\
f(t_n,\mathcal{T}_{rn}) &= \beta\mathcal{T}_{rn} + \varrho\mathcal{T}_{rn} + \gamma\mathcal{U}_n\mathcal{T}_{rn} - \nu\mathcal{E}_n\mathcal{T}_{rn} - \eta_{\mathcal{T}_r}\mathcal{T}_{rn} - \Upsilon\mathcal{T}_{rn}, \\
f(t_n,\mathcal{C}_n) &= \sigma\mathcal{C}_n - \eta_c\mathcal{C}_n - \Upsilon\mathcal{C}_n, \\
f(t_n,\mathcal{U}_n) &= \eta_\mathcal{L}\mathcal{L}_n + \eta_\mathcal{E}\mathcal{E}_n + \eta_s\mathcal{S}_n + \eta_{\mathcal{T}_r}\mathcal{T}_{rn} \\
&\quad + \eta_c\mathcal{C}_n - \varsigma\mathcal{U}_n\mathcal{L}_n - \rho\mathcal{U}_n\mathcal{E}_n - \gamma\mathcal{U}_n\mathcal{T}_{rn} - \Upsilon\mathcal{U}_n,
\end{aligned}
$$

for each n. We refer to the numerical scheme in eqn (11.4.20) the two-step Adam-Bashford fractional numerical method [5] for the dynamics presented in eqn (11.2.5).

Remark 11.4.1. When $\delta \equiv 1$, then eqn (11.4.20) becomes a classical Adam-Bashford numerical method [5], which yields the solution of the dynamics reported in equation (11.2.4).

Theorem 11.4.1. The solution for the dynamics in eqn (11.2.5) and Theorem 11.3.1 imply that the numerical solution for the dynamics presented in (11.2.5) is

$$
\left.
\begin{aligned}
\mathcal{L}(t_{n+1}) &= \mathcal{L}(t_n) + f(t_n, \mathcal{L}_n)\Psi_n + f(t_{n-1}, \mathcal{L}_{n-1})\Psi_{n-1} + R_{\mathcal{L}\delta}, \\[4pt]
\mathcal{E}(t_{n+1}) &= \mathcal{E}(t_n) + f(t_n, \mathcal{E}_n)\Psi_n + f(t_{n-1}, \mathcal{E}_{n-1})\Psi_{n-1} + R_{\mathcal{E}\delta}, \\[4pt]
\mathcal{S}(t_{n+1}) &= \mathcal{S}(t_n) + f(t_n, \mathcal{S}_n)\Psi_n + f(t_{n-1}, \mathcal{S}_{n-1})\Psi_{n-1} + R_{\mathcal{S}\delta}, \\[4pt]
\mathcal{T}_r(t_{n+1}) &= \mathcal{T}_r(t_n) + f(t_n, \mathcal{T}_{rn})\Psi_n + f(t_{n-1}, \mathcal{T}_{rn-1})\Psi_{n-1} + R_{\mathcal{T}_r\delta}, \\[4pt]
\mathcal{C}(t_{n+1}) &= \mathcal{C}(t_n) + f(t_n, \mathcal{C}_n)\Psi_n + f(t_{n-1}, \mathcal{C}_{n-1})\Psi_{n-1} + R_{\mathcal{C}\delta}, \\[4pt]
\mathcal{U}(t_{n+1}) &= \mathcal{U}(t_n) + f(t_n, \mathcal{U}_n)\Psi_n + f(t_{n-1}, \mathcal{U}_{n-1})\Psi_{n-1} + R_{\mathcal{U}\delta},
\end{aligned}
\right\}
$$

where

$$
\begin{aligned}
R_{\mathcal{L}\delta} &= \frac{\delta}{\mathrm{AB}(\delta)\Gamma(\delta)}\left(\int_a^{t_{n+1}}(\tau^2 - (t_n + t_{n-1})\tau + t_{n-1}t_n)(t_{n+1} - \tau)^{\tau-1}\left.\frac{\partial^2 f(t_n,\mathcal{L})}{\partial x^2}(\tau, y(\tau))\right|_{\tau=\varsigma} d\tau\right) \\
&\quad - \frac{\delta}{\mathrm{AB}(\delta)\Gamma(\delta)}\left(\int_a^{t}(\tau^2 - (t_n + t_{n-1})\tau + t_{n-1}t_n)(t_n - \tau)^{\delta-1}\left.\frac{\partial^2 f(t_n,\mathcal{L})}{\partial x^2}(\tau, y(\tau))\right|_{\tau=\varsigma} d\tau\right), \\[6pt]
R_{\mathcal{E}\delta} &= \frac{\delta}{\mathrm{AB}(\delta)\Gamma(\delta)}\left(\int_a^{t_{n+1}}(\tau^2 - (t_n + t_{n-1})\tau + t_{n-1}t_n)(t_{n+1} - \tau)^{\tau-1}\left.\frac{\partial^2 f(t_n,\mathcal{E})}{\partial x^2}(\tau, y(\tau))\right|_{\tau=\varsigma} d\tau\right) \\
&\quad - \frac{\delta}{\mathrm{AB}(\delta)\Gamma(\delta)}\left(\int_a^{t}(\tau^2 - (t_n + t_{n-1})\tau + t_{n-1}t_n)(t_n - \tau)^{\delta-1}\left.\frac{\partial^2 f(t_n,\mathcal{E})}{\partial x^2}(\tau, y(\tau))\right|_{\tau=\varsigma} d\tau\right), \\[6pt]
R_{\mathcal{S}\delta} &= \frac{\delta}{\mathrm{AB}(\delta)\Gamma(\delta)}\left(\int_a^{t_{n+1}}(\tau^2 - (t_n + t_{n-1})\tau + t_{n-1}t_n)(t_{n+1} - \tau)^{\tau-1}\left.\frac{\partial^2 f(t_n,\mathcal{S})}{\partial x^2}(\tau, y(\tau))\right|_{\tau=\varsigma} d\tau\right) \\
&\quad - \frac{\delta}{\mathrm{AB}(\delta)\Gamma(\delta)}\left(\int_a^{t}(\tau^2 - (t_n + t_{n-1})\tau + t_{n-1}t_n)(t_n - \tau)^{\delta-1}\left.\frac{\partial^2 f(t_n,\mathcal{S})}{\partial x^2}(\tau, y(\tau))\right|_{\tau=\varsigma} d\tau\right), \\[6pt]
R_{\mathcal{T}_r\delta} &= \frac{\delta}{\mathrm{AB}(\delta)\Gamma(\delta)}\left(\int_a^{t_{n+1}}(\tau^2 - (t_n + t_{n-1})\tau + t_{n-1}t_n)(t_{n+1} - \tau)^{\tau-1}\left.\frac{\partial^2 f(t_n,\mathcal{T}_r)}{\partial x^2}(\tau, y(\tau))\right|_{\tau=\varsigma} d\tau\right) \\
&\quad - \frac{\delta}{\mathrm{AB}(\delta)\Gamma(\delta)}\left(\int_a^{t}(\tau^2 - (t_n + t_{n-1})\tau + t_{n-1}t_n)(t_n - \tau)^{\delta-1}\left.\frac{\partial^2 f(t_n,\mathcal{T}_r)}{\partial x^2}(\tau, y(\tau))\right|_{\tau=\varsigma} d\tau\right), \\[6pt]
R_{\mathcal{C}\delta} &= \frac{\delta}{\mathrm{AB}(\delta)\Gamma(\delta)}\left(\int_a^{t_{n+1}}(\tau^2 - (t_n + t_{n-1})\tau + t_{n-1}t_n)(t_{n+1} - \tau)^{\tau-1}\left.\frac{\partial^2 f(t_n,\mathcal{C})}{\partial x^2}(\tau, y(\tau))\right|_{\tau=\varsigma} d\tau\right) \\
&\quad - \frac{\delta}{\mathrm{AB}(\delta)\Gamma(\delta)}\left(\int_a^{t}(\tau^2 - (t_n + t_{n-1})\tau + t_{n-1}t_n)(t_n - \tau)^{\delta-1}\left.\frac{\partial^2 f(t_n,\mathcal{C})}{\partial x^2}(\tau, y(\tau))\right|_{\tau=\varsigma} d\tau\right), \\[6pt]
R_{\mathcal{U}\delta} &= \frac{\delta}{\mathrm{AB}(\delta)\Gamma(\delta)}\left(\int_a^{t_{n+1}}(\tau^2 - (t_n + t_{n-1})\tau + t_{n-1}t_n)(t_{n+1} - \tau)^{\tau-1}\left.\frac{\partial^2 f(t_n,\mathcal{U})}{\partial x^2}(\tau, y(\tau))\right|_{\tau=\varsigma} d\tau\right) \\
&\quad - \frac{\delta}{\mathrm{AB}(\delta)\Gamma(\delta)}\left(\int_a^{t}(\tau^2 - (t_n + t_{n-1})\tau + t_{n-1}t_n)(t_n - \tau)^{\delta-1}\left.\frac{\partial^2 f(t_n,\mathcal{U})}{\partial x^2}(\tau, y(\tau))\right|_{\tau=\varsigma} d\tau\right),
\end{aligned}
$$

and

$$
\left.\begin{aligned}
|R_{\mathcal{L}\delta}| &< \infty, \\
|R_{\mathcal{E}\delta}| &< \infty, \\
|R_{\mathcal{S}\delta}| &< \infty, \\
|R_{\mathcal{T}_r\delta}| &< \infty, \\
|R_{\mathcal{C}\delta}| &< \infty, \\
|R_{\mathcal{U}\delta}| &< \infty.
\end{aligned}\right\}
$$

Proof. The proof of Theorem 11.4.1 has been already established in [4]. □

Lemma 11.4.1. The two-step Adam–Bashford fractional numerical method in eqn (11.4.20) is stable if

$$
\left.\begin{aligned}
\|f(t_n, \mathcal{L}_n) - f(t_{n-1}, \mathcal{L}_{n-1})\|_\infty &\to 0, \\
\|f(t_n, \mathcal{E}_n) - f(t_{n-1}, \mathcal{E}_{n-1})\|_\infty &\to 0, \\
\|f(t_n, \mathcal{S}_n) - f(t_{n-1}, \mathcal{S}_{n-1})\|_\infty &\to 0, \\
\|f(t_n, \mathcal{T}_{rn}) - f(t_{n-1}, \mathcal{T}_{rn-1})\|_\infty &\to 0, \\
\|f(t_n, \mathcal{C}_n) - f(t_{n-1}, \mathcal{C}_{n-1})\|_\infty &\to 0, \\
\|f(t_n, \mathcal{U}_n) - f(t_{n-1}, \mathcal{U}_{n-1})\|_\infty &\to 0,
\end{aligned}\right\}
$$

as $n \to \infty$.

Proof. The results of Theorem 11.3.1 and the proof in [4] conclude the proof. □

11.5 Numerical Results and Discussion

Let $a = 0$, $b = 1$ and $N = 100$. Then the numerical scheme in eqn (11.4.20) is implemented in MATLAB R2021a. Thus, for the parameter values under the impact of COVID-19 ($\varpi \equiv 0$, and $\rho \equiv 0$), are presented in Table 11.1, whereas the parameter values without the impact of COVID-19 are presented in Table 11.2. Based on the unwelcome record listed at [29], let the initial conditions for equation in (11.4.20) be $\mathcal{L}_a = 0.6$, $\mathcal{E}_a = 0.08$, $\mathcal{S}_a = 0.2$, $\mathcal{T}_{ra} = 0.02$, $\mathcal{C}_a = 0.07$, $\mathcal{U}_a = 0.03$. To illustrate the main aim of this chapter, it suffices to present two cases emanating from the derived dynamics in eqn (11.2.1) through to the extended dynamics in eqn (11.2.5). These cases are assumed to be the plight of the youths under the impact of natural disaster such as the outbreak of COVID-19 and plight of the youths when the impact of natural disaster such as COVID-19 is non-existence. Therefore, going forward respectively, the two cases are presented in Figures 11.2 and Figure 11.3, respectively.

In Figure 11.2, the plight of the youths is as follows. The number of learners rises well above 63%, (see Figure 11.2 (a)), whereas, those opting for employment before tertiary education (see Figure 11.2 (b)), the chances of getting a job is very slim, because the employment decreases roughly

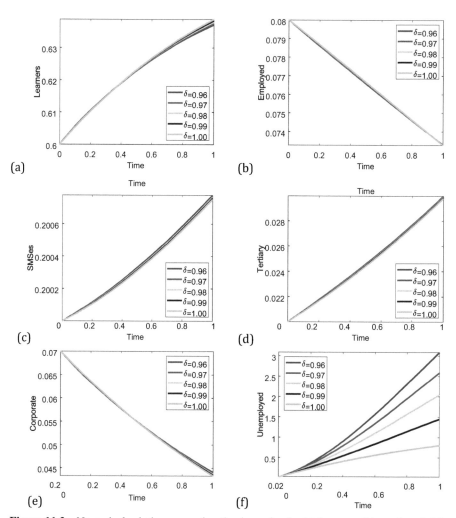

Figure 11.2   Numerical solution presenting the dynamics for (a) learners, (b) employed, (c) SMEs, (d) tertiary, (e) corporate and (f) unemployed youths under the impact of COVID-19.

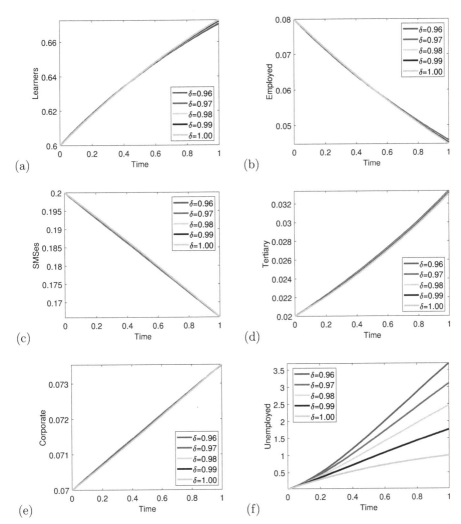

Figure 11.3 Numerical solution presenting the dynamics for (a) learners, (b) employed, (c) SMEs, (d) tertiary, (e) corporate, and (f) unemployed youths, with no COVID-19.

to 0.02%, before it reaches its saturation point [1], which in turns implies persistence of higher unemployment rates (see Figure 11.2 (f)). However, the youths opting for transforming their SMEs into successful corporate, the situation looks bleak (see Figure 11.2 (c)), as there is no growth for any corporate due to the impact of COVID-19 (see Figure 11.2 (c) in conjunction with Figure 11.2 (e)). However, those opting for tertiary education before

Table 11.1   Parameter values for equation in (11.4.20) for Figure 11.2.

$\Upsilon = 0.5$	$\varsigma = 0.4$	$\omega = 0.4$	$\kappa = 0.51$	$\beta = 0.9$	$\eta_{\mathcal{L}} = 0.0007$
$\nu = 0.01$	$\rho = 0.6$	$\varpi = 0.0$	$\varrho = 0.0$	$\eta_{\mathcal{E}} = 0.0007$	$\sigma = 0.02$
$\eta_{\mathcal{S}} = 0.0007$	$\gamma = 0.02$	$\eta_{T_r} = 0.0007$	$\eta_{\mathcal{C}} = 0.0007$	$\Gamma(\delta) = 1.00$	

Table 11.2   Parameter values in equation (11.4.20) for Figure 11.3.

$\Upsilon = 0.5$	$\varsigma = 0.9$	$\omega = 0.0094$	$\kappa = 0.51$	$\beta = 0.9$	$\eta_{\mathcal{L}} = 0.0007$
$\nu = 0.005$	$\rho = 0.9$	$\varpi = 0.02$	$\varrho = 0.1$	$\eta_{\mathcal{E}} = 0.0007$	$\sigma = 0.55$
$\eta_{\mathcal{S}} = 0.0007$	$\gamma = 0.2$	$\eta_{T_r} = 0.0007$	$\eta_{\mathcal{C}} = 0.0007$	$\Gamma(\delta) = 1.00$	

employment, their number is expected to rise well above 0.028% of the intake (see Figure 11.2 (d)). The chances for a youth to transform their SMEs look bleak as no growth at all for the transformed corporate, see Figure 11.2 (e) in conjunction with Figure 11.2 (c).

In Figure 11.3, the plight of the youths as follows. The number of learners rises well above 66% (see Figure 11.3 (a)), whereas, those opting for employment before tertiary education (see Figure 11.3 (b)), the chances of getting a job has not improved at all, because the employment opportunities are becoming fewer and fewer, as reported in [1]. This implies persistence of higher unemployment rates as it can be seen in Figure 11.3 (f) that it has increased well above 3.5%. Figure 11.3(c) presents that, the youths opting for establishing their SMEs are doing well, because their entities are transforming into corporate at an exponential rate, see Figure 11.3 (c) in conjunction Figure 11.3 (e). However, despite the good transformation of SMEs into corporate entities, persistence of higher unemployment rates remains the main challenge see Figure 11.3 (f).

11.6 Conclusion

In this chapter, the dynamics of unemployed youths particularly the graduates are proposed, presented in terms of compartments, as well as a systems of nonlinear ordinary differential equations (ODEs). The dynamics are extended to fractional derivative of Atangana–Baleanu fractional derivatives, for further analysis. Thus, the solution to the proposed dynamics exists, unique, positive, and bounded. Consequently, the asymptotic stability conditions are established as well. This enabled the author to present the desirable two cases. Since the exact solution of the dynamics is not available, a numerical method is derived, analysed and implemented to present the proposed dynamics for the unemployed youths and/or LETSCU. Therefore, the Atangana–Baleanu fractional operator enabled to present various suitable solution for $\delta \in (0, 1]$. Thus, the solution is presented for two cases, to present the main aim of the chapter. The first case (see Figure 11.2) aimed at presenting the fact that indeed the global economy has suffered detrimentally due to the impact COVID-19 pandemic, particularly in the low and middle-income countries. Hence, leading to higher number of unemployed youths across many developing and/or low- and middle-income countries. The second case (see Figure 11.3) is aimed at presenting the fact that even in the absence of natural disaster such as COVID-19 pandemic, the plight of the youths remains at a receiving end, contrary to the current call for a young graduate to become an employer rather than an employee. Therefore, in this chapter, the focus is indeed to portray the impact of COVID-19 on the plight of many youths, as well as to dispose the perception that the existing system currently in place, is not opting well for the graduates. However, contrary to that, as it can be seen in both scenarios of the proposed dynamics that, the persistence of unemployment rates continues to rise, irrespective of the situation on the ground, simply because the existing system in place, calls for an urgent redesigned and/or improved of such system as depicted in Figure 11.1. Therefore, our future research direction is to propose an improved system, through mathematical formulations. Thus, the author believe that the newly designed system will indeed implore more youths engagements with the aim of enabling most of the graduates to become employers rather than employees. In doing so, the author believe this will drastically reduce unemployment rates among youths, which in turn will increase revenues. Increased revenues imply better living conditions for the locals, including women and children. Hence, help accelerate sustainable developments and corporations among all developing and developed nations as urged by the United Nations' sustainable development goals (SDGs) [30].

Declaration

Availability of Data and Materials

Data sharing not applicable to this article as no datasets were generated or analysed during the current study.

Funding

Not funded.

Competing interests

The author declare that they have no competing interests.

Acknowledgement

The author would like to thank the University of Namibia.

References

[1] S. Al-Sheikha, R. Al-Maalwia and H. A. Ashia, Mathematical model of unemployment with the effect of Limited Jobs, *Comptes Rendus Mathématique* **359(3)** 283-290, 2021.

[2] A. Atangana, and D. Baleanu, New fractional derivatives with non-local and non-singular kernel: theory and application to heat transfer model, *Thermal Science* **20(2)** 763-785, 2016.

[3] A. Atangana, and Z. Hammouch, Fractional calculus with power law: The cradle of our ancestors, *European Physical Journa Plus* **134** 429, 2019.

[4] A. Atangana and K. M. Owolabi, Corrigendum: New numerical approach for fractional differential equations, *Mathematical Modelling of Natural Phenomena* **16** 47, 2021.

[5] R. L. Burden and J. D. Faires, *Numerical Analysis*, Brooks/Cole, USA, 2011.

[6] R. Diab-Bahman, and A. Al-Enzi, The impact of COVID-19 pandemic on conventional work settings, *International Journal of Sociology and Social Policy* **(40)9/10** 909-927, 2020.

[7] A. Galindro and D. F. M. Torres, A simple mathematical model for unemployment: a case study in Portugal with optimal control, *Statistics, Optimization & Information Computing* **6(1)** 116-129, 2018.

[8] L. Harding and M. Neamtu, A dynamic model of unemployment with migration and delayed policy intervention, *Computational Economics* DOI: 10.1007/s10614-016-9610-3, 1990.

[9] I. Jaradat, M. Alquran, T. A. Sulaiman and A. Yusuf, Analytic simulation of the synergy of spatial-temporal memory indices with proportional time delay, *Chaos, Solitons & Fractals* **156** 111818, 2022.

[10] J. Kevorkian, *Partial differential equation and analytically techniques*, Wadsworth & Brooks/Cole, Mathematics Series, Springer, USA, 1990.

[11] J-H Kim, W. Su and Y. J. Song, On stability of a polynomial, *Journal of Applied Mathematics & Informatics* **36(3-4)** 231-236, 2018.

[12] J. Kongson, W. Sudsutad and C. Thaiprayoon, On analysis of a nonlinear fractional system for social media addiction involving Atangana–Baleanu–Caputo derivative, *Advances in Difference Equations* **2021** 356, 2021.

[13] G. Lozada-Cruz, Some variants of Cauchy's mean value theorem, *International Journal of Mathematical Education in Science and Technology* **51(7)** 1155-1163, 2020.

[14] A. K. Misra and A. K. Singh, A mathematical model for unemployment, *Nonlinear Analysis Real World Applications* **12(1)** 128-136, 2011.

[15] A. K. Misra and A. K. Singh, A delay mathematical model for the control of unemployment, *Differential Equations and Dynamical Systems* **21(3)** 291-307, 2013.

[16] S. B. Munoli and S. Gani, Optimal control analysis of a mathematical model for unemployment, *Optimal Control Applications and Methods* **37(4)** 798-806, 2016.

[17] C. V. Nikolopoulos and D. E. Tzanetis, A model for housing allocation of a homeless population due to a natural disaster, *Nonlinear Analysis Real World Applications* **4(4)** 561-579, 2003.

[18] S. M. Nuugulu, A. Shikongo, D. Elago, A. T. Salom and K. M. Owolabi, Fractional SEIR model for modelling the spread of COVID-19 in Namibia, *Mathematical Analysis for Transmission of COVID-19* Springer, Singapore, 2021.

[19] Z. M. Odibat and N. T. Shawagfeh, Generalized Taylor's formula, *Applied Mathematics and Computation* **186(1)** 286-293, 2007.

[20] E. C. de Oliveira and J. A. T. Machado, A Review of definitions for fractional derivatives and integral, *Mathematical Problems in Engineering* **2014** 6, 2014.

[21] G. Pathan and P. H. Bhathawala, Mathematical model for unemployment-taking an action without delay, *Advances in Dynamical Systems and Applications*, **12(1)** 41-48, 2017.

[22] T. Petaratip and P. Niamsup, Stability analysis of an unemployment model with time delay, *AIMS Mathematics* **6(7)** 7421-7440, 2021.

[23] Albert Shikongo, *Robust numerical methods to solve differential equations arising in cancer modeling*, University of the Western Cape, http://hdl.handle.net/11394/7250, 2020.

[24] A. Shikongo, S. M. Nuugulu, D. Elago, A. T. Salom and K. M. Owolabi, Fractional derivative operator on quarantine and isolation

principle for COVID-19, chapter in *Advanced Numerical Methods for Differential Equations*, CRC Press, Boca Raton, 2021.

[25] N. Sırghi, M. Neamtu and D. S. Deac, A dynamic model for unemployment control with distributed delay, *Mathematical Methods in Finance and Business Administration* 42-48, 2014.

[26] T. A. Sulaiman, M. Yavuz, H. Bulut and H. M. Baskonus, Investigation of the fractional coupled viscous Burgers' equation involving Mittag-Leffler kernel, *Physica A: Statistical Mechanics and its Applications* **527** 121126, 2019.

[27] M. Yavuz, T. A. Sulaiman, A. Yusuf and T. Abdeljawad, The Schrödinger-KdV equation of fractional order with Mittag-Leffler non-singular kernel, *Alexandria Engineering Journal* **60(2)** 2715-2724, 2021.

[28] A. Yusuf, U. T. Mustapha, T. A. Sulaiman, E. Hincal and M. Bayram, Modeling the effect of horizontal and vertical transmissions of HIV infection with Caputo fractional derivative, *Chaos, Solitons & Fractals* **145** 110794, 2021.

[29] https://africa.businessinsider.com/local/markets/south-africa-namibia-nigeria-have-the-highest-unemployment-rates-in-the-world-report/6zh1b63.

[30] https://sdgs.un.org/goals.

12

A Fractional-order SVIR Model with Two Infection Classes for COVID-19 in India

Nita H. Shah, Kapil Chaudhary, and Ekta Jayswal

Department of Mathematics, Gujarat University, India

Abstract

This chapter describes the fractional-order transmission dynamics of the two variants of SARS-CoV-2, for delta and omicron variants are chosen. An *SVIR*-model is proposed with two infection classes corresponding to the delta and omicron variants. The equilibrium points of the model are determined, and the next-generation matrix method is utilized to compute the corresponding basic reproduction number (R_0). The local stability conditions of the proposed model are investigated around the equilibrium points using the eigenvalue method. The result shows the model is locally stable if $R_0 < 1$ and unstable otherwise. The numerical L1 scheme is used to study the memory effect of the virus variants in the suggested model. Further, numerical simulations are provided for a better insight into the proposed model, and relevant findings are displayed graphically. The memory effect graph shows that as memory increases, there is a relative decline in infection cases of each virus variant. The result also shows early recovery in the case of the latter variant, and the intensity of the recovery relatively increases as the number of cases increases.

Keywords: Caputo Fractional-Order Derivative, COVID-19, Omicron, Next-Generation Matrix Method, L1 Scheme, Memory Effect.

MSC: 37Nxx, 37M05, 26A33.

12.1 Introduction

Regularly throughout history, infectious diseases with pandemic potential have evolved and spread. In the past, the world faced severe infectious diseases, like Spanish flu, smallpox, tuberculosis and many more. Recently, the world has been facing a novel coronavirus disease known as COVID-19. The disease appeared first in Wuhan city of China and subsequently elsewhere. The causing virus SARS-CoV-2, like other viruses, changes its behaviour over time due to mutations and a combination of mutations. World Health Organization named this new variant B.1.1.529 of virus SARS-CoV-2 omicron and referred to it as a variant of concern (VOC) [3]. The first case of this variant was reported in the last week of November 2021 in South Africa [3]. Dr. Angelique Coetzee described the variant infection symptoms as mild compared to delta variants. Further, the first case of the omicron variant in India was reported to the union health ministry of India from Karnataka state on 2 December 2021.

One of the essential things in epidemic research is predicting future patterns, such as how many people will be infected every day, when epidemics turn endemic and so on. Further, in the particular case of COVID-19, the additional questions about the omicron or any new variant are as follows [2]:

- To what extent do the omicron or any new variant bypass vaccination-induced immunity and natural immunity induced from recovery?
- Several questions related to which of the two variants are more infectious and severe.

Background: In the literature, various dynamical methods are used to forecast epidemic patterns and to make government policies to limit the spread. The basic compartmental SIR model has been widely used to describe the disease pattern. From then, numerous integer-order models were obtained by varying basic SIR models [20]. However, coronavirus, like all viruses, evolves throughout time. Most alterations have little or no impact on the virus's properties [21]. The system of fractional-order equations better represents this kind of natural phenomenon due to its hereditary properties. Further, the fractional order improves the consistency of the model with actual data and observations as it has a degree of freedom to fit the actual data compared to integer-order models [11, 17].

Literature Survey: Many scholars from various fields have researched the prediction, study and creation of crucial epidemic-fighting strategies. Grassly and Fraser discussed the importance of developing models that can capture key features of the spread of infection [10]. Yang et al. [20] discussed many integer-order models obtained by varying basic SIR model. It is observed that epidemic models based on fractional-order derivatives are a very effective way to study the dynamics of diseases like COVID-19 as fractional-order derivatives are influenced by past events and the history of the topic under study the immediate surroundings [5]. Koziol et al. [12] presented the generalized basic compartmental SIR epidemic model with fractional-order derivatives for predicting the spread of the COVID-19 disease. The time-domain model implementation was based on the fixed-step method using the nabla fractional-order difference defined by the Grünwald–Letnikov formula. Özköse et al. [21] proposed a fractional-order model of the COVID-19 omicron variant containing the heart attack effect with real data from the United Kingdom. Niak [14] investigated a non-linear fractional-order SIR epidemic model and applied the L1 scheme in a fractional-order disease model. Teka et al. [19] suggested a fractional-order model based on spiking activities of transmitting information in the brain through neurons and used the L1 scheme for its memory trace effects emerging from the past activities of neuro. Otunuga [15] in his chapter proposed an epidemic SEIRS model with vital dynamics for COVID-19 and estimated its epidemiological parameters. Shah et al. [18] applied the optimal control theory in the COVID-19 fractional-order model to pretend the impact of various intervention strategies.

This chapter is arranged in the following manner. Section 12.2 introduces preliminary definitions and results to be used in this chapter. In Section 12.3, a model is formulated, and its equilibrium points are calculated. In Section 12.4, the next-generation matrix method is utilized to compute the basic reproduction number. Local stability around the equilibrium points is discussed in Section 12.5. Memory trace and heredity trait are discussed in Section 12.6, and numerical simulation is done in Section 12.7.

12.2 Preliminaries

Definition 1. The fractional-order ϕ-derivative of the function f in Caputo sense on the interval $(0, t)$ is defined by

$$
{}_0^C D_t^\phi f(t) = \frac{1}{\Gamma(n-\phi)} \int_0^t (t-s)^{n-\phi-1} f^{(n)}(s) ds,
$$

where $n = \lceil \phi \rceil$ is the least integer greater than or equal to ϕ and $\Gamma(x)$ denotes the gamma function.

As $\phi \in (0,1]$, the fractional-order ϕ-derivative of function f in Caputo's sense can be written as:

$$\begin{smallmatrix} C \\ 0 \end{smallmatrix} D_t^\phi f(t) = \frac{1}{\Gamma(1-\phi)} \int_0^t \frac{f'(s)}{(t-s)^\phi} ds.$$

Definition 2. The Laplace transform of the function f is defined as

$$\mathcal{L}\{f(t)\}(s) = F(s) = \int_0^\infty f(t)e^{-st} dt.$$

The Laplace transform of the Caputo fractional ϕ-derivative of the function f is defined as:

$$\mathcal{L}\{\begin{smallmatrix} C \\ 0 \end{smallmatrix} D_t^\phi f(t)\}(s) = s^\phi L(f(t))(s) - \sum_{k=0}^{n-1} s^{\phi-k-1} f^{(k)}(0).$$

Definition 3. The Mittag–Leffler function $E_{\phi,\eta}$ is a complex function depending upon two parameters ϕ and η is defined as,

$$E_{\phi,\eta}(z) = \sum_{k=0}^\infty \frac{z^k}{\Gamma(\phi k + \eta)}.$$

For the particular case of $\eta = 1$, Mittag–Leffler function E_ϕ in one parameter ϕ can be described as

$$E_\phi(z) = E_{\phi,1}(z) = \sum_{k=0}^\infty \frac{z^k}{\Gamma(\phi k + 1)}. \tag{12.1}$$

Some well-known results based on the Mittag–Leffler function and its Laplace transform are given below, which will be used further in this chapter.

$$\mathcal{L}\{t^{\eta-1} E_{\phi,\eta}(-\lambda t^\phi)\}(s) = \frac{s^{\phi-\eta}}{s^\phi + \lambda}, \tag{12.2}$$

where $\mathrm{res}(s) > |\lambda|^{1/\phi}$ and

$$\mathcal{L}^{-1}\left\{\frac{1}{s(s^\phi + \lambda)}\right\}(t) = t^\phi E_{\phi,\phi+1}(-\lambda t^\phi)$$

(12.3)

$$= \frac{1}{\lambda}\left[1 - E_\phi(-\lambda t^\phi)\right],$$

where $\mathrm{res}(s) > |\lambda|^{1/\phi}$.

12.3 Formulation of Model

In this epidemic model, the total population denoted by $n(t)$ is divided into five compartments s, v, i_0, i_δ and r, where each compartment denotes the specific stage of the disease transmission. In this model, i_0 and i_δ are two classes representing the omicron and delta variants of viral COVID-19 infection, respectively.

To understand the system and to reduce the number of variables, the new variables S, V, I_0, I_δ and R are introduced by dividing each compartment respectively by total population $n(t)$ (see Table 12.1 for respective meaning). The flow diagram of the model is given in Figure 12.1. Table 12.2 provides the value and significance of the parameters used in this model.

Assumptions for the proposed model:

(a) The infected population is homogeneously mixed with susceptible people.

Table 12.1   Compartments used in the proposed model.

Notation	Meaning
S	Fraction of individuals who are vulnerable to getting the infection and are not vaccinated.
V	Fraction of individuals who are vulnerable to getting the infection and are vaccinated.
I_0	Fraction of individuals responsible for spreading the omicron variant among the population.
I_δ	Fraction of individuals responsible for spreading the delta variant among the population.
R	Fraction of individuals who have recovered.

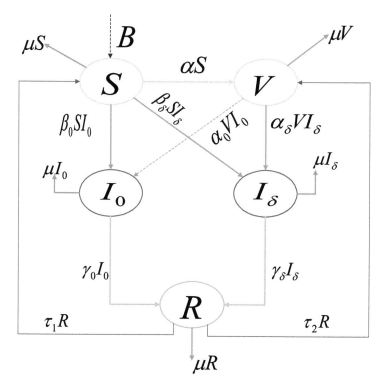

Figure 12.1 The flow diagram of the proposed model.

(b) The recovered individuals may have temporary immunity but are vulnerable to catching the infection again.

(c) The vaccinated individuals are prone to both disease variants.

(d) The newly infected people are immediately contagious.

The governing system of the fractional-order non-linear differential equations that describe the proposed epidemic model is as follows:

$$
\begin{aligned}
{}_0^C D_t^\phi S &= B - \beta_0 S I_0 - \beta_\delta S I_\delta - (\mu + \alpha)S + \tau_1 R, \\
{}_0^C D_t^\phi V &= \alpha S - \mu V - \alpha_0 V I_0 - \alpha_\delta V I_\delta + \tau_2 R, \\
{}_0^C D_t^\phi I_0 &= \beta_0 S I_0 + \alpha_0 V I_0 - (\mu + \mu_0 + \gamma_0)I_0, \\
{}_0^C D_t^\phi I_\delta &= \beta_\delta S I_\delta + \alpha_\delta V I_\delta - (\mu + \mu_\delta + \gamma_\delta)I_\delta, \\
{}_0^C D_t^\phi R &= \gamma_0 I_0 + \gamma_\delta I_\delta - (\mu + \tau_1 + \tau_2)R
\end{aligned}
\tag{12.4}
$$

Table 12.2 Values and meaning of parameters used in the proposed model.

Parameter	Meaning	Value	Reference
B	Birth rate of the population	0.000052705	Calculated
α	Vaccination rate	0.060925	Estimated
β_0	Transmission rate of non-vaccinated getting the omicron	0.35	Estimated
β_δ	Transmission rate of non-vaccinated getting the delta variant	0.35	Estimated
α_0	Transmission rate of vaccinated getting the omicron variant	0.25	Estimated
α_δ	Transmission rate of vaccinated getting the delta variant	0.25	Estimated
γ_0	Recovery rate from the omicron variant	0.14	Calculated
γ_δ	Recovery rate from the delta variant	0.14	Calculated
μ_0	Death rate due to the omicron variant	0.005	Calculated
μ_δ	Death rate due to the delta variant	0.0001	Calculated
μ	Death rate	0.000019135	Calculated
τ_1	Rate of recovered individuals moving to non-vaccinated susceptible	0.1	Estimated
τ_2	Rate of recovered individuals moving to vaccinated susceptible	0.9	Estimated

with non-negative initial conditions $S(0) = S_0$, $V(0) = V_0$, $I_0(0) = I_{00}$, $I_\delta(0) = I_{\delta 0}$, $R(0) = R_0$.

Note: The parameters used in the system (12.4) are non-negative.

Theorem 12.1. The feasible region of the system (12.4) is given by

$$\Omega = \left\{ (S, V, I_0, I_\delta, R) \in \mathbb{R}_+^5 \mid S + V + I_0 + I_\delta + R \leq N(0) + \frac{B}{\mu} \right\}.$$

Proof. Let us assume that $N(t) = S(t) + V(t) + I_0(t) + I_\delta(t) + R(t)$. Adding all the equations in the system (12.4),

$$_0^C D_t^\phi N(t) = B - \mu N(t) - \mu_0 I_0(t) - \mu_\delta I_\delta(t) \leq B - \mu N(t). \qquad (12.5)$$

Using Laplace transform both sides into inequality (12.5),

$$s^\phi \mathcal{L}\{N(t)\} - s^{\phi-1} N(0) \leq \frac{B}{s} - \mu \mathcal{L}\{N(t)\}.$$

On solving,

$$\mathcal{L}\{N(t)\} \leq \frac{B}{s(s^\phi + \mu)} + N(0)\frac{s^{\phi-1}}{s^\phi + \mu}.$$

Taking inverse Laplace transform, and using equations (12.1)–(12.3),

$$N(t) \le Bt^\phi E_{\phi,\phi+1}(-\mu t^\phi) + N(0)E_{\phi,1}(-\mu t^\phi)$$

$$= B/\mu(1 - E_\phi(-\mu t^\phi)) + N(0)E_\phi(-\mu t^\phi).$$

Since $0 \le E_\phi(-\mu t^\phi) \le 1$, Thus,

$$S(t) + V(t) + I_0(t) + I_\delta(t) + R(t) = N(t) \le \frac{B}{\mu} + N(0).$$

□

12.3.1 Existence and uniqueness of the solution

Lemma 12.1. (13). The fractional-order system ${}^C_0 D^\phi_t(X(t)) = F(t, X(t))$ such that $X(0) = X_0$ has unique solution if the following holds:

(i) $F(t, X(t))$ and $(\partial F/\partial X)(X)$ are continuous functions.
(ii) $\|F(X)\| \le K_1 + K_2\|X\|$, where K_1 and K_2 are positive constants.

Theorem 12.2. The solution of the fractional order system (12.4) exists and is unique.

Proof. The system (12.4) of fractional order differential equations can be written as follows:

$${}^C_0 D^\phi_t(X(t)) = F(X) = M_1 + M_2 X + SM_3 X + VM_4 X, \qquad (12.6)$$

where

$$X(t) = \begin{bmatrix} S(t) \\ V(t) \\ I_0(t) \\ I_\delta(t) \\ R(t) \end{bmatrix} \text{ and } M_1 = \begin{bmatrix} B \\ 0 \\ 0 \\ 0 \\ 0 \end{bmatrix} \text{ are column vectors}$$

and the rest M_2, M_3 and M_4 are square matrices:

$$M_2 = \begin{bmatrix} -(\mu + \alpha) & 0 & 0 & 0 & \tau_1 \\ \alpha & -\mu & 0 & 0 & \tau_2 \\ 0 & 0 & -(\mu + \mu_0 + \gamma_0) & 0 & 0 \\ 0 & 0 & 0 & -(\mu + \mu_\delta + \gamma_\delta) & 0 \\ 0 & 0 & \gamma_0 & \gamma_\delta & -(\mu + \tau_1 + \tau_2) \end{bmatrix},$$

$$M_3 = \begin{bmatrix} 0 & 0 & -\beta_0 & -\beta_\delta & 0 \\ 0 & 0 & 0 & 0 & 0 \\ 0 & 0 & \beta_0 & \beta_\delta & 0 \\ 0 & 0 & 0 & 0 & 0 \\ 0 & 0 & 0 & 0 & 0 \end{bmatrix} \text{ and } M_4 = \begin{bmatrix} 0 & 0 & 0 & 0 & 0 \\ 0 & 0 & -\alpha_0 & -\alpha_\delta & 0 \\ 0 & 0 & \alpha_0 & \alpha_\delta & 0 \\ 0 & 0 & 0 & 0 & 0 \\ 0 & 0 & 0 & 0 & 0 \end{bmatrix}.$$

Now, using the norm on both sides in equation (12.6) and the properties of norm on $C^*[0,T]$

$$\|F(X)\| = \|M_1 + M_2 X + S M_3 X + V M_4 X\|$$

$$\leq \|M_1\| + \|M_2 X\| + \|S M_3 X\| + \|V M_4 X\|$$

$$\leq \|M_1\| + (\|M_2\| + \|S M_3\| + \|V M_4\|) \|X\|$$

$$\leq K_1 + K_2 \|X\|,$$

where $K_1 = \|M_1\|$ and $K_2 = (\|M_2\| + \|S M_3\| + \|V M_4\|)$ are positive constants. Hence, using Lemma (12.1), system (12.4) has a unique solution☐

12.3.2 Equilibrium points

In this section, the equilibrium points of the system (12.4) are evaluated. The points are the steady-state solution of the system (12.4). There are two equilibrium points of the proposed system to be analyzed for the proposed model. The disease-free equilibrium point E^0 is given by:

$$E^0 = \left(S = \frac{B}{\mu + \alpha}, V = \frac{B\alpha}{\mu(\mu + \alpha)}, I_\delta = 0, I_0 = 0, R = 0 \right).$$

The endemic equilibrium point E^1 is given by

$$E^1 = (S^*, V^*, I_\delta^*, I_0^*, R^*)$$

where

$$S^* = \frac{\alpha_\delta(\mu + \mu_0 + \gamma_0) - \alpha_0(\mu + \mu_\delta + \gamma_\delta)}{\beta_0 \alpha_\delta - \alpha_0 \beta_\delta},$$

$$V^* = \frac{\beta_0(\mu + \mu_\delta + \gamma_\delta) - \beta_\delta(\gamma_0 + \mu + \mu_0)}{\beta_0 \alpha_\delta - \alpha_0 \beta_\delta}.$$

Further, $I_0^* = A_1/A_4$, $I_\delta^* = A_2/A_4$, $R^* = A_3/A_4$ where each A_1, A_2, A_3 and A_4 can be written in terms of S^*, V^* and the disease parameters as follows:

$$A_1 = -\beta_\delta\alpha(\mu + \tau_1 + \tau_2)S^{*2} + ((\mu + \tau_1 + \tau_2)((-\mu - \alpha)\alpha_\delta + \mu\beta_\delta)V^*$$

$$+ \gamma_\delta(\mu\tau_2 + \alpha(\tau_1 + \tau_2)))S^* + (B(\mu + \tau_1 + \tau_2)\alpha_\delta - \gamma_\delta\mu\tau_1)V^*$$

$$- B\tau_2\gamma_\delta,$$

$$A_2 = \alpha\beta_0(\mu + \tau_1 + \tau_2)S^{*2} + (-(\mu + \tau_1 + \tau_2)((-\mu - \alpha)\alpha_0 + \beta_0\mu)V^*$$

$$- \gamma_0(\mu\tau_2 + \alpha(\tau_1 + \tau_2)))S^* + (-B(\mu + \tau_1 + \tau_2)\alpha_0 + \gamma_0\mu\tau_1)V^*$$

$$+ B\gamma_0\tau_2,$$

$$A_3 = -\alpha(-\beta_0\gamma_\delta + \gamma_0\beta_\delta)S^{*2} + V^*(-\gamma_0(\mu + \alpha)\alpha_\delta + ((\mu + \alpha)\alpha_0 -$$

$$\beta_0\mu)\gamma_\delta + \gamma_0\beta_\delta\mu)S^* + BV^*(-\alpha_0\gamma_\delta + \gamma_0\alpha_\delta),$$

$$A_4 = (-(\mu + \tau_1 + \tau_2)(\alpha_0\beta_\delta - \beta_0\alpha_\delta)V^* + \tau_2(-\beta_0\gamma_\delta + \gamma_0\beta_\delta))S^*$$

$$- V^*\tau_1(-\alpha_0\gamma_\delta + \gamma_0\alpha_\delta).$$

12.4 Calculation of Basic Reproduction Number

The next-generation matrix method [15, 7] is utilized to compute the basic reproduction number [9] of the system. Consider the vectors $\vec{x} = (I_0, I_\delta)^T$ and $\vec{y} = (S, V, R)^T$,

$$_0^C D_t^\phi \vec{x}(t) = \vec{f}(t) - \vec{v}(t).$$

Let F and V be the corresponding Jacobian matrix of $\vec{f}(t)$ and $\vec{v}(t)$, respectively at the disease-free equilibrium point E^0,

$$F(E^0) = \frac{B}{\mu(\mu + \alpha)}\begin{bmatrix} \mu\beta_0 + \alpha\alpha_0 & 0 \\ 0 & \mu\beta_\delta + \alpha\alpha_\delta \end{bmatrix},$$

$$V(E^0) = \begin{bmatrix} \gamma_0 + \mu + \mu_0 & 0 \\ 0 & \mu + \mu_\delta + \gamma_\delta. \end{bmatrix}$$

The matrix V is non-singular. The basic reproduction number R_0 is the spectral radius of the matrix FV^{-1}:

$$R_0 = \rho(FV^{-1}) = \sup \left[\frac{B(\alpha\alpha_0 + \beta_0\mu)}{\mu(\mu + \alpha)(\gamma_0 + \mu + \mu_0)}, \frac{B(\alpha\alpha_\delta + \mu\beta_\delta)}{\mu(\mu + \alpha)(\mu + \mu_\delta + \gamma_\delta)} \right],$$

The basic reproduction number (R_0) represents the typical number of secondary infections brought on by an infected person in a susceptible community. In the above context, the first term in sup expression of R_0 denotes the basic reproduction number corresponding to the omicron infection, and the second term denotes the reproduction number for the delta variant.

12.5 Stability

In this section, the stability conditions of the model are investigated about equilibrium points.

12.5.1 Local stability of the disease-free equilibrium E^0

Lemma 12.2. [8, 16] Consider the fractional-order system

$$_0^C D_t^\phi x = Ax, x(0) = x_0,$$

where A is the arbitrary matrix and $\phi \in (0, 1)$

- The trivial solution is asymptotically stable if and only if all the eigenvalues λ_j of matrix A satisfy $|arg(\lambda_j)| > \phi\pi/2$ for $j = 1, 2, 3, \cdots, n$.
- The trivial solution is stable if and only if all the eigenvalues of matrix A satisfy $|arg(\lambda_j)| \geq \phi\pi/2$ and eigenvalues with $|arg(\lambda_j)| = \phi\pi/2$ have the same algebraic and geometric multiplicity for $j = 1, 2, 3, \cdots, n$.

Theorem 12.3. The disease-free equilibrium point E^0 of the system (12.4) is locally asymptotically stable if $R_0 < 1$ and unstable if $R_0 > 1$.

Proof. The Jacobian matrix J of the system (12.4) is given by

$$J = \begin{pmatrix} -\beta_0 I_0 - \beta_\delta I_\delta \\ -\alpha - \mu & 0 & -\beta_0 S & -\beta_\delta S & \tau_1 \\[1em] \alpha & \begin{array}{c} -\alpha_0 I_0 - \alpha_\delta I_\delta \\ -\mu \end{array} & -\alpha_0 V & -\alpha_\delta V & \tau_2 \\[1em] \beta_0 I_0 & \alpha_0 I_0 & \begin{array}{c} \beta_0 S + \alpha_0 V \\ -\gamma_0 - \mu - \mu_0 \end{array} & 0 & 0 \\[1em] \beta_\delta I_\delta & \alpha_\delta I_\delta & 0 & \begin{array}{c} \beta_\delta S + \alpha_\delta V \\ -\mu - \gamma_\delta - \mu_\delta \end{array} & 0 \\[1em] 0 & 0 & \gamma_0 & \gamma_\delta & -\mu - \tau_1 - \tau_2 \end{pmatrix}.$$

At disease-free equilibrium point E^0,

$$J(E^0) = \begin{pmatrix} -\alpha - \mu & 0 & \frac{-\beta_0 B}{\mu+\alpha} & \frac{-B\beta_\delta}{\mu+\alpha} & \tau_1 \\[1em] \alpha & -\mu & \frac{-\alpha_0 B\alpha}{\mu(\mu+\alpha)} & \frac{-\alpha_\delta B\alpha}{\mu(\mu+\alpha)} & \tau_2 \\[1em] 0 & 0 & \begin{array}{c} \frac{(B\mu\beta_0 + B\alpha\alpha_0)}{\mu(\mu+\alpha)} \\ -\gamma_0 - \mu - \mu_0 \end{array} & 0 & 0 \\[1em] 0 & 0 & 0 & \begin{array}{c} \frac{(B\mu\beta_\delta + B\alpha\alpha_\delta)}{\mu(\mu+\alpha)} \\ -\mu - \gamma_\delta - \mu_\delta \end{array} & 0 \\[1em] 0 & 0 & \gamma_0 & \gamma_\delta & -\mu - \tau_1 - \tau_2. \end{pmatrix}$$

The eigenvalues of the matrix $J(E^0)$ are as follows,

$$\lambda_1 = -\mu,$$

$$\lambda_2 = -\mu - \tau_1 - \tau_2,$$

$$\lambda_3 = -\mu - \alpha,$$

$$\lambda_4 = \frac{B\alpha\alpha_\delta, + B\mu\beta_\delta - \alpha\mu^2 - \alpha\mu\gamma_\delta - \alpha\mu\mu_\delta - \mu^3 - \mu^2\gamma_\delta - \mu^2\mu_\delta}{\mu(\mu+\alpha)}$$

$$\lambda_5 = \frac{B\alpha\alpha_0 + B\mu\beta_0 - \alpha\mu^2 - \alpha\mu\gamma_0 - \alpha\mu\mu_0 - \mu^3 - \mu^2\gamma_0 - \mu^2\mu_0}{\mu(\mu+\alpha)}.$$

It can be observed that λ_1, λ_2 and λ_3 are negative and

$$\lambda_4 < 0 \iff \frac{B(\alpha\alpha_\delta + \mu\beta_\delta)}{\mu(\mu + \alpha)(\mu + \mu_\delta + \gamma_\delta)} < 1$$

Similarly,

$$\lambda_5 < 0 \iff \frac{B(\alpha\alpha_0 + \beta_0\mu)}{\mu(\mu + \alpha)(\gamma_0 + \mu + \mu_0)} < 1$$

Consequently,

$$R_0 = \sup\left[\frac{B(\alpha\alpha_0 + \beta_0\mu)}{\mu(\mu + \alpha)(\gamma_0 + \mu + \mu_0)}, \frac{B(\alpha\alpha_\delta + \mu\beta_\delta)}{\mu(\mu + \alpha)(\mu + \mu_\delta + \gamma_\delta)}\right] < 1$$

$$\updownarrow$$

eigenvalues λ_4 and λ_5 are negative.

Therefore, all the eigenvalues of the matrix $J(E^0)$ have a negative real part if and only if $R_0 < 1$. Thus, whenever $R_0 < 1$, the argument of each eigenvalue has absolute value π.

$$|arg(\lambda_j)| = \pi > \frac{\phi\pi}{2} \qquad \text{for each } j = 1, 2, 3, 4, 5.$$

Hence, the disease-free equilibrium point E^0 of the system (12.4) is locally asymptotically stable if and only if $R_0 < 1$.

Whenever $R_0 > 1$, the eigenvalues λ_4 and λ_5 are positive. So, the disease-free equilibrium point E^0 of the system (12.4) is unstable. $\qquad\square$

Next, the stability conditions of the endemic equilibrium point are observed.

12.5.2 Local stability of the endemic equilibrium point E^1

Theorem 12.4. The endemic point E^1 of the system (12.4) is locally asymptotically stable if the following condition holds,

(a)
$$(\beta_0 + \beta_\delta)S^* + (\alpha_0 + \alpha_\delta)V^* < (\alpha_0 + \beta_0)I_0^* + (\alpha_\delta + \beta_\delta)I_\delta^* + \alpha + 4\mu$$
$$+ \mu_\delta + \mu_0 + \gamma_\delta + \gamma_0.$$

Proof. The Jacobian of the system (12.4) at endemic equilibrium point E^1 is given by

$$J^* = \begin{pmatrix} -\beta_0 I_0^* - \beta_\delta I_\delta^* \\ -\alpha - \mu & 0 & -\beta_0 S^* & -\beta_\delta S^* & \tau_1 \\[2mm] \alpha & -\alpha_0 I_0^* - \alpha_\delta I_\delta^* \\ -\mu & -\alpha_0 V^* & -\alpha_\delta V^* & \tau_2 \\[2mm] \beta_0 I_0^* & \alpha_0 I_0^* & \beta_0 S^* + \alpha_0 V^* \\ -\gamma_0 - \mu - \mu_0 & 0 & 0 \\[2mm] \beta_\delta I_\delta^* & \alpha_\delta I_\delta^* & 0 & \beta_\delta S^* + \alpha_\delta V^* \\ -\mu - \gamma_\delta - \mu_\delta & 0 \\[2mm] 0 & 0 & \gamma_0 & \gamma_\delta & -\mu - \tau_1 - \tau_2. \end{pmatrix}$$

The characteristic polynomial of matrix J^* is as follows,

$$P(\lambda) = (\lambda + \mu + \tau_1 + \tau_2)(\lambda^4 + k_3\lambda^3 + k_2\lambda^2 + k_1\lambda + k_0), \qquad (12.7)$$

where the coefficients k_0, k_1, k_2 and k_3 are defined as follows,

$$k_3 = (\alpha_0 + \beta_0)I_0^* - (\beta_0 + \beta_\delta)S^* - (\alpha_0 + \alpha_\delta)V^* + (\alpha_\delta + \beta_\delta)I_\delta^* + \alpha + \gamma_0$$
$$+ 4\mu + \mu_0 + \gamma_\delta + \mu_\delta,$$

$$k_2 = 6\mu^2 + 3\mu((\beta_\delta + \alpha_\delta)I_\delta^* + (\beta_0 + \alpha_0)I_0^* - (\beta_\delta + \beta_0)S^* - (\alpha_\delta V^* + \alpha_0)$$
$$V^* + (\mu_\delta + \gamma_\delta) + (\mu_0 + \alpha + \gamma_0)) + I_\delta^{*2}\beta_\delta\alpha_\delta + ((\alpha_0\beta_\delta + \beta_0\alpha_\delta)I_0^*$$
$$+ (-\alpha_0 V^* + (-S^* - V^*)\beta_\delta - \beta_0 S^* + \mu_\delta + \gamma_\delta + \mu_0 + \alpha + \gamma_0)\alpha_\delta$$
$$- \beta_\delta(\beta_0 S^* + \alpha_0 V^* - \gamma_0 - \mu_0 - \gamma_\delta - \mu_\delta))I_\delta^* + I_0^{*2}\alpha_0\beta_0 + (-V^*($$
$$\beta_0 + \alpha_0)\alpha_\delta + (-\beta_\delta S^* + (-S^* - V^*)\beta_0 + \mu_\delta + \gamma_\delta + \mu_0 + \alpha + \gamma_0)$$
$$\alpha_0 - \beta_0(\beta_\delta S^* - \gamma_0 - \mu_0 - \gamma_\delta - \mu_\delta))I_0^* + V^*(\beta_0 S^* + \alpha_0 V^* - \alpha -$$
$$\gamma_0 - \mu_0)\alpha_\delta + V^*(\beta_\delta S^* - \alpha - \gamma_\delta - \mu_\delta)\alpha_0 + S^*(\beta_0 S^* - \alpha - \gamma_0 -$$
$$\mu_0)\beta_\delta - S^*(\mu_\delta + \gamma_\delta + \alpha)\beta_0 + (\mu_\delta + \gamma_\delta + \mu_0 + \gamma_0)\alpha + (\mu_0 + \gamma_0)($$
$$\mu_\delta + \gamma_\delta),$$

$$k_0 = \mu^4 + ((\beta_\delta + \alpha_\delta)I_\delta^* + (\beta_0 + \alpha_0)I_0^* - \beta_\delta S^* - \beta_0 S^* - \alpha_\delta V^* - \alpha_0 V^*$$

$$+ \mu_\delta + \gamma_\delta + \mu_0 + \alpha + \gamma_0)\mu^3 + (I_\delta^{*2}\beta_\delta\alpha_\delta + ((\alpha_0\beta_\delta + \beta_0\alpha_\delta)I_0^* +$$

$$(-\alpha_0 V^* + (-S^* - V^*)\beta_\delta - \beta_0 S^* + \mu_\delta + \gamma_\delta + \mu_0 + \alpha + \gamma_0)\alpha_\delta$$

$$- \beta_\delta(\beta_0 S^* + \alpha_0 V^* - \gamma_0 - \mu_0 - \gamma_\delta - \mu_\delta))I_\delta^* + I_0^{*2}\alpha_0\beta_0 + (-V^*$$

$$(\beta_0 + \alpha_0)\alpha_\delta + (-\beta_\delta S^* + (-S^* - V^*)\beta_0 + \mu_\delta + \gamma_\delta + \mu_0 + \alpha +$$

$$\gamma_0)\alpha_0 - \beta_0(\beta_\delta S^* - \gamma_0 - \mu_0 - \gamma_\delta - \mu_\delta))I_0^* + V^*(\beta_0 S^* + \alpha_0 V^* -$$

$$\alpha - \gamma_0 - \mu_0)\alpha_\delta + V^*(\beta_\delta S^* - \alpha - \gamma_\delta - \mu_\delta)\alpha_0 + S^*(\beta_0 S^* - \alpha -$$

$$\gamma_0 - \mu_0)\beta_\delta - S^*(\mu_\delta + \gamma_\delta + \alpha)\beta_0 + (\mu_\delta + \gamma_\delta + \mu_0 + \gamma_0)\alpha + (\mu_0$$

$$+ \gamma_0)(\mu_\delta + \gamma_\delta))\mu^2 + (-\beta_\delta\alpha_\delta(\beta_0 S^* + \alpha_0 V^* - \gamma_0 - \mu_0 - \gamma_\delta -$$

$$\mu_\delta)I_\delta^{*2} + (((-V^*(\beta_\delta + \beta_0)\alpha_0 - \beta_0(\beta_\delta S^* - \gamma_0 - \mu_0 - \gamma_\delta - \mu_\delta))$$

$$\alpha_\delta - \beta_\delta\alpha_0(\beta_0 S^* - \gamma_0 - \mu_0 - \gamma_\delta - \mu_\delta))I_0^* + (((S^* + V^*)\beta_\delta - \mu_\delta$$

$$- \gamma_\delta - \alpha)V^*\alpha_0 + (\beta_0 S^* - \gamma_0 - \mu_0)(S^* + V^*)\beta_\delta - S^*(\mu_\delta + \gamma_\delta$$

$$+ \alpha)\beta_0 + (\mu_\delta + \gamma_\delta + \mu_0 + \gamma_0)\alpha + (\mu_0 + \gamma_0)(\mu_\delta + \gamma_\delta))\alpha_\delta - \beta_\delta($$

$$\mu_\delta + \gamma_\delta)(\beta_0 S^* + \alpha_0 V^* - \gamma_0 - \mu_0))I_\delta^* - \beta_0\alpha_0(\beta_\delta S^* + \alpha_\delta V^* -$$

$$\gamma_0 - \mu_0 - \gamma_\delta - \mu_\delta)I_0^{*2} + ((((S^* + V^*)\beta_0 - \mu_0 - \alpha - \gamma_0)\alpha_0 -$$

$$\beta_0(\mu_0 + \gamma_0))V^*\alpha_\delta + (((S^* + V^*)\beta_0 - \mu_0 - \alpha - \gamma_0)S^*\beta_\delta$$

$$- (\mu_\delta + \gamma_\delta)(S^* + V^*)\beta_0 + (\mu_\delta + \gamma_\delta + \mu_0 + \gamma_0)\alpha + (\mu_0 + \gamma_0)($$

$$\mu_\delta + \gamma_\delta))\alpha_0 - \beta_0(\mu_0 + \gamma_0)(\beta_\delta S^* - \gamma_\delta - \mu_\delta))I_0^* + \alpha(\beta_0 S^* +$$

$$\alpha_0 V^* - \gamma_0 - \mu_0)(\beta_\delta S^* + \alpha_\delta V^* - \gamma_\delta - \mu_\delta))\mu - \beta_\delta\alpha_\delta(\mu_\delta + \gamma_\delta)$$

$$(\beta_0 S^* + \alpha_0 V^* - \gamma_0 - \mu_0)I_\delta^{*2} + (((-((\mu_0 + \gamma_0)\beta_\delta + \beta_0(\mu_\delta + \gamma_\delta))$$

$$V^*\alpha_0 - \beta_0(\mu_0 + \gamma_0)(\beta_\delta S^* - \gamma_\delta - \mu_\delta))\alpha_\delta - \beta_\delta\alpha_0(\mu_\delta + \gamma_\delta)(\beta_0 S^*$$

$$- \gamma_0 - \mu_0))I_0^* - \alpha_\delta\alpha(\mu_\delta + \gamma_\delta)(\beta_0 S^* + \alpha_0 V^* - \gamma_0 - \mu_0))I_\delta^* -$$

$$I_0^*\alpha_0(\mu_0 + \gamma_0)(I_0^*\beta_0 + \alpha)(\beta_\delta S^* + \alpha_\delta V^* - \gamma_\delta - \mu_\delta),$$

$$k_1 = 4\mu^3 + 3((\beta_\delta + \alpha_\delta)I_\delta^* + (\beta_0 + \alpha_0)I_0^* - \beta_\delta S^* - \beta_0 S^* - \alpha_\delta V^* - \alpha_0 V^*$$

$$+ \mu_\delta + \gamma_\delta + \mu_0 + \alpha + \gamma_0)\mu^2 + 2(I_\delta^{*2}\beta_\delta\alpha_\delta + ((\alpha_0\beta_\delta + \beta_0\alpha_\delta)I_0^*$$

$$+ (-\alpha_0 V^* + (-S^* - V^*)\beta_\delta - \beta_0 S^* + \mu_\delta + \gamma_\delta + \mu_0 + \alpha + \gamma_0)$$

$$\alpha_\delta - \beta_\delta(\beta_0 S^* + \alpha_0 V^* - \gamma_0 - \mu_0 - \gamma_\delta - \mu_\delta))I_\delta^* + I_0^{*2}\alpha_0\beta_0 + (-$$

$$V^*(\beta_0 + \alpha_0)\alpha_\delta + (-\beta_\delta S^* + (-S^* - V^*)\beta_0 + \mu_\delta + \gamma_\delta + \mu_0 + \alpha +$$

$$\gamma_0)\alpha_0 - \beta_0(\beta_\delta S^* - \gamma_0 - \mu_0 - \gamma_\delta - \mu_\delta))I_0^* + V^*(\beta_0 S^* + \alpha_0 V^* -$$

$$\alpha - \gamma_0 - \mu_0)\alpha_\delta + V^*(\beta_\delta S^* - \alpha - \gamma_\delta - \mu_\delta)\alpha_0 + S^*(\beta_0 S^* - \alpha -$$

$$\gamma_0 - \mu_0)\beta_\delta - S^*(\mu_\delta + \gamma_\delta + \alpha)\beta_0 + (\mu_\delta + \gamma_\delta + \mu_0 + \gamma_0)\alpha + (\mu_0 +$$

$$\gamma_0)(\mu_\delta + \gamma_\delta))\mu - \beta_\delta\alpha_\delta(\beta_0 S^* + \alpha_0 V^* - \gamma_0 - \mu_0 - \gamma_\delta - \mu_\delta)I_\delta^{*2} +$$

$$(((-V^*(\beta_\delta + \beta_0)\alpha_0 - \beta_0(\beta_\delta S^* - \gamma_0 - \mu_0 - \gamma_\delta - \mu_\delta))\alpha_\delta - \beta_\delta\alpha_0$$

$$(\beta_0 S^* - \gamma_0 - \mu_0 - \gamma_\delta - \mu_\delta))I_0^* + (((S^* + V^*)\beta_\delta - \mu_\delta - \gamma_\delta - \alpha)$$

$$V^*\alpha_0 + (\beta_0 S^* - \gamma_0 - \mu_0)(S^* + V^*)\beta_\delta - S^*(\mu_\delta + \gamma_\delta + \alpha)\beta_0 + ($$

$$\mu_\delta + \gamma_\delta + \mu_0 + \gamma_0)\alpha + (\mu_0 + \gamma_0)(\mu_\delta + \gamma_\delta))\alpha_\delta - \beta_\delta(\mu_\delta + \gamma_\delta)(\beta_0 S^*$$

$$+ \alpha_0 V^* - \gamma_0 - \mu_0))I_\delta^* - \beta_0\alpha_0(\beta_\delta S^* + \alpha_\delta V^* - \gamma_0 - \mu_0 - \gamma_\delta - \mu_\delta$$

$$)I_0^{*2} + ((((S^* + V^*)\beta_0 - \mu_0 - \alpha - \gamma_0)\alpha_0 - \beta_0(\mu_0 + \gamma_0))V^*\alpha_\delta +$$

$$(((S^* + V^*)\beta_0 - \mu_0 - \alpha - \gamma_0)S^*\beta_\delta - (\mu_\delta + \gamma_\delta)(S^* + V^*)\beta_0 +$$

$$(\mu_\delta + \gamma_\delta + \mu_0 + \gamma_0)\alpha + (\mu_0 + \gamma_0)(\mu_\delta + \gamma_\delta))\alpha_0 - \beta_0(\mu_0 + \gamma_0)$$

$$(\beta_\delta S^* - \gamma_\delta - \mu_\delta))I_0^* + \alpha(\beta_0 S^* + \alpha_0 V^* - \gamma_0 - \mu_0)(\beta_\delta S^* + \alpha_\delta V^*$$

$$- \gamma_\delta - \mu_\delta).$$

Using Routh–Hurwitz criteria [4, 6], the characteristic polynomial (12.7) has eigenvalues with negative real part if $k_0, k_1, k_3 > 0$ and $k_1 k_2 k_3 > k_1^2 + k_3^2 k_0$ that gives the required stability conditions around E^1. $\qquad\square$

12.6 Memory Trace and Hereditary Trait

In this section, a numerical solution is obtained using the L1 scheme [14, 19]. The numerical approximation of the Caputo fractional-order derivative of the function $f(t)$ is described as:

$$
{}^{C}_{0}D^{\phi}_{t}f(t) \approx \frac{(dt)^{-\phi}}{\Gamma(2-\phi)} \left[\sum_{k=0}^{T-1} (f(t_{k+1}) - f(t_k)) \left((T-k)^{1-\phi} \right. \right.
$$

$$
\left. \left. -(T-1-k)^{1-\phi} \right) \right]
$$

Now, the above approximation is made for each equation of the fractional-order system (12.4),

$$
{}^{C}_{0}D^{\phi}_{t}S(t) \approx \frac{(dt)^{-\phi}}{\Gamma(2-\phi)} \left[\sum_{k=0}^{T-1} (S(t_{k+1}) - S(t_k)) \left((T-k)^{1-\phi} \right. \right.
$$

$$
\left. \left. -(T-1-k)^{1-\phi} \right) \right],
$$

$$
{}^{C}_{0}D^{\phi}_{t}V(t) \approx \frac{(dt)^{-\phi}}{\Gamma(2-\phi)} \left[\sum_{k=0}^{T-1} (V(t_{k+1}) - V(t_k)) \left((T-k)^{1-\phi} \right. \right.
$$

$$
\left. \left. -(T-1-k)^{1-\phi} \right) \right],
$$

$$
{}^{C}_{0}D^{\phi}_{t}I_0(t) \approx \frac{(dt)^{-\phi}}{\Gamma(2-\phi)} \left[\sum_{k=0}^{T-1} (I_0(t_{k+1}) - I_0(t_k)) \left((T-k)^{1-\phi} \right. \right.
$$

$$
\left. \left. -(T-1-k)^{1-\phi} \right) \right],
$$

$$
{}^{C}_{0}D^{\phi}_{t}I_\delta(t) \approx \frac{(dt)^{-\phi}}{\Gamma(2-\phi)} \left[\sum_{k=0}^{T-1} (I_\delta(t_{k+1}) - I_\delta(t_k)) \left((T-k)^{1-\phi} \right. \right.
$$

$$
\left. \left. -(T-1-k)^{1-\phi} \right) \right],
$$

$$\begin{aligned} {}^{C}_{0}D^{\phi}_{t}R(t) \approx \frac{(dt)^{-\phi}}{\Gamma(2-\phi)} & \left[\sum_{k=0}^{T-1} (R(t_{k+1}) - R(t_k)) \left((T-k)^{1-\phi} \right. \right. \\ & \left. \left. -(T-1-k)^{1-\phi} \right) \right]. \end{aligned}$$

Using the definition of Caputo derivative and the above equations, the numerical solution of each equation is given by the difference of Markov term and Memory trace.

- **For the omicron variant (I_0) :**

$$I_0(t_T) \approx \text{Markov term of } I_0 - \text{Memory trace of } I_0$$

Where Markov term and Memory trace of I_0 is given as follows:

$$\text{Markov term} = dt^{\phi}\Gamma(2-\phi)F(X) + I_0(t_{T-1})$$

$$\text{Memory trace} = \sum_{k=0}^{T-2} (I_0(t_{k+1}) - I_0(t_k)) \left((T-k)^{1-\phi} \right.$$
$$\left. - (T-1-k)^{1-\phi} \right)$$

- **For the delta variant (I_δ) :**

$$I_\delta(t_T) \approx \text{Markov term of } I_\delta - \text{Memory trace of } I_\delta$$

Where Markov term and Memory trace of I_δ is given as follows:

$$\text{Markov term} = dt^{\phi}\Gamma(2-\phi)F(X) + I_\delta(t_{T-1})$$

$$\text{Memory trace} = \sum_{k=0}^{T-2} (I_\delta(t_{k+1}) - I_\delta(t_k)) \left((T-k)^{1-\phi} \right.$$
$$\left. -(T-1-k)^{1-\phi} \right)$$

The term memory trace combines all previous activities and records of the long-term history of the system. In contrast, the Markov term refers to a memory-less term where the system's future state does not depend upon the past state of the system. Notice that whenever $\phi = 1$, the memory trace is zero and $0 < \phi < 1$ gives non-zero memory trace.

12.7 Numerical Simulation

In this section, the numerical simulation is performed using MATLAB software for the proposed model. The parametric values for the numerical simulation are taken from the daily case data being updated at [1]. The initial values $(S_0, V_0, I_{00}, I_{\delta 0}, R_0)$ are assumed to be $(0.78, 0.02, 0.05, 0.05, 0.1)$.

Figure 12.2 represents the transmission patterns of the each compartment of the proposed model over a period of time (days). In the initial days, the omicron variant and the delta variant have similar transmission patterns, but after a week, omicron has a slight decline in cases. Figure 12.3 represents the variation in delta variant I_δ over different values of ϕ. Smaller values of fractional-order result in a rapid decrement in values of I_δ over time. It can be observed that as the memory effect increases, there is a sharp decrease in I_δ values.

Figure 12.4 represents the variation in omicron variant I_0 over different values of fractional-order ϕ. Similar to I_δ, It can be observed that as the

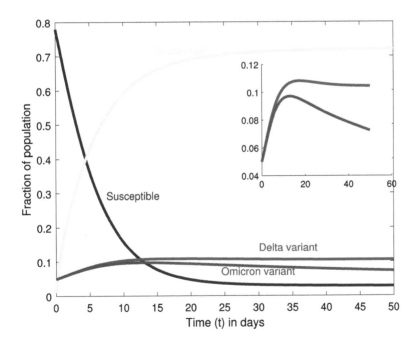

Figure 12.2 Variation in S, V, I_0 and I_δ over a period of time in days.

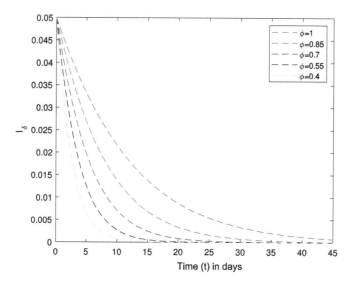

Figure 12.3   Changes in I_δ for different values of ϕ over a period of time in days.

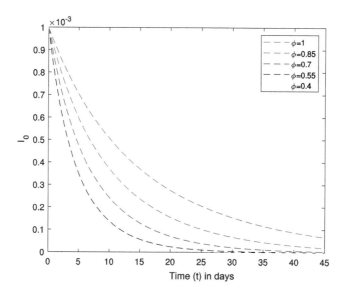

Figure 12.4   Changes in I_0 for different values of ϕ over a period of time in days.

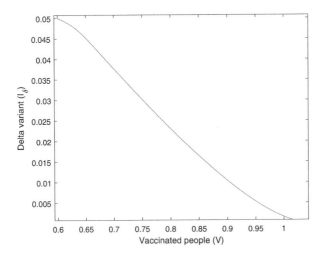

Figure 12.5 Variation in I_δ over V

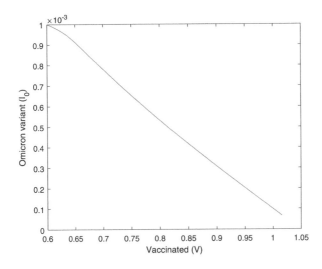

Figure 12.6 Variations in V with respect to I_0.

memory effect increases, there is a sharp decrease in I_0 values as well. Figure 12.5 represents the variation in delta variant I_δ over vaccinated susceptibles V Increment in the value of vaccinated susceptible V results in to decrement in the values of I_δ.

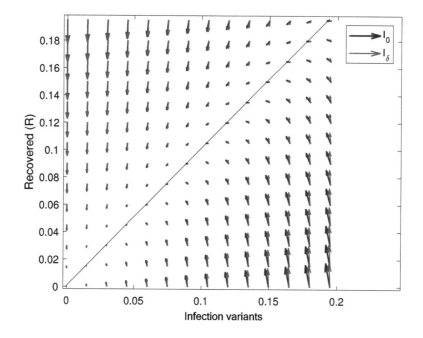

Figure 12.7   Quiver of I_δ, I_0 with respect to R.

Figure 12.6 represents the variation in the omicron variant I_0 over vaccinated susceptibles V Increment in the values of vaccinated susceptible (V) results in increment in the values of the omicron infectious individuals (I_0). From figure 12.7, it is observed that individuals infectious from the delta and the omicron variants are getting recovered, and are again vulnerable to catching the infection. Further, the intensity of recovery also increases as the number of infections (irrespective of variant) increases. As compared to red arrows representing I_δ), blue ones I_0 are slightly tilted towards recovered R representing early recovery of latter variants cases.

12.8 Conclusion

In this chapter, a fractional-order *SVIR* model is proposed for COVID-19 with two infection classes related to pre- and post-evolution of the SARS-CoV-2. Two important equilibrium points are obtained, disease-free E^0 and endemic E^1. The basic reproduction number R_0 is obtained using the next-generation matrix method and is found to be the maximum of the basic reproduction numbers corresponding to the delta and the omicron variants. The equilibrium point E^0 is found to be locally asymptotically stable only when $R_0 < 1$, and E^1 is found to be asymptotically stable with certain conditions, as mentioned in Section 12.5.2. Numerical simulations are performed using MATLAB software. It is clear from the numerical simulations that the vaccination drive is a very effective intervention measure, and the latter variant of the virus is not much severe and infectious. Further, the memory effect graph shows that as it increases, there is a decline in infectious cases of each variant.

Declaration of Competing Interest

The authors declare that they have no known competing financial interests or personal relationships that could have appeared to influence the work reported in this chapter.

Credit Authorship Contribution Statement

Shah Nita: gave the concept.
Chaudhary Kapil: analysed model mathematically.
Jayswal Ekta: carried out numerical simulation.

Acknowledgements

The authors acknowledge the valuable technical support given by DST-FIST to the department. Chaudhary Kapil is supported by CSIR-JRF (Council of Scientific and Industrial Research) File No. 09/0070(13467)/2022-EMR-I. The authors are thankful to everyone who gave valuable suggestions to improve the quality of this manuscript.

References

[1] Ministry of Health and Family Welfare, Goverment of India.
[2] Omicron is more transmissible, but is it really milder? | doctor's note | al jazeera.
[3] Who coronavirus (covid-19) dashboard | who coronavirus (covid-19) dashboard with vaccination data.

[4] E. Ahmed and A. S. Elgazzar. On fractional order differential equations model for nonlocal epidemics. *Physica a*, 379:607, 6 2007.

[5] O. Balatif, L. Boujallal, A. Labzai, and M. Rachik. Stability analysis of a fractional-order model for abstinence behavior of registration on the electoral lists. *International Journal of Differential Equations*, 2020, 2020.

[6] C.F. Beards. Automatic control systems. *Engineering Vibration Analysis with Application to Control Systems*, pages 171–279, 1995.

[7] O. Diekmann, J. A.P. Heesterbeek, and J. A.J. Metz. On the definition and the computation of the basic reproduction ratio r0 in models for infectious diseases in heterogeneous populations. *Journal of Mathematical Biology*, 28:365–382, 1990.

[8] Kai Diethelm. The analysis of fractional differential equations. 2004, 2010.

[9] P. Van Den Driessche and James Watmough. Reproduction numbers and sub-threshold endemic equilibria for compartmental models of disease transmission. *Mathematical biosciences*, 180:29–48, 2002.

[10] Nicholas C. Grassly and Christophe Fraser. Mathematical models of infectious disease transmission. *Nature Reviews Microbiology 2008 6:6*, 6:477–487, 5 2008.

[11] Hesam Khajehsaeid. A comparison between fractional-order and integer-order differential finite deformation viscoelastic models: Effects of filler content and loading rate on material parameters. *https://doi.org/10.1142/S1758825118500990*, 10, 1 2019.

[12] Kamil Kozioł, Rafał Stanisławski, and Grzegorz Bialic. Fractional-order sir epidemic model for transmission prediction of covid-19 disease. *Applied Sciences 2020, Vol. 10, Page 8316*, 10:8316, 11 2020.

[13] Wei Lin. Global existence theory and chaos control of fractional differential equations. *Journal of Mathematical Analysis and Applications*, 332:709–726, 8 2007.

[14] Parvaiz Ahmad Naik. Global dynamics of a fractional-order sir epidemic model with memory. *https://doi.org/10.1142/S1793524520500710*, 13, 8 2020.

[15] Olusegun M. Otunuga. Estimation of epidemiological parameters for covid-19 cases using a stochastic seirs epidemic model with vital dynamics. *Results in Physics*, 28:104664, 9 2021.

[16] Prabir Panja. Stability and dynamics of a fractional-order three-species predator–prey model. *Theory in Biosciences*, 138:251–259, 11 2019.

[17] Karthikeyan Rajagopal, Navid Hasanzadeh, Fatemeh Parastesh, Ibrahim Ismael Hamarash, Sajad Jafari, and Iqtadar Hussain. A fractional-order model for the novel coronavirus (covid-19) outbreak. *Nonlinear Dynamics*, 101:711–718, 7 2020.

[18] Nita H. Shah, Ankush H. Suthar, and Ekta N. Jayswal. Control strategies to curtail transmission of covid-19. *International Journal of Mathematics and Mathematical Sciences*, 2020, 2020.

[19] Wondimu W. Teka, Ranjit Kumar Upadhyay, and Argha Mondal. Spiking and bursting patterns of fractional-order izhikevich model. *Communications in Nonlinear Science and Numerical Simulation*, 56:161–176, 3 2018.

[20] Wuyue Yang, Dongyan Zhang, Liangrong Peng, Changjing Zhuge, and Liu Hong. Rational evaluation of various epidemic models based on the covid-19 data of china. *Epidemics*, 37:100501, 12 2021.

[21] Fatma Özköse, Mehmet Yavuz, M. Tamer Şenel, and Rafla Habbireeh. Fractional order modelling of omicron sars-cov-2 variant containing heart attack effect using real data from the United Kingdom. *Chaos, Solitons & Fractals*, 157:111954, 4 2022.

13

Forecasting of a COVID-19 Model using LSTM

Kopila Chhetri, Himanshi, Sidhanth Karwal, and Sudipa Chauhan

Amity Institute of Applied Science, Amity University,
Amity Institute of Science and Technology, Amity University,

Abstract

The COVID-19 pandemic is still wreaking havoc on health care infrastructure, the economy, and agriculture. This pandemic impacted masses at physical, mental and financial levels. Several of the fastest-growing economies in the world are struggling as a result of the epidemic's intensity and contagiousness. Forecasting the number of infected COVID-19 patients could be helpful in preparing future hospital resources and planning due to the expanding diversity of cases and the resulting load on healthcare providers and the government.

Due to the complexity of virus propagation, even the most well-known computational and mathematical models have been shown to be incorrect. This paper focuses on developing and implementing artificial intelligence (AI) algorithms to predict COVID-19 propagation using available time-series data. We have focused on 10 hotspot countries (China, Germany, Iran, Italy, Spain, United States, France, Turkey, United Kingdom and India). To forecast COVID-19 confirmed, active, recovered and deaths, we used a suggested model based on long short-term memory (LSTM) which has an input layer that is followed by three LSTM layers and ten hidden units (neurons).

Keywords: COVID-19, 10 countries, LSTM, Prediction, Forecast.

13.1 Introduction

The 2019-nCoV, also known as the '2019 novel coronavirus' or 'COVID-19', is the culprit behind the current pneumonia outbreak, which started in early December 2019 in the area around Wuhan City in Hubei Province, China. The abrupt halt in certain countries caused a slew of issues, particularly for low-income groups and migratory workers. More than 534 million cases have been documented over the world as of today (10 June 2022), resulting in more than 6.31 million deaths. A large percentage of those who have been rehabilitated have long-COVID health problems, which can last anywhere from months to a lifetime. In terms of vaccines, more than 5.17 billion people worldwide have received a dose of a COVID-19 vaccine, which is about 67.4% of the world's population as per June 2022 update.

Due to the intricacy of infection propagation, popular computational and statistical models for COVID-19 predictions are incorrect. Due to a paucity of data, developing a model that takes into account population density, the impact of lock-downs, the impact of viral mutations and variations like the delta variant, logistics and travel, and qualitative social elements like culture and lifestyle is difficult.

Culture and lifestyle, on the other hand, are examples of variables of interest that cannot be quantified. Modelling attempts have been generally inaccurate due to the qualitative character of key factors of interest, such as lifestyle, and lack of data collecting. We need to take a fresh look at the issue using the most up-to-date data and the most extensive forecasting algorithms.

COVID-19 has been labelled a global pandemic by the WHO on 11 March 2020. A pandemic is defined by the WHO as the widespread of a new illness against which the majority of people do not have a good immune system. India is one of the largest populated countries reported its first COVID-19 case on 30 January 2020, when a medical student returned to his home in Kerela from Wuhan. On 23 March, the first lockdown in India was announced in Kerela, followed by the rest of the country on 25 March.

Then the deadly second wave known as Delta variant lasted from February till June 2021, where on the 9 April the greatest number of COVID cases $(144, 829)$ were detected in India. Globally the second wave which has hit the countries hard are Belgium, South Korea, Iran, Germany, Spain and the Czech Republic. It was first detected in India in October 2021 and has rapidly surpassed the other countries. It wreaked havoc on the Indian health system, with India having the highest daily infection rate in the world at the time of its peak. On the plus side, India boasts one of the world's quickest recovery rates, ranking ninth in the world while being second in total cases [22].

The third wave Omicron variant $(B.1.1.529)$ is a mutated variant of the SARS-CoV. This variant has many mutations compared to the reference/ancestral variant [1]. It was the first detected in South Africa and has been listed as a variant of interest by WHO [2]. Since then many other countries including India have seen an increased number of COVID-19 cases primarily of the omicron variant in contrast to the Delta variant of the second wave. It reached India on 2 December 2021, characterized by a subdued Omicron-driven outbreak. The pandemic of coronavirus disease 2019 (COVID-19) has put an enormous burden on health systems, public health infrastructure, and the economies of many countries. Many countries were obliged to restrict their borders and impose a partial or full lockdown as a result of the COVID-19 epidemic, which had a disastrous effect on the global economy. Agriculture, which is a key source of income for the population in rural areas, particularly in developing countries, has also had a significant impact. The abrupt halt in certain countries caused a slew of issues, particularly for low-income groups and migratory workers. A large percentage of those who have been rehabilitated have long-COVID health problems, which can last anywhere from months to a lifetime. In terms of vaccines, 61.1% world's population has got at least one shot, provided globally as of 1 June 2022. Furthermore, in low-income nations, about 2.4% of the population has received at least one dose.

About 534 million people are infected by COVID-19 globally and 6.31 million deaths as of June 10, 2022. The United States has the highest number of cases followed by India. In comparison, there are 2,197 fatalities per million people in the United States and 2,013 deaths per million people in the United Kingdom. Due to the intricacy of infection propagation, popular computational and statistical models for COVID-19 predictions are incorrect. Due to a paucity of data, developing a model that takes into account population density, the impact of lock downs, the impact of viral mutations and variations like the delta variant, logistics and travel, and qualitative social elements like culture and lifestyle is difficult. Modelling attempts have been generally inaccurate due to the qualitative character of key factors of interest, such as lifestyle, and a lack of data collecting. We need to take a fresh look at the issue using the most up-to-date data and the most extensive forecasting algorithms. Furthermore, there are several other constraints, including noisy or incorrect data from active cases, shifting death rates as a result of various variations, and asymptotic carriers. According to sources, the models contain a variety of flaws and have failed in many scenarios. Despite these obstacles, it has been demonstrated that country-based mitigating variables such as lockdown level and monitoring have a significant influence on infection rates.

The question today is how to predict the peak of a pandemic while taking into account all of the efforts undertaken in all directions. In many organizations' most modern applications and systems, artificial intelligence and machine learning are deployed. Deep learning is a type of machine learning that generates correct results and outperforms other machine learning approaches, especially as data scales up. There is a need to analyse the most recent deep-learning models for COVID-19 forecasting in hotspot countries. When compared to ordinary neural networks, deep learning models like recurrent neural networks (RNNs) are better at representing spatiotemporal sequences and dynamical systems. Long short-term memory networks have solved the problem of training RNNs for long-term dependencies in data sequences that span hundreds or thousands of time steps (LSTMs). In China, LSTMs have been employed to forecast COVID-19 with good results when compared to epidemic models. In Canada, LSTMs have also been used to forecast COVID-19. Convolutions neural networks (CNNs) and other deep learning models have recently shown promise for time series forecasting. In this article, we use LSTM models to anticipate the development of COVID infections in the world's 10 hotspot countries. We examine the performance of univariate and multivariate time series forecasting algorithms for short-term forecasting.

COVID-19 is spreading and goes through four stages: first, the import stage of the disease, then the contact stage, where the disease begins to spread, then community transmission, and eventually, the entire thing is labelLed a pandemic. As a result, we will use LSTM to anticipate COVID-19 to control further spread and transmission. Many researchers have pointed out that deep learning improves prediction accuracy. To evaluate the risk of infection, prediction models that integrate many features have been developed. These are intended to aid medical personnel around the world in triaging patients,especially in light of the scarce healthcare resources. We developed a machine-learning algorithm that was trained on the records of all the 10 countries' confirmed cases from the very beginning of the pandemic till 8 June 2022.

In recent years, AI technology has made inroads into clinical and public health applications, such as early warning of epidemics and intelligent

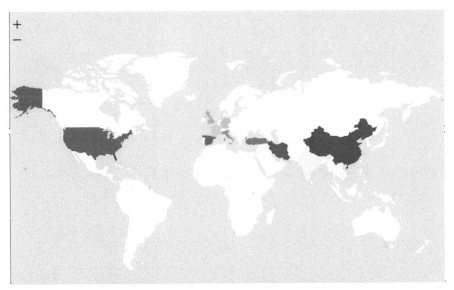

Figure 13.1 Top 10 hotspot countries of COVID-19 cases

analysis of large amounts of medical data. AI had substantially enhanced our diagnosis, prediction, and treatment level in the COVID-19 battle.

This paper focuses on the prediction of COVID-19 with the available data from 10 countries namely: China, Germany, Iran, Italy, Spain, United States, France, Turkey, United Kingdom and India. Its shown in the map in Figure 1.

The remainder of the study is divided into two sections: Sections 2 and 3 focus on LSTM methodology and data preparation respectively, while Section 4 reflects on a proposed LSTM model. We also have discussed root mean square error in Section 5, and lastly Sections 6 and 7 for employing anticipated results for selected countries and conclusions, respectively. In the data analysis carried out for the infection and the mortality rate, it was found that the symptoms of the virus infection are moderate in about 82% of the reported case and the other group of individual show severe and critical symptoms and was fatal in terms of those group [25].

The *confirmed, active, recovered* and *deaths* of the top 10 hotspot countries with the most COVID-19 cases are shown in Figure 2.

Similarly, the graphs show that the highest number of *active cases* are in United States and India, whereas China has the lowest number of cases among them.

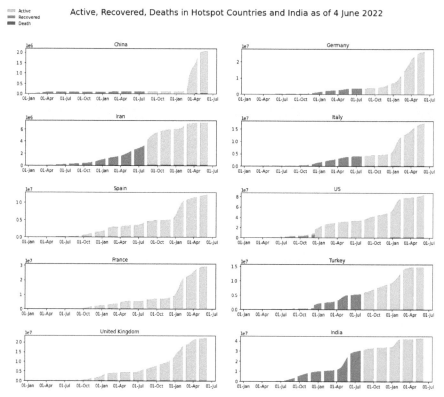

Figure 13.2   Active, Recovered and Deaths in Hotspot Countries as of 4 June 2022.

Even in the third wave (Omicron) is the muted variant of the SARS-CoV. As per WHO, over 514 million confirmed cases and over 6 million deaths have been reported globally as of 8 May 2022, but the cases and deaths have continued to decline since the end of March 2022. The world is caught in a never-ending cycle of disease breakouts and is dangerously unprepared for such disasters. Then third wave(Omicron) which is the muted variant of the SARS-CoV. It was first detected in South Africa and it hit India on 2 December 2021. Although the first wave took many lives in India, the third wave resulted in fewer deaths. Dr Padmanabha Shenoy, a rheumatologist and medical director at CARE Hospital in Kochi, showed that hybrid immunity is more effective than two vaccine doses in limiting the spread of the new type. Data showed that individuals with hybrid immunity had an immunological response 30 times stronger than those who had two vaccination doses, according to Dr Shenoy [3].

With each wave of the coronavirus pandemic, there has been a collection of established statistical data from WHO, domestic medical organizations and medical journals on various aspects of the Omicron variant of SARS-Cov 19 such as its infection rate morbidity and mortality [1].

With the help of these data and multiple statistical and LSTM, many models have been used in India to forecast death as seen in ref [3]. The author used Transfer Learning in LSTM networks to forecast new COVID cases and deaths. Models processed with data from early COVID-infected countries like Italy and the United States were used in their model to forecast the spread in other countries [23]. Single and multi-step forecasting were performed from these models. The results from these models were also tested against various data from Germany, France, Brazil, India and Nepal to check the validity of the method. The results stipulated that new cases and deaths could be forecasted pretty well with their proposed models. Complex patterns of a steep rise, spikes, and flattening effects of new cases and deaths could be learned from mature datasets, that is data from countries infected earlier. This approach was more useful for countries that are in their early phase of virus spread, to forecast based on other countries' models.

These LSTM models can also be used to predict the trends of new COVID-19 infections. As seen in this ref [4, 5] a study by Bedi et al. a modified SEIRD (susceptible exposed infected recovered deceased) model was proposed for predicting the COVID-19 trend and peak in India's four worst-affected states. The modified SEIRD model is based on the SEIRD model, which likewise uses an asymptomatic but infectious exposed population to make predictions. In their research, they also used a deep learning-based long short-term memory (LSTM) model for trend prediction. Predictions of LSTM were compared with the predictions obtained from the proposed modified SEIRD model for the next 30 days.

The use of LSTM models are not only limited to forecasting morbidity and mortality pattern and can also be used to associate certain factors which may play a role in predicting outcomes. In a study in India [5], The association between meteorological factors and COVID-19 instances was investigated, and a forecasting model based on long short-term memory (LSTM), a deep learning model, was developed. In many geographic regions across India, the study discovered that specific humidity has a strong positive correlation, whereas maximum temperature has a negative correlation and minimum temperature has a positive correlation. Univariate and multivariate LSTM time series forecast models were optimized using weather data and COVID-19-verified case data. The optimized models were used to forecast daily COVID-19 instances for a period of one month with a lead time of 1-âĂŞ14 days. The results showed that the univariate LSTM model was reasonably good for forecasting COVID-19 instances in the short term (1 day ahead of time) (relative error $< 20\%$). In addition, after taking weather parameters into account, the multivariate LSTM model enhanced the accuracy of medium-range forecasts ($1 - 7$ days advance). According to the study, India's west and northwest regions saw the greatest improvement in forecasting skill due to the particular humidity. Similar to this, utilizing new or changing current models, like the weather prediction model, the temperature played a vital part in model improvement in the Southern and Eastern areas of India.

With Omicron variant being a variant of concern and the implications for control of the pandemic to reduce the burden of the healthcare system, with the help of such LSTM Models many state or country-specific models can be constructed for forecasting Omicron variant cases and such forecasts could be of great help to all policymakers.

13.2 Methodology

13.2.1 Long short-term memory (LSTM)

Machine learning is the use of artificial intelligence to teach computers to learn from data and make decisions on their own, much like humans do. Deep learning, sometimes known as deep Structured learning, is a subcategory of machine learning algorithms. Deep learning extracts high-level features from the data it is given using a multi-layer approach.

Long short-term memory networks or LSTM, is an augmentation of simple RNN cell. It is a type of recurrent neural network (RNN) that was developed to deal with instances where RNNs failed. It is a unique form of network that can handle and retain long-term information and being able to forecast the current occurrences. Hochreiter & Schmidhuber (1997) introduced them, and numerous individuals developed and popularised them in subsequent work [26]. For situations involving sequence prediction, LSTM are frequently employed and have shown to be incredibly successful. The reason they work tremendously well on a large variety of problems is that LSTM can store past important information and forget the information that is not required. LSTM also provides us with a large range of parameters such as learning rates, and input and output biases. They don't have to work hard to remember knowledge for lengthy periods.

An LSTM is a sort of recurrent neural network that performs numerous math operations to improve its memory rather than just passing its results into the following component of the network. For forecasting the likely numbers of COVID-19 cases, the deep learningÂătechnique comprises recurrent neural network (RNN) and long short-term memory (LSTM) networks. The LSTM models had a precision accuracy of 98.58%, while the RNN models had a precision accuracy of 93.45%.

LSTM was outlined to reduce the vanishing and exploding gradient problem. Apart from the hidden state vector, each LSTM cell maintains a cell state vector, and each time step the next LSTM can choose to read from it, write to it or reset the cell using an explicit gate mechanism. This model consists of three parts, and each part performs an individual function.The first section determines whether the information from the preceding timestamp needs to be remembered or can be ignored. In the second phase the cell attempts to learn new data obtained from the input to this cell. Lastly in the third part, the cell transmits the revised data from the current timestamp to the next timestamp. A series of "gates" used by LSTMs control how data in a sequence enters, is stored, and leaves the network. The *forget gate, input gate*, and *output gate* are the three gates in a standard LSTM [7]. All these gates are connected with a Sigmoid dense layer as filters and they filter the information like what information to pass through and what not to.

13.2.2 How does LSTM work

We can visualize well in Figure 3 that the *cell state* C_t, the key to LSTM is the horizontal line that runs through the top of the diagram. The cell is in a state resembling a conveyor belt. With only a few tiny linear interactions, it flows straight down the entire chain. It is incredibly easy for data to simply follow it without being altered. It flows quite nicely through LSTM and updated through two points $\otimes$ which is element-wise and $\oplus$ sum messages. With these two operations, we can stabilize the gradient. The function $\otimes$ decides what to forget and $\oplus$ decides what new information to remember. Similarly, the bottom horizontal line is the *hidden state* h_t which is the same as the output and the rectangular boxes show the dense layers which contain σ and tanh. The next step is to determine what additional data will be stored in the cell state. There are two components to this. The input gate layer, a sigmoid layer, determines which values to update first. A tanh layer then generates a C_t vector of new candidate values that could be added to the state. We will combine these two in the next step to make a state update. Here σ is the

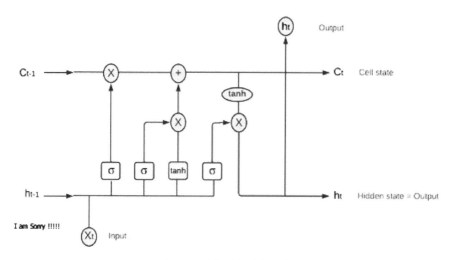

Figure 13.3   LSTM block.

activation LSTM function and its output ranges from 0 to 1. Here 0 does not allow any data to flow through while 1 allows everything to flow.

It's time to switch from the old cell state C_t to the new cell state C_t. We already know what to do because of the previous steps; now we just have to perform it.

We magnify the previous state by foot, forgetting the items we had previously agreed to forget. After that, we sum up $i_t * C_t$ This is the updated candidate values, scaled by how much each state value was updated. Finally, we must determine what we will produce. This output will be filtered and will be depending on our cell condition. To identify which features of the cell state will be output, we first run a sigmoid layer.The cell state is then multiplied by the sigmoid gate output, which only outputs the parts we wish to output, after being passed through tanh (to force the values to be between ($1 and 1$).

In Figure 3 each sign and symbol has its important function to play. It is explained in detail.

Ⓧ
= Cell state gets updated and decides what to forget from these long-term memory state vectors.

⊕ = It is the sum between two matrices. It tries to keep track of the most important information. It drops the less important things and adds very important things.

`tanh` = tanh is the dense layer and activation function. Tanh is an activation function that is non-linear. It manages the network's values, keeping them between −1 and 1. A function whose second derivative may persist for longer is required to avoid information fading. Some values may increase to huge proportions, rendering them meaningless. This will assist the network in determining which data can be ignored and which data must be retained.

(X_t) = It is the Input Vector.

(h_t) = It is the output of the cell which is connected with a hidden state which is the same thing as the output.

σ = The sigmoid function is a non-linear activation function. The gate keeps it contained. Sigmoid, unlike tanh, keeps the values between 0 and 1. It assists the network in updating or erasing data. If the multiplication yields a zero, the data are considered lost. Similarly, if the value is 1, the information remains.

tanh = It is not a dense layer it is just the function itself.

$$f_t = \sigma(W_f[h_{t-1}, x_t] + b_f).$$

Here, $\mathbf{f_t}$ is the forget matrix for time $\mathbf{t}$, and $\mathbf{C_{t-1}}$ is responsible for long-term memory. $\sigma(W_f)$ is the weight matrix for the dense layer between forget gate and input gate and C_{t-1} is the cell state from the previous time step and we perform element-wise multiplication with f_t matrix, while b_f is connection bias at f.

$$C_t^f = C_{t-1} * f_t.$$

When multiplied by 1, remembers to take the value, while 0, filters.

$$C_{t-1} = [1, 2, 4],$$

$$f_t = [1, 0, 1].$$

So we have taken one example here. When 1 and 4 are multiplied by 1, we get the same values as 1 and 4 which is remembered, and keeping the information in as f_t is equal to 1 but 2 multiplied by 0, it filters as 0 and not 2

$$C_t^f = [1, 2, 4] * [1, 0, 1] = [1, 0, 4],$$

$$i_t = \sigma(W_i[h_{t-1}, x_t] + b_i),$$

tanh is the dense layer which is a simple **RNN** cell which is already present in **RNN**.

$$C_t^i = \tanh\left(W_c[h_{t-1}, x_t] + b_c\right),$$

$$C_t^i = C_t^i * i_t,$$

C_t is the cell state and C_t^f symbolizes what to forget from pre-state (long-term memory) while C_t^i tells what is important to add as new information and the result is the cell state C_t. Along with b, which is the bias, **W** and **U** are the weight matrices that are changed throughout learning. It is worth noting that all of the gates are the same size and it is the intermediate cell state, C_t is the current cell memory, and d_h is the size of the hidden state. $C_0 = 0$ and $h_0 = 0$ are the initial values at t = 0:

$$C_t = C_t^f + C_t^i.$$

O_t is the output filter

$$O_t = \sigma(W_o[h_{t-1}, x_t] + b_o),$$

$$h_t = O_t * \tanh(C_t).$$

13.3 Data Preparation

Time series forecasting using deep learning and a comparative case study of COVID-19 confirmed cases in 10 different countries. These data were prepared and turned into appropriate interpretation and coding in order to fit into the LSTM model for predicting and forecasting issues. The data were split such that 66% of the data were reserved for training and 34% of the data were reserved for testing.

1. The COVID-19 time series data for the 10 countries were downloaded from the beginning of 2020 till 2022, from https://raw.githubuser content.com

2. Based on the number of confirmed cases and death cases, we employed the long short-term memory (LSTM) as a regression model to predict and forecast the COVID-19 spread till date.

13.4 The Proposed LSTM Model

For modelling and forecasting COVID-19 in various regions of the world, a variety of machine learning and statistical models have been applied. In this paper, We have implemented a straightforward long short-term memory (LSTM),, which consists of an input layer, a single hidden layer, and an output layer.

We chose RNN over other models for time series prediction because LSTMs are better at discovering complicated pattern logic from data by memorizing what is valuable and discarding what is not.

A simple long short-term memory (LSTM) model comprising an input layer, a single hidden layer, and an output layer that is utilized to make a prediction has been built. The input layer contains neurons that correspond to seven sequence steps. An LSTM layer with 10 hidden units (neurons) and a rectified linear unit (ReLU) as an activation function makes up the hidden layer. A dense tanh layer with 1 unit was used to forecast the output in the output layer. The model was then fitted to the data to make a forecast. Because of the stochastic nature of the LSTM model, the generated results may vary; as a result, we ran it numerous times. Finally, we arrive at the final stage.

13.5 Root Mean Square Error (RMSE)

The root mean square error (RMSE) is the residuals' standard deviation (prediction errors). The RMSE is a measure of how evenly distributed the residuals are. In other words, it indicates how tightly the data are clustered around the line of best fit:

$$RMSE = \sqrt{\frac{SSE}{N}} = \sqrt{\frac{\sum_{i=1}^{N}(x_i - \widehat{x}_i)^2}{N}}, \qquad (13.1)$$

$$R^2 = 1 - \frac{\sum(y - \widehat{y})^2}{\sum(y - \bar{y})^2}, \qquad (13.2)$$

where x_i is the actual observations time series, $\widehat{x}_i$ is the estimated or forecasted time series and N is the number of non-missing data points. We have

	Country	Cases	Train RMSE	Test RMSE	R^2
0	China	660066	1582.28	135057.82	0.9683
1	Germany	26493235	21530.77	3412640.58	0.8219
2	Iran	7232678	15588.38	746231.82	-0.1605
3	Italy	17490451	20294.14	338270.28	0.9951
4	Spain	12403245	34464.06	1091760.06	0.8724
5	USA	84748884	129661.53	1420474.52	0.9928
6	France	28794888	47900.67	1498058.24	0.9721
7	Turkey	15072747	38556.31	666382.21	0.9573
8	UK	22305893	42487.39	4172867.72	0.4720
9	India	43176817	98666.11	1331959.92	0.9034

Figure 13.4   Train and test RMSE with R^2.

used LSTM to train the model basically and have also evaluated the degree of error by calculating R^2.

The root mean square error (RMSE) of the LSTM model proposed in this paper was used to assess the model fit. Lower the RMSE value, the better the performance. The model fit to the training data is used for the test RMSE, which is then tested on the omitted test data. Not just how well the model fits the data used to train it, as is the case with train RMSE, but also how well the fitted model will predict in general. The R^2 was used to assess the models' robustness; the closer the R^2 was to 1, the better the model worked. The error rate was utilized to measure forecast accuracy; the closer to 0, the more accurate the prediction. We could see in figure 13.4 that the R^2 of United States, Italy, France and Turkey are closer to 1, which shows that the data fitted perfectly while for Iran we got a negative R^2 value, which means the regression line is making more errors than mean line.

13.6 Experimental Results

The dataset is made up of daily unique instances that are normalized by taking into consideration the maximum number of daily cases over the complete dataset. To smooth the original data, we took the data from the beginning of 2020 and continue until the first week of June 2022. The creation of a series of prediction and forecasting models based on time-series data to determine when the situation in 10 countries (China, Germany, Iran, Italy, Spain, United States, France, Turkey, United Kingdom and India) will be under control was simple straightforward. By analysing the forecasted LSTM graph in figure 13.5 it became evident that there will be slightly rise of COVID cases in China and Italy while Iran, Spain, France, Turkey and United Kingdom show a more rise in cases but Germany and India shows a decline. United States has a unique curve which shows neither rise nor fall in cases.

Through observation and study of the various factors such as the mass awareness resulting in precautionary steps taken on an individual basis and assuring proper social distancing, wearing of mask in public places and use of sanitizer flattened the curve and that resulted in the breakdown of the transmission of infection in most of the countries.

Several effective COVID-19 vaccines have so far been created by renowned businesses worldwide. But the three most effective ways for preventing COVID-19 are still masked, following official recommendations for safety precautions, and social distance. The most important parameter and the reason are the vaccination drive carried out which resulted in the mass vaccination of the population and enhanced the body's defense mechanism against the virus and improved immunity among the society, another most important reason that impacted the virus transmission was the lifestyle changes adopted by the society at large that flattened the curve accompanied by mass awareness that encouraged people to take preventive steps taken to curtail the diseases' that contributed in the rapid decline of fresh COVID-19 cases (Barman, Rahman, Bora, & Borgohain, 2020).

13.7 Conclusions and Future Work

In recent years, Artificial Intelligence technology has progressed from the lab to clinical and public health applications, such as early epidemic warnings and intelligent analysis of large amounts of medical data. Through large data of clinical cases, AI could study the epidemiological, clinical, and therapy impacts of COVID-19. AI was also utilized to guide diagnosis by doing

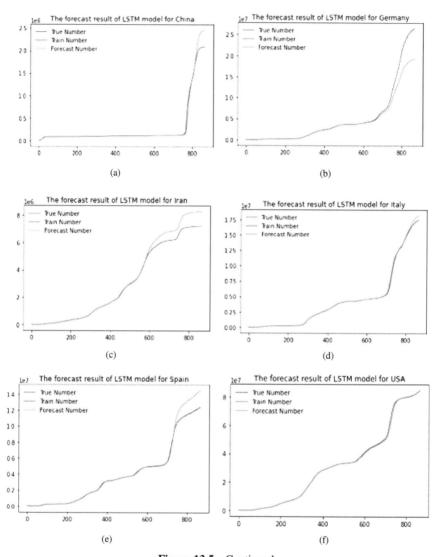

Figure 13.5   Continued.

quantitative digital analysis of medical imagery. The model was strengthened by the expert's knowledge system. The analysis' accuracy improved as the number of learning samples grew. As a result, it may be used to help identify and treat COVID-19 patients. Based on large data, scientists created several new COVID-19 prediction models.

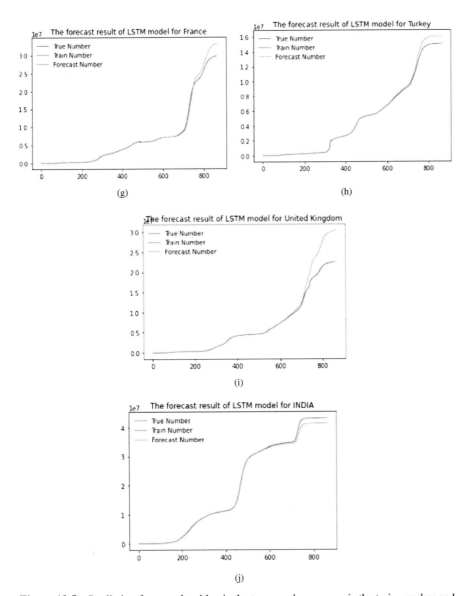

Figure 13.5 Prediction for next day, blue is the true number, orange is the train number and green is the forecast number. (a) China, (b) Germany, (c) Iran, (d) Italy, (e) Spain, (f) United States, (g) France, (h) Turkey, (i) United Kingdom and (j) India.

The use of artificial intelligence-based technologies to model the transmission of this infectious disease has been emphasized as a viable alternative to clinical procedures. Time-series prediction using machine learning was successfully accomplished in modelling and projecting the virus's final status, particularly in COVID-19. The purpose of this research is to employ Artificial Intelligence (AI) algorithms to forecast COVID-19 using time-series data in 11 different countries.

The further we study the more we have an understanding of the virus' behaviour and it becomes easier to predict its impact in the future. This behaviour will help us understand the long-term impact of COVID-19 and its impacts on the immune system at a deeper level.

This recent pandemic as stated earlier impacted the human population on a massive scale making it a medical emergency to prevent the spread of the disease and the solution came in the form of vaccine. That did not discourage the medical teams of various countries who worked round the clock to deliver the vaccine. This became possible only after the intensive study and research done by various organizations of different countries that focused mainly on understanding the behaviour of the virus and its impact. Even before the development of vaccination took clinical-stage it was in the pre-clinical stage which mainly focused on its outbreak, and its impact both long term and short term. There were various methods/tools used to study and one of the ways of studying has been presented in this dissertation. This kind of outbreak can happen anytime considering the lifestyle changes and the way we mould the natural environment as per our needs. However, tools like LSTM can help us understand and arrest such an outbreak before it can cause large-scale loss of life. The last two years have been challenging for humans as a race and this fight against COVID-19 was made successful only after the collective work.

This paper's work can be expanded to a higher level in the future, such as a predictive model (LSTM) for COVID-19 for all 10 countries in four parameters *(confirmed, active, recovered and death)*. Here we have shown forecasting graphs only for confirmed cases. Furthermore, the future forecast study can be expanded, resulting in a more precise prediction of the overall number of countries. We also came across challenges in plotting the graphs as our data were very less initially. We have learned that for results and clean graphs, we need to have more datasets.

References

[1] Amar LA, Taha AA, Mohamed MY. Prediction of the final size for COVID-19 epidemic using machine learning: A case study of Egypt. Infect Dis Model. 2020;5:622-634. doi: 10.1016/j.idm.2020.08.008.

[2] Bedi P, Dhiman S, Gole P, Gupta N, Jindal V. Prediction of COVID-19 Trend in India and Its Four Worst-Affected States Using Modified SEIRD and LSTM Models. SN Comput Sci. 2021;2(3):224. doi: 10.1007/s42979-021-00598-5.

[3] Bhimala KR, Patra GK, Mopuri R, Mutheneni SR. Prediction of COVID-19 cases using the weather integrated deep learning approach for India. Transbound Emerg Dis. 2022 May;69(3):1349-1363. doi: 10.1111/tbed.14102.

[4] Bragazzi NL, Dai H, Damiani G, Behzadifar M, Martini M, Wu J. How Big Data and Artificial Intelligence Can Help Better Manage the COVID-19 Pandemic. Int J Environ Res Public Health. 2020 May 2;17(9):3176. doi: 10.3390/ijerph17093176.

[5] Chandra R, Jain A, Singh Chauhan D. Deep learning via LSTM models for COVID-19 infection forecasting in India. PLoS One. 2022 Jan 28;17(1):e0262708. doi: 10.1371/journal.pone.0262708.

[6] Chandra R. and Kapoor A,(2019). Bayesian neural multi-source transfer learning. Neurocomputing. 378. 10.1016/j.neucom.2019.10.042.

[7] Chandra R., Jain K., Deo R. V, and Cripps S,(2019), 2019. Langevin-gradient parallel tempering for Bayesian neural learning. Neurocomput. 359, C (Sep 2019), 315âĂŞ326. https://doi.org/10.1016/j.neucom.2019.05.082

[8] Chowdhry A,(September 13, 2021) COVID-19 Prediction using LSTM.

[9] Callaway E,(December 2, 2021) Heavily Mutated Omicron Variant Puts Scientists on Alert, Nature 600, no. 7887 10.1038/d41586-021-03552-w.

[10] Olah, C. (2015). Understanding lstm networks.

[11] Dolphin R, (October 21, 2020), LSTM Networks, A detailed explanation.

[12] Ghany KKA, Zawbaa HM, Sabri HM. COVID-19 prediction using LSTM algorithm: GCC case study. Inform Med Unlocked. 2021;23:100566. doi: 10.1016/j.imu.2021.100566.

[13] A. Gumaei, M. Al-Rakhami, M. Mahmoud Al Rahhal, F. Raddah H. Albogamy, E. Al Maghayreh et al., "Prediction of covid-19 confirmed cases using gradient boosting regression method," Computers, Materials & Continua, vol. 66, no.1, pp. 315âĂŞ329, 2021.

[14] Van Houdt, G., Mosquera, C., & Nápoles, G. (2020). A review on the long short-term memory model. Artificial Intelligence Review, 53(8), 5929-5955. https://doi.org/10.1007/s10462-020-09838-1.

[15] Ma, R., Zheng, X., Wang, P. et al. The prediction and analysis of COVID-19 epidemic trend by combining LSTM and Markov method. Sci Rep 11, 17421 (2021). https://doi.org/10.1038/s41598-021-97037-5

[16] Hoffmann M, Krüger N, Schulz S, Cossmann A, Rocha C, Kempf A, Nehlmeier I, Graichen L, Moldenhauer AS, Winkler MS, Lier M, Dopfer-Jablonka A, Jäck HM, Behrens GMN, Pöhlmann S. The Omicron variant is highly resistant against antibody-mediated neutralization: Implications for control of the COVID-19 pandemic. Cell. 2022 Feb 3;185(3):447-456.e11. doi: 10.1016/j.cell.2021.12.032.

[17] Mittal S., 2020, An Exploratory Data Analysis of COVID-19 in India, INTERNATIONAL JOURNAL OF ENGINEERING RESEARCH & TECHNOLOGY (IJERT) Volume 09, Issue 04 (April 2020).

[18] Narin, A., Kaya, C. & Pamuk, Z. Automatic detection of coronavirus disease (COVID-19) using X-ray images and deep convolutional neural networks. Pattern Anal Applic 24, 1207âĂŞ1220 (2021). https://doi.org/10.1007/s10044-021-00984-y

[19] Rani V. and Jakka A.,(2020). Forecasting COVID-19 cases in India Using Machine Learning Models. 466-471. 10.1109/ICSTCEE49637.2020.9276852.

[20] Saxena S,(March 16, 2021)Introduction to Long Short Term Memory (LSTM).

[21] Shastri S, Singh K, Kumar S, Kour P, Mansotra V. Time series forecasting of Covid-19 using deep learning models: India-USA comparative case study. Chaos Solitons Fractals. 2020 Nov;140:110227. doi: 10.1016/j.chaos.2020.110227. Epub 2020 Aug 20. PMID: 32843824; PMCID: PMC7440083.

[22] Barman MP, Rahman T, Bora K, Borgohain C. COVID-19 pandemic and its recovery time of patients in India: A pilot study. Diabetes Metab Syndr. 2020 Sep-Oct;14(5):1205-1211. doi: 10.1016/j.dsx.2020.07.004. Epub 2020 Jul 9. PMID: 32673841; PMCID: PMC7343664.

[23] Gautam Y,(January 4 2021), Transfer Learning for COVID-19 Cases and Deaths Forecast Using LSTM Network, ISA Transactions, S0019-0578(20)30576-0.

[24] Zhao Z,Nehil-Puleo K, and Zhao Y,(May 15,2020),How well can we forecast the COVID-19 pandemic with curve fitting and recurrent neural network?

[25] Ilbeigipour S, Albadvi A, (18 April 2022). Supervised learning of COVID-19 patients' characteristics to discover symptom patterns and improve patient outcome prediction.

14

Simulation of COVID-19 Cases in India using AR and ANN Models

Kokila Ramesh[1], Radha Gupta[2], and Nethravathi N[3]

[1]Department of Mathematics, FET, Jain (Deemed-to-be-University), India
[2,3]Department of Mathematics, Dayananda Sagar College of Engineering, Banglore, India
E-mail: r.kokila@jainuniversity.ac.in; radha.gaurav.gupta@gmail.com; nandukowshik@gmail.com

Abstract

Coronavirus infections are not going to be obsolete but will remain a part of life in the coming years in the form of different waves. These waves exhibit special effects on human lives with varying degrees of freedom. With this in context, a statistical model has been developed to predict the number of cases in July. This wave is known as the fourth wave of COVID-19 infections. This is achieved by conducting an extensive statistical analysis of the confirmed, active, recovered and death cases for the period of five months, that is from January 2022 to June 2022. Detailed analysis of data enables us to understand inter- and intra-relationships between the cases. The modelling and prediction of the confirmed, active, and recovered cases will provide information to the human community at large and in a way, it can be prevented. Therefore, an effort has been made in this chapter for all India levels to model and predict these cases using a new artificial neural network (ANN) model with the help of a backpropagation algorithm. In this network, the variability between the confirmed, active and recovered cases has been incorporated along with the intra-variability of each case. Data have been divided into training and testing periods and the efficiency in the training and testing periods for active cases is found to be more than 95% and 85%, respectively. Further with this efficiency, active cases have been predicted for July 2022.

Keywords: COVID-19; Auto-regression Model; ANN Model; Validation and Prediction
MSC Code: 60G25, 62J05, 62M10

14.1 Introduction

Over the past three years, coronavirus disease (COVID-19) has been found to be infectious all over the world. Initially, in the first wave, the infection was severe, having common symptoms, such as fever, dry cough or wet cough, runny nose, tiredness, sore throat and body aches. In some cases, diarrhoea was also a common symptom of infection. This was continued in the subsequent waves of the outbreak of COVID-19 with added symptoms, such as headache and nausea. This is considered to be a pandemic situation, even though the death rate is as little as 1.21%.

As a part of precautions, the Indian government has taken several measures to prevent the outburst of COVID-19's fourth wave. The effect of this wave is not severe as it does not need hospitalization, but the infections still spread to a large extent, and also the number of active cases is increasing day by day. The most affected state in India is Maharashtra and slowly, in all the other states, the situation is becoming the same. With this in view, an exercise of statistical analysis of different cases has been carried out in this chapter for further prediction.

It has been observed that various methods are carried out to model and predict the number of active cases all over the world. Specifically, in India, researchers are trying to model and forecast using dynamical and statistical models. In this chapter, the authors have used data that are publicly available on the website referenced in the data section. These data consist of four cases, such as confirmed, active, recovered and death cases, which contribute to the study of understanding and modelling. In the research work done here, data have been procured from primary sources, and exhaustive data analysis has been performed on the same. It is seen that the different cases are intra- and interconnected. As a part of the study, the connections have been utilized to construct a new ANN model for inter- and intra-relationship between cases. This relationship has given better results compared to the previous work of the authors. The model has been verified in the training and testing periods and found to be effective.

14.2 Literature

Zeb et al. (2020) developed a mathematical model to study the behaviour of COVID-19 by including an isolation class. The local and global stability of the proposed model depends on the basic reproduction number of the virus. The numerical solution to this problem is obtained using a non-standard finite difference (NSFD) scheme and the Runge–Kutta method of fourth order. This is followed by Biswas et al. (2020) who formulated a deterministic compartmental model to identify the influential parameters and suggest prevention strategies to reduce the outbreak size. Estimation of the basic reproduction number was found using actual data based on reported cases.

Further, a toolkit has been developed by Overton et al. (2020) that goes beyond the SIR differential equation model, which is considered to be a basic one. They preferred statistical and mathematical models to investigate the early stages of infectious disease outbreaks. Also Torrealba-Rodriguez, Conde-Gutiérrez and Hernández-Javier (2020) used mathematical models of Gompertz and applied the computational model of inverse Artificial Neural Network to predict the number of cases of Coronavirus in Mexico. The data are considered from 27 February to 8 May 2022 to understand the pattern and estimate the projection till the epidemic ends. Furthermore, Takele (2020) emphasized the need for continuous efforts in the direction of monitoring the spread of Coronavirus (COVID-19). They applied the Autoregressive Integrated Moving Average (ARIMA) approach to predict the spreading rate of COVID in East Africa. Initially (Tiwari, Deyal, and Bisht 2020) focused on the spread of SARS-CoV-2 in South Asian countries, including India, up to July 2020. A seven-compartment mathematical model was created by Shah et al. (2020) to curtail the COVID-19 spread by recommending the BCG vaccine nationwide. They introduced two types of reproduction numbers to study the local stability along with sensitivity analysis.

In continuation, epidemiologic mathematical models with five compartments have been developed to investigate the progression of the disease, its predicted time and its impact on India. Kumar et al. (2021) also suggested a simple mathematical model for predicting the expansion of the coronavirus by December 2021. The data collected for this study are those of 2020. The finding from predicted empirical results has a close match with the observed data.

The case study presented by Lewis and Al Mannai (2021) speaks about the lasting period (i.e. the size and duration) of the COVID-19 pandemic. They focused on the curve-fitting solution of the Bass model and later on compared the results with the Kermack-McKendrick SIR and Tsallis-Tirnakli model. It is further seen that Gumel et al. (2021) formulated, analysed and simulated mathematical models to know the dynamics of COVID-19. They introduced an endemic model with only one compartment, for homogeneous and heterogeneous populations. The idea was to assess the impact of the COVID-19 vaccine on this population.

Gupta et al. (2021) have made a statistical model for 4 dominant cases (positive, active, death and recovered cases) in 4 metropolitan cities of India to predict the spread of COVID-19 spread. An auto-regressive model was formulated with 10 lag days to predict the number of active cases in four metropolitan cities of India by considering the data from 26 April to 31 July 2020. The model was tested in August 2020. Also, an Artificial Neural Network (ANN) model was developed using a backpropagation algorithm for active cases in India and particularly Bangalore to study the comparison between the two models. In this model, inter-relationship between the cases was missing.

Bandekar and Ghosh (2022) considered the classic SIR model as the base model and developed a model for India and its most affected states, Maharashtra, Karnataka and Tamil Nadu. The model deals with ten compartments to discuss the dynamics of coronavirus. Some of the factors included here are the effectiveness of face masks, contact tracing, quarantine and isolation, along with testing. The model has been fitted to estimate the transmission rate of the disease and the rate of detection for undetected asymptomatic and symptomatic individuals.

Sinha, Namdev and Shende (2022) established a mathematical model to show the spread of the coronavirus in India and across the globe. They considered physical and financial factors as key reasons why population density is more likely to interact and spread the disease. They found the infected cases, infected fatality rate and recovery rate of coronavirus using simulation and validated the model using the rough set method. Tsetimi, Ossaiugbo and Atonuje (2022) used a deterministic SEIR model to understand the transmission dynamics of the Coronavirus. The effects of some sensitive parameters of the basic reproduction number of the COVID-19 disease are simulated in the numerical solutions of the model.

Looking at the gaps and the importance of research studies in COVID-19, an effort has been made for all India levels to model and predict the Coronavirus cases in July 2022, using a new artificial neural network (ANN) model with the help of a backpropagation algorithm.

14.3 Data

The data at all India's levels have been collected from the following links: (National Current Weather | AccuWeather n.d.) and (Coronavirus Delhi, India – live map tracker from Microsoft Bing n.d.). These data comprise all states cumulative and a broad region which contributes to the total population. The major cases considered in the study are death, recovered, active and confirmed cases. These cases are interrelated and contribute to the statistical analysis of each other. The basic statistics of COVID-19 cases in India are shown in Table 14.1.

The data have large variation, skewness, and kurtosis, which indicates that the data are highly non-stationary. The data have been plotted for the period of five months (from January to June 2022) for confirmed, active, recovered and death cases and are shown in Figure 14.1.

The descriptive statistics of the data describe the non-stationarity of the data, which indicates that data are highly non-linear in nature. If the data have to be Gaussian in nature, then its skewness has to be 0 and its kurtosis has to be 3. Hence, it is non-Gaussian. The data have been non-dimensionalized using its long-term average and long-term deviation, to understand the data on both sides of the axes of reference. Even after non-dimensionalizing the data, skewness and kurtosis remain the same. The relative frequency histograms of the modified data depict the distribution of the same in the real line and are shown in Figure 14.2. These relative frequency histograms help one identify the underlying distribution pattern, in turn, identifying the distribution function or probabilistic model. The probabilistic model can be used for one-step prediction ahead. The distribution encourages the construction of a non-Gaussian model for all the cases observed in the present study. Thus, in the next section, a model has been developed to model these cases and capture the relations existing between the cases.

Table 14.1   The descriptive statistics of all the four cases, such as death, recovered, active and confirmed at India level.

Statistics		Average	Standard deviation	Skewness	Kurtosis
India	Confirmed cases	51498	92554	1.833	4.968
	Active cases	381060	652060	1.7629	4.6668
	Recovered cases	51649	90590	1.7725	5.0220
	Death cases	256	442	4.7515	37

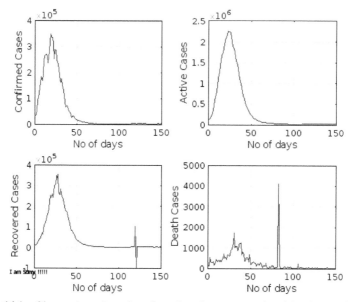

Figure 14.1 Observed number of confirmed, active, recovered and death cases in India.

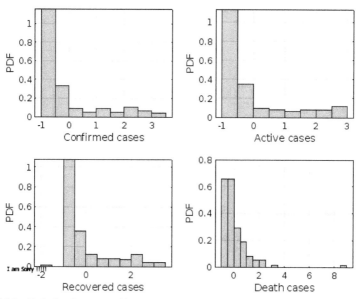

Figure 14.2 Relative frequency histogram plot of COVID-19 confirmed cases, active cases, recovered cases and death cases at all India level.

14.4 Methodology

Modelling of any statistical data is always based on the exhaustive analysis done within and between the variables. The variables considered here are confirmed, active, recovered and death cases at all Indian levels of the COVID-19 situation. Even though this situation is not new, it is a variant of the first three waves. Hence, its infections are also new to people. Therefore, in this chapter, an attempt has been made to model such a situation to help the community at large to understand the pattern in the new COVID-19 fourth wave. Data have been divided into training and testing periods. The training period (from 1 January to 1 June) has been used for modelling the data based on the relationship between the variables and also within the variables. The relationship between the variables and within the variables is shown in Figure 14.3.

The two groups of data known as training and testing periods have been taken into account for modelling and validation. If the number of parameters

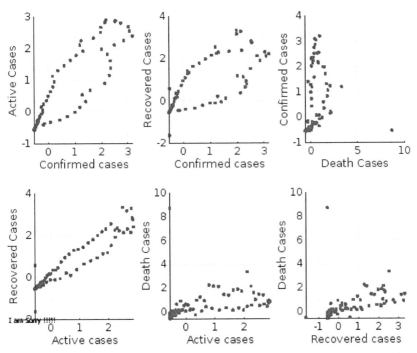

Figure 14.3   Scatter plot between the confirmed, active, recovered and death cases to observe the relationship between the variables.

for training the data is more than 50%, then the result obtained may be spurious. This is because it can be considered to be a polynomial fit which is based on the degree of the equation. The validation of the model in the training period is done based on parameters, such as coefficient of correlation, coefficient of determination and root mean square error. These measures will be accepted if they fall in the confidence level, otherwise, they will be rejected. The same procedure has been adopted for the validation period also.

Model 1: (Auto-regressive Model)

It is observed that there is a lag relationship specifically in the active cases; to visualize the same autocorrelation plot is shown in Figure 14.4. In this plot, it can be easily observed that, except for active cases, no other cases depict the lag relationship. Hence, an Auto-regressive (AR) model with ten lag days as an input has been done on the active cases to predict the cases for future days. For a sample size considered for modelling purposes apart

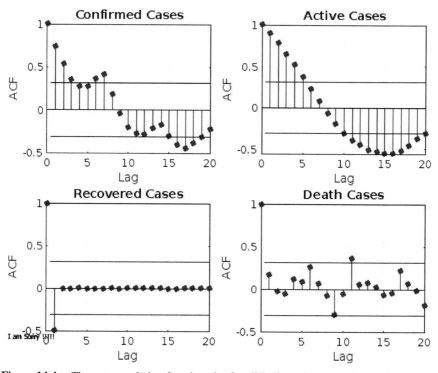

Figure 14.4   The autocorrelation function plot for all India regions to understand the intra-relationship within the variables.

from 10 days' lag, the remaining lags become ineffective and start going towards the negative side. With this in view, the following equation has been considered as the auto-regression model for prediction:

In eqn (14.1) A_t represents number of active cases at time t. The parameters in eqn (14.1) for all India level during training period of the model are given in Table 14.2

$$A_t = C_O + (C_1 * A_{t-1}) + (C_2 * A_{t-2}) + (C_3 * A_{t-3}) + (C_4 * A_{t-4}) \tag{14.1}$$

$$+(C_5 * A_{t-5}) + (C_6 * A_{t-6}) + (C_7 * A_{t-7}) + (C_8 * A_{t-8})$$
$$+(C_9 * A_{t-9}) + (C_{10} * A_{t-10})\dots\dots$$

A comparison between the observed data and the model 1 fit is shown in Figure 14.5. Visually, it seems to be a good fit, but to compare the same, descriptive statistics for both observed and model fit have been found and are compared in Table 14.3. In this average, standard deviation has been matched with the model, and the measures, such as correlation coefficient (CC), have been found and are listed in Table 14.3. Since a negligible relationship exists between the model fit and the error, CC is found to be 0.99.

Model 2:

Apart from the active cases, no other cases show the lag linear relationship with the variables, and also from the histogram plot, it is clearly seen that there is a nonlinear relationship between the variables. Therefore, ignoring the nonlinearity in the data will not be a good decision. Hence, a new Artificial Neural Network (ANN) model has been presented in this chapter with 5 days' lags of confirmed, active and recovered cases as 15 inputs in

Table 14.2 The unknowns of the regression (1) for all India.

Place	C0	C1	C2	C3	C4	C5	C6	C7	C8	C9	C10
India	−0.0018	0.0590	0.0897	−0.5936	0.8506	−0.5900	0.3602	−0.4443	0.3982	−1.0782	1.9436

Table 14.3 The parameters depicting the model performance with the observed data for the training period from 11 January 2022 to 1 June 2022.

Parameters		India
Average	Actual	407233
	Model fit	406889
Standard deviation	Actual	684649
	Model fit	684299
Correlation coefficient	Between observed data and fit	0.99

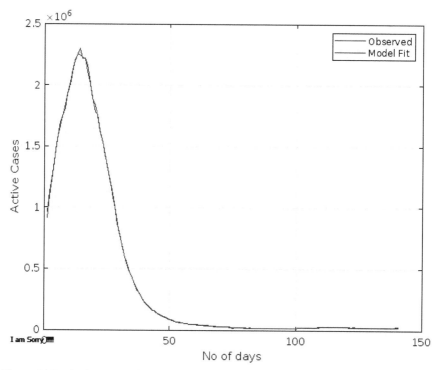

Figure 14.5 Active cases of all India from 11 January to 1 June have been compared between observed (blue line) and model fit (red line).

the input layer having one hidden layer having six neurons and three outputs in the output layer. This network has been finalized by comparing different networks with variations in the inputs starting from six inputs with the error obtained from the observed data and model fit comparison. The network has been trained using a backpropagation algorithm in the training period using the sigmoid function. There are a total of 117 weights used while training on the network. The network has experimented with only the India region for active, confirmed and recovered case data. The comparison observed in the actual data and the network model for these confirmed, active and recovered cases is shown in Figure 14.6.

Model comparison with the observed data for its performance is shown in Table 14.4 for the ANN model. It is seen that the model and observed data agree well even though only 5 lag information was used in the network for training the entire length of data using all three cases, such as confirmed,

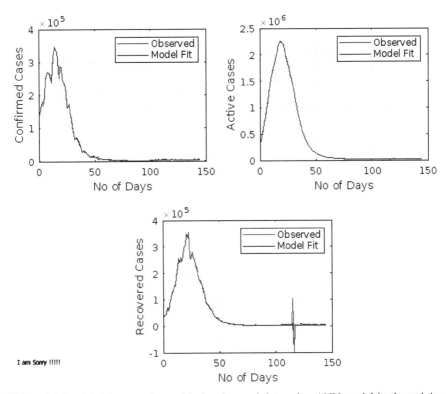

I am Sorry !!!!!

Figure 14.6   Model comparison with the observed data using ANN model in the training period.

Table 14.4   Model comparison with the observed data statistics during the training period.

Parameters		India		
		Confirmed	Active	Recovered
Average	Actual	54715	416430	56827
	Model fit	54715	416430	56827
Standard deviation	Actual	96687	678290	94399
	Model fit	96665	678270	94040
Correlation coefficient (CC)	Between observed data and fit	0.99	0.99	0.99
Performance parameter (PP)	Between observed data and fit	0.99	0.99	0.99

active and recovered. Both the Model 1 (AR model) and the Model 2 (ANN model) performs well in the first part of the data, which is known as the training period. This needs to be verified in the testing period. Now the observation will be in the second part of the data, which is the testing period for the validation of the model and can be used for forecasting.

14.5 Results and Discussion

It is observed that the two models considered in the present study have performed almost similarly in the first part of the data (training period). This discussion has to be precise in the testing period; hence, both the models have been tested in the second part (testing period) of the data. Both AR and ANN models have their disadvantages and advantages. With this in view, verification with the same parameters as in the training period is used to extend the model for the testing period to understand the performance of the models used in this chapter for the period of 30 days (1 June to 30 June). The model performance of Model 1 and Model 2 is shown in Table 14.5.

The comparison between the model prediction by AR and ANN models day wise is listed in Table 14.5 for the number of active cases. This helps one choose the appropriate model for forecasting the same for July. This is a crude way of comparison and it is represented by the reference of values. Further, the same has been compared using the RMSE obtained from observed data and forecast by AR and ANN models during the testing period. The CC obtained from observed data and the forecast by two models and the performance parameter (PP) obtained from the observed data and forecast are given in Table 14.6. These parameters measure the efficiency of the models in the best way.

The training and testing periods play an important role in deciding the efficiency of the forecasting model. It may be seen that both the models developed for the present study perform well in the training period as the weights try to adjust the data using the minimum mean square method. But the actual efficiency can be gauged only in the testing period as with the same weights the forecasting has been done. One has to clearly understand that the first and second parts of the data will not behave similarly, as the measures considered to find the efficiency of the model will be different. But in the present case, the AR model for active cases has been developed with ten lag relationship, and the ANN model is developed for Active cases using confirmed, active and recovered cases relationship. Hence, compared to previous work by the authors, this method provides a better result. These

Table 14.5 Day wise comparison of the testing of a model for the period of 30 days (1 June to 30 June 2022) using AR and ANN methods with observed data and forecast for the month of July.

Day	India			India		
	Observed data	Model 1 (AR)	Model 2 (ANN)	Day	Model 1 (AR)	Model 2 (ANN)
01-06-2022	26,152	26,408	27,129	01-07-2022	1,14,151	1,19,151
02-06-2022	27,506	27,854	28,465	02-07-2022	1,15,984	1,05,743
03-06-2022	28,781	29,374	29,481	03-07-2022	1,18,419	1,14,567
04-06-2022	30,196	30,615	31,655	04-07-2022	1,20,734	1,20,056
05-06-2022	31,940	32,337	32,989	05-07-2022	1,23,525	1,22,567
06-06-2022	33,000	33,911	33,622	06-07-2022	1,25,703	1,23,458
07-06-2022	35,788	34,546	35,540	07-07-2022	1,28,714	1,29,514
08-06-2022	39,430	39,209	40,126	08-07-2022	1,30,889	1,31,264
09-06-2022	43,363	43,142	44,127	09-07-2022	1,33,220	1,33,510
10-06-2022	47,582	47,741	47,454	10-07-2022	1,35,643	1,34,231
11-06-2022	51,343	51,910	52,748	11-07-2022	1,37,483	1,36,587
12-06-2022	54,523	55,836	55,581	12-07-2022	1,39,640	1,37,523
13-06-2022	56,942	57,629	56,042	13-07-2022	1,41,331	1,40,870
14-06-2022	61,320	59,775	59,763	14-07-2022	1,43,160	1,42,574
15-06-2022	65,480	66,111	67,391	15-07-2022	1,44,852	1,45,687
16-06-2022	70,246	69,387	71,644	16-07-2022	1,46,207	1,45,543
17-06-2022	75,055	75,505	75,349	17-07-2022	1,47,734	1,47,658
18-06-2022	79,645	79,681	82,811	18-07-2022	1,48,904	1,47,456
19-06-2022	83,438	84,699	85,955	19-07-2022	1,50,076	1,50,005

Table 14.5   Continued.

20-06-2022	85,663	86,590	85,558	20-07-2022	1,51,095	1,50,457
21-06-2022	89,282	88,573	87,549			
22-06-2022	90,585	92,944	92,987			
23-06-2022	95,065	91,710	95,672			
24-06-2022	98,683	1,00,016	1,00,410			
25-06-2022	98,307	1,01,414	1,01,200			
26-06-2022	1,00,468	99,192	1,01,130			
27-06-2022	1,02,784	1,01,918	1,06,550			
28-06-2022	1,06,006	1,05,545	1,01,920			
29-06-2022	11,0153	1,08,380	1,10,230			
30-06-2022	1,12,366	1,14,151	1,12,560			

Table 14.6   Model performance during the testing period of 30 days (from 1 June to 30 June 2022) using the parameters, such as RMSE, CC and PP, between the observed data and the forecast.

Place	Model	RMSE	CC	PP
India	AR	1.2723e+003	0.99	0.99
	ANN	1.2485e+003	0.99	0.99

models have been validated for the parameters within the confidence bands. Hence, these models can be accepted for modelling as well as forecasting purposes of COVID-19 data, specifically for active cases. The work has been extended to forecast COVID-19 active cases for July month, which is given in Table 14.5. The forecast for the period of 20 days shown in Table 14.5 is obtained based on the model parameters.. Further, based on the data collection in future, if the efficiency of the model continues to be well within the confidence limits, it can be extended to forecast for some more days.

14.6 Conclusions

The research conducted in this chapter is based on the COVID-19 data in the fourth wave, which is found to be increasing in nature. Both AR and ANN models have been developed upon the relationship between the variables for active cases. Any model is developed only by conducting extensive statistical analysis and has been done in this chapter to find the underlying patterns in the data which are both inter- and intra-related. It is understood that there is a 10-year lag relationship specifically in the active cases, and hence, an AR model is developed both for training and testing periods. Further, it is extended to forecast the active cases for 20 days in July. Even though three outputs were given in the ANN model, only active cases could capture the non-linear relationship between the variables and be able to forecast the same five-year lag as the inputs of confirmed, active and recovered cases. Using the ANN model, the forecast for July has been done and found to be similar to the AR model. With this in view, any model can be considered for modelling and forecasting the active cases. But it cannot be the same for the remaining cases. In the future, a better model for other cases will be developed.

References

[1] Bandekar, Shraddha Ramdas, and Mini Ghosh. 2022. "Mathematical Modeling of COVID-19 in India and Its States with Optimal Control." Modeling Earth Systems and Environment 8 (2): 2019–34.

[2] Biswas, Sudhanshu Kumar, Jayanta Kumar Ghosh, Susmita Sarkar, and Uttam Ghosh. 2020. "COVID-19 Pandemic in India: A Mathematical Model Study."

[3] "Coronavirus Delhi, India - Live Map Tracker from Microsoft Bing." https://www.bing.com/covid/local/delhi_india?vert=graph(August7,2022).

[4] Gumel, Abba B., Enahoro A. Iboi, Calistus N. Ngonghala, and Elamin H. Elbasha. 2021. "A Primer on Using Mathematics to Understand COVID-19 Dynamics: Modeling, Analysis and Simulations." Infectious Disease Modelling 6: 148–68.

[5] Gupta, Radha, Kokila Ramesh, N. Nethravathi, and B. Yamuna. 2021. "Impact of COVID-19 in India and Its Metro Cities: A Statistical Approach." In Mathematical Engineering, Springer Science and Business Media Deutschland GmbH, 185–201.

[6] Kumar, Harish et al. 2021. "A Simple Mathematical Model to Predict and Validate the Spread of Covid-19 in India." In Materials Today: Proceedings, Elsevier Ltd., 3859–64.

[7] Lewis, Ted G., and Waleed I. Al Mannai. 2021. "Predicting the Size and Duration of the COVID-19 Pandemic." Frontiers in Applied Mathematics and Statistics 6: 72. https://www.frontiersin.org/articles/10.3389/f ams.2020.611854/full(August7,2022).

[8] "National Current Weather I AccuWeather." https://www.accuweather. com/en/in/india-weather(August7,2022).

[9] Overton, Christopher E. et al. 2020. "Using Statistics and Mathematical Modelling to Understand Infectious Disease Outbreaks: COVID-19 as an Example." Infectious Disease Modelling 5: 409–41.

[10] Shah, Nita H. et al. 2020. "Modelling the Impact of Nationwide BCG Vaccine Recommendations on COVID-19 Transmission, Severity, and Mortality." medRxiv: 2020.05.10.20097121.

[11] Sinha, Arvind Kumar, Nishant Namdev, and Pradeep Shende. 2022. "Mathematical Modeling of the Outbreak of COVID-19." Network Modeling Analysis in Health Informatics and Bioinformatics 11(1): 1–19.

[12] Takele, Rediat. 2020. "Stochastic Modelling for Predicting COVID-19 Prevalence in East Africa Countries." Infectious Disease Modelling 5: 598–607.

[13] Tiwari, Vipin, Namrata Deyal, and Nandan S. Bisht. 2020. "Mathematical Modeling Based Study and Prediction of COVID-19 Epidemic Dissemination Under the Impact of Lockdown in India." Frontiers in Physics 8: 443. https://www.frontiersin.org/articles/10.3389/fphy.2020 .586899/full(August7,2022).

[14] Torrealba-Rodriguez, O., R. A. Conde-Gutiérrez, and A. L. Hernández-Javier. 2020. "Modeling and Prediction of COVID-19 in Mexico Applying Mathematical and Computational Models." Chaos, Solitons and Fractals 138: 109946.

[15] Tsetimi, Jonathan, Marcus Ifeanyi Ossaiugbo, and Augustine Atonuje. 2022. "A Mathematical Model and Analysis for the COVID-19 Infection." Journal of Mathematics and Statistics 18(1): 49–64.

[16] Zeb, Anwar, Ebraheem Alzahrani, Vedat Suat Erturk, and Gul Zaman. 2020. "Mathematical Model for Coronavirus Disease 2019 (COVID-19) Containing Isolation Class." BioMed Research International 2020.

Index

A
Air Pollution 66, 67, 77
ANN Model 288, 296, 298
Atangana–Baleanu fractional 215, 236
Auto regression model

B
Basic reproduction number 1, 83, 111, 133
Booster dose 125, 145

C
Caputo derivative 52, 56, 83
Caputo fractional-order derivative 56, 241, 257
Classifying motive of information 168
Co-infection 83, 86, 96
Coronavirus 3, 25, 65, 110
COVID-19 1, 26, 52, 84

D
Dynamical system 14, 125

E
Environment 66, 152, 284
Epidemic violence

F
Forecast 62, 267, 270

Forward bifurcation 83, 105

H
Hopf Bifurcation 197, 206, 209

L
L1 scheme 241, 243
LSTM 267, 271, 274

M
Mathematical model 20, 27, 85, 111
Medical Wastage 66
Memory effect 3, 83, 105
Misinformation spread dynamics 168

N
Next-generation matrix method 27, 122, 178, 241
Numerical simulation 13, 36, 60, 118

O
Omicron 127, 241, 259
Optimal control 11, 109, 122, 243

P
Pandemic 3, 65, 84, 151
Pattern identification of misinformation spread 168
Pneumonia 84, 96
Prediction 242, 268, 273
Prey-Predator Model 199

S

Simulation 26, 53, 60, 96
Social gatherings 109, 120
Square root functional responses 197
Stability Analysis of Equilibrium points 197

T

Threshold 109, 122, 125
Transmissible diseases in predator 197

U

Unemployment 215, 236

V

Vaccination 42, 51, 60, 281
Validation and Prediction 288

W

Water Pollution 67

About the Editors

Dr. Mandeep Mittal started his career in the education industry in 2000 with the Amity Group. Currently, he is working as Head and Professor in the Department of Mathematics, Amity Institute of Applied Sciences, Amity University Uttar Pradesh, Noida. He earned his post doctorate from Hanyang University, South Korea, 2016, Ph.D. (2012) from the University of Delhi, India, and postgraduation in Applied Mathematics from IIT-Roorkee, India (2000). He has published more than 70 research papers in International Journals and International conferences. He authored one book with Narosa Publication on C language and edited five research books with IGI Global and Springer. He is a series editor of Inventory Optimization, Springer Singapore Pvt. Ltd. He was awarded Best Faculty Award by the Amity School of Engineering and Technology, New Delhi for the year 2016–2017. He guided four Ph.D. scholars, and six students working with him in the area inventory control and management. He also served as Dean of Students Activities at Amity School of Engineering and Technology, Delhi, for nine years, and worked as Head, Department of Mathematics in the same institute for one year. He is a member of editorial boards of Revista Investigacion Operacional, Journal of Control and Systems Engineering and Journal of Advances in Management Sciences and Information Systems. He has actively participated as a core member of organizing committees of international conferences in India and outside India.

Professor Nita H. Shah received her Ph.D. in Statistics from Gujarat University in 1994. She is Head of Department of the Department of Mathematics in Gujarat University, India. She is Postdoctoral Visiting Research Fellow of the University of New Brunswick, Canada. Professor Nita's research interests include inventory modeling in supply chain, robotic modeling, mathematical modeling of infectious diseases, image processing, dynamical systems and its applications. She has completed 3 UGC sponsored projects and has published 13 monographs, 5 textbooks and 475+ peer-reviewed research papers. Five edited books have been prepared for IGI Global and Springer with co-editor

305

Dr. Mandeep Mittal. Her papers are published in high impact Elsevier, Inderscience, Springer, and Taylor and Francis journals. Google Scholar, shows citations of over 4041 and the maximum number of citations for a single paper is over 214. The H-index is 28 up to April 23, 2023 and i-10 index is 101. She has guided 30 Ph.D. students and 15 M.Phil. students. She has travelled in USA, Singapore, Canada, South Africa, Malaysia and Indonesia to give talks.